VOLUME ONE HUNDRED AND SEVEN

Advances in
PROTEIN CHEMISTRY AND STRUCTURAL BIOLOGY

Chromatin Proteins and Transcription Factors as Therapeutic Targets

VOLUME ONE HUNDRED AND SEVEN

ADVANCES IN PROTEIN CHEMISTRY AND STRUCTURAL BIOLOGY

Chromatin Proteins and Transcription Factors as Therapeutic Targets

Edited by

ROSSEN DONEV
Biomed Consult Ltd
Swansea, United Kingdom

ACADEMIC PRESS
An imprint of Elsevier
elsevier.com

Academic Press is an imprint of Elsevier
50 Hampshire Street, 5th Floor, Cambridge, MA 02139, United States
525 B Street, Suite 1800, San Diego, CA 92101-4495, United States
The Boulevard, Langford Lane, Kidlington, Oxford OX5 1GB, United Kingdom
125 London Wall, London, EC2Y 5AS, United Kingdom

First edition 2017

ISBN: 978-0-12-812390-4
ISSN: 1876-1623

For information on all Academic Press publications
visit our website at https://www.elsevier.com/books-and-journals

Publisher: Zoe Kruze
Acquisition Editor: Alex White
Senior Editorial Project Manager: Helene Kabes
Production Project Manager: Surya Narayanan Jayachandran
Senior Cover Designer: Christian J. Bilbow

Typeset by SPi Global, India

CONTENTS

CONTRIBUTORS

A.S. Baldwin
Lineberger Comprehensive Cancer Center, University of North Carolina, Chapel Hill, NC, United States

Y. Ben-David
The Key Laboratory of Chemistry for Natural Products of Guizhou Province and Chinese Academy of Sciences, Guizhou Medical University, Guiyang, Guizhou, PR China

A.R. Brasier
University of Texas Medical Branch; Sealy Center for Molecular Medicine; Institute for Translational Sciences, University of Texas Medical Branch, Galveston, TX, United States

E. Cesaro
University of Naples Federico II, Naples, Italy

L.A. Cordova
Stanford University, Stanford, CA, United States; Faculty of Dentistry, University of Chile, Santiago, Chile

P. Costanzo
University of Naples Federico II, Naples, Italy

C.G.P. Doss
School of Biosciences and Technology, VIT University, Vellore, Tamil Nadu, India

J.K. Durand
Curriculum in Genetics and Molecular Biology; Lineberger Comprehensive Cancer Center, University of North Carolina, Chapel Hill, NC, United States

S.B. Goodman
Stanford University, Stanford, CA, United States

M. Grosso
University of Naples Federico II, Naples, Italy

Z. Jiang
Toronto General Research Institute, University Health Network, Toronto, ON, Canada

B. Li
Key Laboratory of Molecular Virology and Immunology, CAS Center for Excellence in Molecular Cell Science, Unit of Molecular Immunology, Institut Pasteur of Shanghai, Shanghai Institutes for Biological Sciences, Chinese Academy of Sciences; Shanghai Institute of Immunology, Shanghai JiaoTong University School of Medicine, Shanghai, PR China

T.-h. Lin
Stanford University, Stanford, CA, United States

J.C. Liu
Toronto General Research Institute, University Health Network, Toronto, ON, Canada

L. Lu
Stanford University, Stanford, CA, United States

A. Lupo
University of Sannio, Benevento, Italy

G. Montano
BioMedical Center, Lund University, Lund, Sweden

A. Nabeshima
Stanford University, Stanford, CA, United States

J. Pajarinen
Stanford University, Stanford, CA, United States

E. Pasheva
"Roumen Tsanev" Institute of Molecular Biology, Bulgarian Academy of Sciences, Sofia, Bulgaria

J. Ren
Key Laboratory of Molecular Virology and Immunology, CAS Center for Excellence in Molecular Cell Science, Unit of Molecular Immunology, Institut Pasteur of Shanghai, Shanghai Institutes for Biological Sciences, Chinese Academy of Sciences, Shanghai, PR China

M. Shrestha
Toronto General Research Institute, University Health Network; University of Toronto, Toronto, ON, Canada

P. Sneha
School of Biosciences and Technology, VIT University, Vellore, Tamil Nadu, India

G. Sodaro
University of Naples Federico II, Naples, Italy

O.A.E. Sparagano
Coventry University, Coventry, United Kingdom

B. Tian
University of Texas Medical Branch; Sealy Center for Molecular Medicine; Institute for Translational Sciences, University of Texas Medical Branch, Galveston, TX, United States

I. Ugrinova
"Roumen Tsanev" Institute of Molecular Biology, Bulgarian Academy of Sciences, Sofia, Bulgaria

L. Xia
The Key Laboratory of Chemistry for Natural Products of Guizhou Province and Chinese Academy of Sciences, Guizhou Medical University, Guiyang, Guizhou, PR China

J. Yang
University of Texas Medical Branch; Sealy Center for Molecular Medicine, University of Texas Medical Branch, Galveston, TX, United States

Y. Yao
The Key Laboratory of Chemistry for Natural Products of Guizhou Province and Chinese Academy of Sciences, Guizhou Medical University, Guiyang, Guizhou, PR China

Z. Yao
Stanford University, Stanford, CA, United States

E. Zacksenhaus
Toronto General Research Institute, University Health Network; University of Toronto, Toronto, ON, Canada

PREFACE

Targeting proteins that regulate chromatin structure and gene transcription for treatment of different diseases is an exciting but very challenging concept. Various stimuli modulate either specific or broad signaling pathways in cells, resulting in the expression of effector proteins and transcripts that prompt biological response. Activated transcription factors and chromatin proteins usually function as part of multiprotein complexes that bind directly to DNA and regulate gene expression. Although a very large number of studies have focused on targeting these classes of nuclear proteins, so far there has not been much of success in applying these targeting strategies in clinical practice. There are many reasons for this relatively low success, including the complexity of the cellular response, delivery of the therapeutic molecules into the nuclei, preventing the immune system from reaction against nuclear proteins to which it has not been exposed, etc. The published promising results targeting transcription factors and chromatin proteins and the outstanding challenges to the therapeutic applications of these strategies justify further in-depth studies in this very promising and dynamic field.

The current volume of the *Advances in Protein Chemistry and Structural Biology* is focused on *Chromatin Proteins and Transcription Factors as Therapeutic Targets*. The first chapter in this volume elucidates protective host responses by triggering activation of the nuclear factor κB (NFκB)/RelA transcription factor and formation of a complex with the positive transcription elongation factor (P-TEFb)/cyclin-dependent kinase (CDK)9 and Bromodomain-containing protein 4 (BRD4) epigenetic reader. So formed RelA·BRD4·P-TEFb complex produces acute inflammation by regulating transcriptional elongation, which produces a rapid genomic response by inactive genes maintained in an open chromatin configuration engaged with hypophosphorylated RNA polymerase II. The overview of the studies in this field suggests that disruption of the RelA·BRD4·P-TEFb pathway and EZH2 methyltransferase has important implications for reversing fibrosis and restoring normal mucosal immunity in chronic inflammatory diseases. The second chapter presents an overview of the major nucleolar and extracellular properties of the high mobility group box 1 protein (HMGB1) and describes its interaction with various molecular partners, such as specific receptors or inhibitors, important for the HMGB1 role as a target in multiple diseases. This paper further explores in details the

HMGB1 pivotal role as a target for cancer treatment. Two different ways of targeting have been discussed: first, in terms of its substantial impact on the repair capacity of cancer cells, thus affecting the effectiveness of chemotherapy with the antitumor drug *cis*-platinum; and second, the possibility of HMGB1 to be targeted by microRNAs influencing different pathways of human diseases and thus making it a promising candidate for a new strategy for therapeutic interventions against cancer and some other pathologies.

The third and fourth chapters are dedicated to the NF-κB family of transcription factors as therapeutic targets in various pathologies. The NF-κB family plays a significant role in regulating the immune system and also promotes cellular proliferation and survival. NF-κB and its activating kinase, IKK, have become appealing therapeutic targets because of their critical roles in the progression of many diseases including chronic inflammation and cancer. The third chapter discusses the conditions that lead to pathway activation, the effects of constitutive activation, and some of the strategies used to inhibit NF-κB signaling. An increasing number of in vitro and in vivo evidences suggest that NF-κB is an important potential therapeutic target for inflammation-associated bone disorders by modulating inflammation and bone remodeling process simultaneously. The challenges of NF-κB targeting therapy in bone disorders are detailed in Chapter 4. Recent developments of novel inhibitors with differential approaches to modulate NF-κB activity, and the controlled release (local) or bone-targeting drug delivery (systemic) strategies, have largely increased the translational application of NF-κB therapy in bone disorders.

In the fifth chapter in this volume of the *Advances in Protein Chemistry and Structural Biology* authors elucidate the plasticity of Tregs and Th17 cells and the related human diseases resulting from the imbalance of them. Additionally, this paper focuses on the mechanisms regulating the plasticity between Tregs and Th17 cells and the potential therapeutic strategies by targeting regulators of the expression and activity of FOXP3 and RORγt or regulators of Treg/Th17 balance in autoimmune diseases, allergy, infection, and cancer.

Next two chapters are focusing on the role of transcription factors in different cancer types. The sixth chapter reviews the Kruppel-associated box zinc finger proteins (KRAB-ZFP) family, and more specifically its member ZNF224 protein, as a target for cancer treatment. Recent findings indicated that ZNF224 can behave as both tumor suppressor and oncogene in different human tumor types. The transcriptional regulatory properties of ZNF224 appear to be complex and influenced by specific sets of interactors.

The functional activity of ZNF224 as a transcription factor and the influence of protein partners on its dual behavior are discussed. The seventh chapter in this volume evaluates multi-platform genomic analyses that have identified 93 frequently altered genes in breast cancer. Of these, as many as 49 genes are found directly or indirectly involved in transcription. These include constitutive and inducible DNA-binding transcription factors (DB-TFs, 13 genes), corepressors/coactivators (14 genes), epigenetic factors (10 genes), and mediator/splicing/rRNA (3 genes) factors. At least 9 additional genes are determined as immediate upstream regulators of transcriptional cofactors. Authors review these diverse oncogenic alterations in transcription factors in breast cancer and discuss implications for therapy.

The next chapter in this volume moves the focus on the transcription factors as a target for vaccination against ticks and mites. Ticks and mites are ectoparasites that can serve as potential vectors for numerous bacteria, viruses, and parasites. To control them, transcription factors have been identified and studied for their role to sustain such arthropod pests. This review summarizes some of the work done on those transcription factors involved during blood feeding, reproduction, or when interacting with their pathogens and symbiont populations. Any transcription factor supporting the equilibrium developed by the ticks/mites could become a potential target for new control methods to prevent some of their key physiological functions.

The final ninth chapter in this thematic volume focuses on the application of a series of computational strategies to analyze the effect of mutations on the structure or function of the Hepatocyte Nuclear Factor 1 beta (HNF1B) protein in forming a protein–DNA complex. The HNF1B is a transcription factor that results in Maturity-onset Diabetes of the Young-5 (MODY-5). Prediction scores obtained from the in silico tools suggest that H153N and A241T are the major nonsynonymous mutations involved in destabilizing this transcription factor. Furthermore, a high-end microscopic computational study, such as molecular dynamics simulations, was used to relate the structural and functional effects upon mutations. Although both mutations were found to cause similar structural differences, authors observed that A241T has higher destabilizing effect on the protein. This work is a demonstration for understanding the genotype–phenotype relationships in transcription factor genes using fast and accurate computational approaches.

The aim of this thematic volume is to promote further basic and translational research, and design of new therapeutics in this very exciting and central field for the molecular control of expression of genes involved in different disorders and diseases. Advancing our knowledge on this essential topic would allow further design of therapeutic strategies to control gene expression in desire activity range in order to treat diseases such as fibrosis, sepsis, vascular diseases, inflammatory-associated bone diseases, and different types of cancer.

Dr. Rossen Donev
Biomed Consult
United Kingdom

CHAPTER ONE

Targeting Chromatin Remodeling in Inflammation and Fibrosis

J. Yang*,†,1, B. Tian*,†,‡,1, A.R. Brasier*,†,‡,2

*University of Texas Medical Branch, Galveston, TX, United States
†Sealy Center for Molecular Medicine, University of Texas Medical Branch, Galveston, TX, United States
‡Institute for Translational Sciences, University of Texas Medical Branch, Galveston, TX, United States
[2]Corresponding author: e-mail address: arbrasie@utmb.edu

Contents

Abstract

Mucosal surfaces of the human body are lined by a contiguous epithelial cell surface that forms a barrier to aerosolized pathogens. Specialized pattern recognition receptors detect the presence of viral pathogens and initiate protective host responses by triggering activation of the nuclear factor κB (NFκB)/RelA transcription factor and formation of a complex with the positive transcription elongation factor (P-TEFb)/cyclin-dependent kinase (CDK)9 and Bromodomain-containing protein 4 (BRD4) epigenetic reader. The RelA·BRD4·P-TEFb complex produces acute inflammation by regulating transcriptional elongation, which produces a rapid genomic response by inactive genes

[1] Equal contributors.

Advances in Protein Chemistry and Structural Biology, Volume 107
ISSN 1876-1623
http://dx.doi.org/10.1016/bs.apcsb.2016.11.001

maintained in an open chromatin configuration engaged with hypophosphorylated RNA polymerase II. We describe recent studies that have linked prolonged activation of the RelA–BRD4 pathway with the epithelial–mesenchymal transition (EMT) by inducing a core of EMT corepressors, stimulating secretion of growth factors promoting airway fibrosis. The mesenchymal state produces rewiring of the kinome and reprogramming of innate responses toward inflammation. In addition, the core regulator Zinc finger E-box homeodomain 1 (ZEB1) silences the expression of the interferon response factor 1 (IRF1), required for type III IFN expression. This epigenetic silencing is mediated by the Enhancer of Zeste 2 (EZH2) histone methyltransferase. Because of their potential applications in cancer and inflammation, small-molecule inhibitors of NFκB/RelA, CDK9, BRD4, and EZH2 have been the targets of medicinal chemistry efforts. We suggest that disruption of the RelA·BRD4·P-TEFb pathway and EZH2 methyltransferase has important implications for reversing fibrosis and restoring normal mucosal immunity in chronic inflammatory diseases.

1. INFLAMMATION INITIATION IN MUCOSAL SURFACES

Vertebrates are distinguished by highly specialized internal respiratory and digestive tracts lined by a contiguous surface of interconnected epithelial cells. In humans, the airways mucosa covers 70 m^2 of surface area and is exposed to over 11,000 L of air daily (Knight & Holgate, 2003; Whitsett & Alenghat, 2015). This epithelial sheet is therefore highly adapted to sense and respond to the presence of inhaled pathogenic invasive microorganisms. To accomplish this, regionally specialized epithelia (Knight & Holgate, 2003) provide a semiimpermeable physical barrier that secretes protective mucopolysaccharides and antimicrobial proteins constituting a mucosal surface (Holt, Strickland, Wikstrom, & Jahnsen, 2008). In addition, airway epithelial cells are armed with an inducible defense mechanism consisting of germline-encoded pattern recognition receptors (PRRs) that trigger a protective antiviral and inflammatory response known as the innate immune response (IIR). The IIR is intended to reduce the spread of a foreign pathogen and activate the adaptive immune response that will serve to provide sterilizing immunity and protection from future infections (Bals & Hiemstra, 2004).

Viral pathogens represent the most common infections of the respiratory tract linked to exacerbations of obstructive lung diseases by triggering inflammation and remodeling/fibrosis (Al-Muhsen, Johnson, & Hamid, 2011; Johnston, 2007). The IIR to viral infections represents a dynamic activation of gene expression programs (Pichlmair & Reis e Sousa, 2007) via the activation of PRRs by pathogen-associated molecular patterns (PAMPs), molecules produced by viral replication that distinguish the presence of microbial pathogens from normal eukaryotic cellular products of metabolism. Examples of PAMPs include dsRNA, 5-phosphorylated RNA, or CpG-DNA, indicating the presence of replicating virus; or lipopeptides, mannans, and flagellins that would indicate the presence of dividing bacteria (Bianchi, 2007).

Two major classes of PRRs that sense the presence of viral PAMPs are the cytoplasmic PRRs known as the retinoic acid-inducible gene (RIG)-I-like RNA helicases (RLHs) and membrane/endosomal-anchored Toll-like receptors (TLRs). Because these PRRs have been extensively studied and reviewed (Loo & Gale, 2011), only salient features are described here.

The primary RLH in airway epithelial cells is RIG-I itself—the major cytoplasmic PRR (Liu et al., 2007) that recognizes double-stranded- and 5-phosphorylated RNAs, representing PAMPs from diverse single-stranded (ss) RNA viruses (Kato et al., 2008, 2006; Loo & Gale, 2011). Mechanistically, the RNA helicase domain of RIG-I binds RNA patterns, an event that induces ATP-dependent conformational changes and its nondestructive polyubiquitylation by Lys63 linkages. Lys63 ubiquitylation is a posttranslational modification involved in innate signaling that promotes protein complex formation with a scaffold integral membrane protein, known as IFNβ promoter stimulator/mitochondrial antiviral signaling (MAVS; Oshiumi, Matsumoto, & Seya, 2012; Skaug, Jiang, & Chen, 2009). Binding of RIG-I to mitochondrial membrane-anchored MAVS induces the complex to aggregate into "prion-like" particles, producing an active intracellular signal transducing complex (Hou, Sun, Zheng, Skaug, Jiang, & Chen, 2011).

The TLR family is a family of membrane-localized PRRs that bind pathogen-associated lipopolysaccharide, lipoteichoic acids, mannans, flagellins, dsRNA, and DNA oligonucleotides (Hopkins & Sriskandan, 2005). TLR3 is the primary receptor for dsRNA and activates innate signaling by recruiting the TRIF adapter. TLR7/8 and 9 bind ssDNA and unmethylated CpGs, respectively. Upon binding its cognate PAMP, TLR signaling is initiated by recruiting specific adapter proteins, including MyD88 and TRIF, to form active signaling complexes (O'Neill & Bowie, 2007).

2. CENTRAL ROLE OF NFκB IN REGULATING THE INNATE IMMUNE RESPONSE

High-density microarrays and next-generation RNA sequencing studies have shown that activation of innate immune pathways triggers waves of expression for ~5000 genes controlling antiviral immunity, chemokine networks, antioxidant responses, and airway remodeling (Tian & Brasier, 2003; Tian et al., 2015, 2002; Yang et al., 2013). This sequential gene expression program is the result of signaling emanating from activated RLH and TLR-type PRRs, whose pathways converge on two primary transcription effector arms—interferon regulatory factor (IRF) and nuclear factor κB (NFκB). NFκB and IRF are preformed transcription factors regulated by signal-induced posttranslational modifications. Consequently, infected cells express and secrete antiviral type I and type III interferons (IFNs) (Garofalo et al., 1994; Jamaluddin et al., 2001; Tian et al., 2013), 17 C-, CXC-, and CC-type chemokines (Bao et al., 2008, 2007; Zhang et al., 2001), and over 570 proteins enriched in microvesicles/exosomes (Bertolusso et al., 2014; Choudhary, Boldogh, & Brasier, 2016; Fang et al., 2015; Zhao, Jamaluddin, et al., 2016). The intracellular signaling pathways of the IIR have been extensively reviewed (Akira, Uematsu, & Takeuchi, 2006; Brasier, 2008a, 2008b; Brasier, Garcia-Sastre, & Lemon, 2008; Hopkins & Sriskandan, 2005).

The NFκB signaling arm of the IIR has been thought to control inflammatory gene expression and be largely independent of IRF-mediated antiviral IFN signaling. However, computational studies of the integrated innate response to viral patterns have shown that NFκB influences IFN regulatory factor (IRF)-IFN signaling by a positive cross talk pathway (Bertolusso et al., 2014; Fang et al., 2015). In this cross talk pathway, NFκB/RelA induces IRF-1 and -7 isoforms. Downstream, RIG-I mRNA expression is amplified by synergistic interaction of RelA and IRF isoforms binding to the RIG-I promoter (Bertolusso et al., 2014). This interaction is required for sustained IFN production (Bertolusso et al., 2014; Fang et al., 2015). RIG-I upregulation further promotes IRF1/7 expression, enhancing type I IFN production (Marie, Durbin, & Levy, 1998), resulting in a positive autoamplification loop. In this way, NFκB is a master regulator of both the inflammatory and antiviral arms of the innate pathway.

NFκB is a cytoplasmic transcription factor complex that functions as an integrator of innate inflammatory signals (Brasier, 2006, 2008a, 2008b). In normal epithelial cells, the major transcriptional activating subunit is the 65 kDa transactivating subunit RelA, a protein maintained in an inactivated state in the cytoplasm by association with IκB proteins. Relief from inhibition is mediated by rate-limiting serine-directed IκB kinases (IKKs) that phosphorylate the NH_2-terminus of IκB within minutes of PRR activation, triggering association with the beta-transducin repeat-containing E3 ubiquitin protein ligase (BTRC/βTrCP), and subsequent proteolytic destruction via the 26S proteasome and calpain pathways (Brasier, 2006). Quantitative analysis of the signaling response to the potent activating molecule TNFα has shown that ~10% of all ~300K molecules of cytoplasmic RelA are translocation competent, and that the major ankyrin repeat-containing inhibitor of RelA is IκBα (Kalita et al., 2011). Consequently, IκB proteolysis represents the rate-limiting step in controlling NFκB entry into the nucleus in the IIR (Fig. 1).

Although NFκB's release from its cytoplasmic stores is necessary for triggering the IIR, it is not sufficient. PAMP signaling stimulates nuclear reactive oxygen stress pathways that are also required for triggering the genomic response (Choudhary et al., 2016). Our recent studies have shown that nuclear reactive oxygen species (ROS)-induced stress activates a PI3 kinase, ataxia telangiectasia mutated (ATM), which plays a critical role in licensing NFκB/RelA transcriptional activity (Choudhary et al., 2011; Fang et al., 2015; Fang et al., 2014). Viral replication is a potent activator of intracellular ROS generation (Hosakote, Liu, Castro, Garofalo, & Casola, 2009;

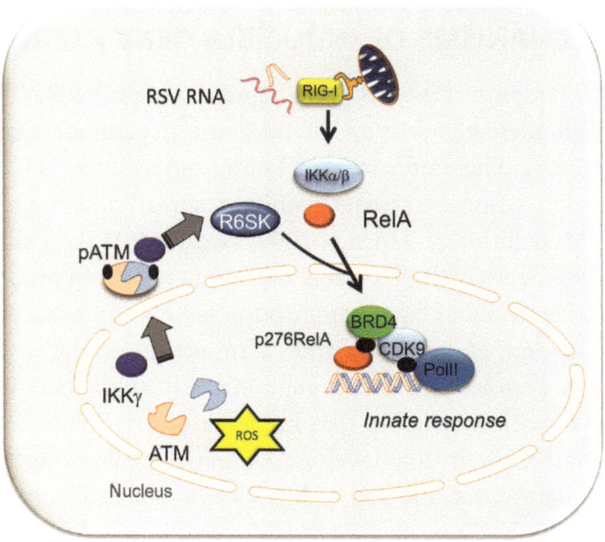

Fig. 1 Activation of NFκB by coupled posttranslational modifications. Two-step activation of NFκB/RelA. NFκB/RelA is released from cytoplasmic stores through IκBα proteolysis. Activation is mediated by Ser276 phosphorylation by nuclear ROS stress initiating ATM signaling via the ribosomal S6 kinases (R6SK). Illustrated is Ser276 complex formation with the BRD4 adapter that interacts with acetylated RelA, recruiting P-TEFb (CDK9) into the complex.

Liu et al., 2004). Nuclear ROS activate ATM by stimulating redox-mediated intermolecular dimerization and autophosphorylation on Ser1981, a hallmark of ATM activation. Phospho-ATM undergoes rapid IKKγ-dependent nuclear export and interacts with the cytoplasmic NFκB pathway at two distinct points in the cytoplasm (Fang et al., 2014). First, cytoplasmic ATM complexes with BTRC to promote rapid proteasomal degradation of phospho-IκBα described earlier. Second, cytosolic phospho-ATM is required for activating ribosomal S6 kinases (RSK), a family of redundant RelA Ser276-directed kinases (Jamaluddin, Tian, Boldogh, Garofalo, & Brasier, 2009; Jamaluddin, Wang, Boldogh, Tian, & Brasier, 2007). Phospho-Ser276 RelA, in turn, is a rate-limiting activation coupled to acetylation of Lys310, mediated by p300/CBP acetyltransferase (Chen et al., 2005). In this way, "outside-in" signaling initiated by PRRs is coordinated with "inside-out" nuclear oxidative stress response pathways and ATM to fully activate the innate pathway (Choudhary et al., 2016). This requirement ensures that the airway cell responds fully to bone fide microbial threats and not to nonreplicating adventitious molecular patterns (Fig. 1).

3. MECHANISMS OF INDUCIBLE GENE EXPRESSION

A critical aspect of the IIR is the rapid activation of RNA polymerase II (Pol II)-dependent gene expression programs to generate intracellular and antiviral effectors. These effectors, including intracellular GTPases, soluble mediators, and exosomes, restrict pathogen spread and trigger protective immunity (Medzhitov & Horng, 2009; Smale, 2012). Transcription of protein-coding genes by RNA Pol II is a highly regulated process involving preinitiation (assembly of basal transcription factors and coactivator recruitment), initiation, elongation, and termination (Chiba, Yamamoto, Yamaguchi, & Handa, 2010). Although virtually all of these steps are regulated, recent work has shown that immediate-early genes in the IIR are primarily controlled at the level of transcriptional elongation (Brasier et al., 2011; Hargreaves, Horng, & Medzhitov, 2009; Tian et al., 2013).

4. TRANSCRIPTIONAL ELONGATION IN RAPID INFLAMMATORY RESPONSES

As a consequence of terminal cellular differentiation steps, preinitiation complexes are formed on the promoters of genes destined to particular in stimulus response. In these latent-inducible genes, paused RNA Pol II accumulates in the promoter and 30–50 nt downstream of the transcription start site (Adelman & Lis, 2012; Kwak & Lis, 2013); the paused polymerase produces only short uncapped, abortive transcripts. Pol II pausing has been observed to be a common regulatory step in global genomic studies, indicating the importance of this regulatory mechanism (Adelman & Lis, 2012; Kwak & Lis, 2013). This paused form of RNA Pol II is hypophosphorylated, containing only phospho-Ser 5 located within the heptad repeats of the carboxyl terminal domain (CTD) of the major regulatory subunit. Additionally, paused Pol II is maintained in a complex with the negative elongation factor (NELF) and DRB sensitivity-inducing factor (DSIF) (Adelman & Lis, 2012; Kwak & Lis, 2013) (Fig. 2). In some cases, nucleosomes may also contribute to pausing (Kwak & Lis, 2013; Li & Gilmour, 2013; Weber, Ramachandran, & Henikoff, 2014).

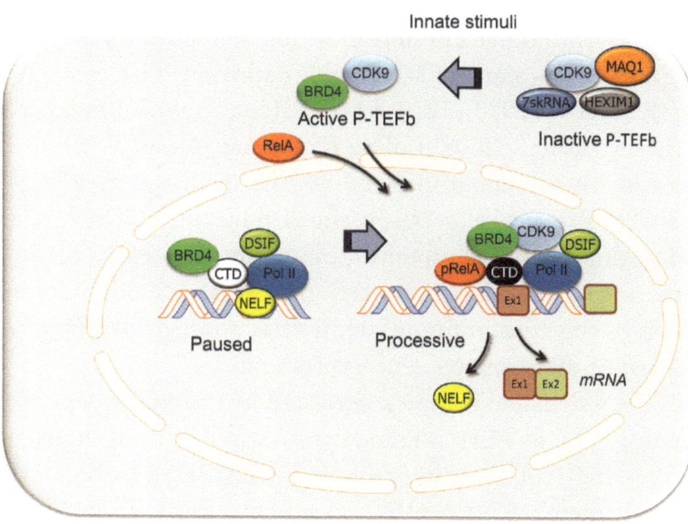

Fig. 2 Transcriptional elongation in the innate immune response. Schematic view of the pause–release mechanism mediated by the RelA·BRD4·CDK9 complex. Innate stimuli induce the inactive P-TEFb complex (bound to 7SK RNA, HEXIM1) to transition into an active complex with BRD4. In latent-inducible promoters, RNA polymerase II (Pol II) is associated with promoters at—and just downstream of—the transcription start sites. Sequence-specific binding of RelA results in BRD4·CDK9 recruitment and formation of a competent preinitiation complex. Mechanistically, P-TEFb kinase activity phosphorylates the negative elongation factor (NELF), DRB sensitivity-inducing factor (DSIF), and the carboxy-terminal domain (CTD) of Pol II on Ser2. DSIF becomes a positive elongation factor after phosphorylation. Consequently full-length, fully spliced RNA transcripts are produced to mediate the innate immune response.

Release of paused Pol II is mediated by the positive transcription elongation factor-b (P-TEFb), a complex containing cyclin-dependent kinase 9 (CDK9) and Cyclin T1 (Peterlin & Price, 2006; Zhou, Li, & Price, 2012). P-TEFb is recruited to paused promoters directly by interacting with acetylated histones or indirectly by interactions with specific transcription factors. The kinase activity of P-TEFb phosphorylates Ser2 in the heptad repeats of the RNA Pol II CTD, as well as other substrates, including NELF (which is evicted from the Pol II complex upon its phosphorylation) and DSIF (which is retained and functions as a positive elongation factor) (Adelman & Lis, 2012; Kwak & Lis, 2013; Peterlin & Price, 2006). Released Pol II acquires a processive function, enabling RNA Pol II to traverse the gene body and express full-length, fully spliced RNA transcripts (Fig. 2). This regulatory mechanism enables terminally differentiated cells to rapidly express developmentally determined latent-inducible genes.

Originally described as a mediator of heat-shock genes (Lis, Mason, Peng, Price, & Werner, 2000), P-TEFb has been shown to be a key rate-limiting regulator of other stimulus-inducible transcription factor networks, including the NFκB- and IRF3-inducible pathways (Tian et al., 2013), IFN signaling (Patel et al., 2013), IL-6–STAT3 (Hou, Ray, & Brasier, 2007), and others (Brasier, 2008a, 2008b). The functional role of P-TEFb has been demonstrated by the application of siRNA-mediated silencing (Nowak et al., 2008; Tian et al., 2013), small-molecule CDK9 and BRD4 (Bromodomain-containing protein 4) inhibitors (Tian et al., 2013), and artificial tethering to paused promoters (Lis et al., 2000). Consequently, P-TEFb has received significant attention as a target for pharmacological manipulation of dysregulated gene expression.

P-TEFb is a dynamic complex regulated in a stimulus-dependent manner. In resting cells, P-TEFb is found in two states—one as an inactive complex with methyl-CpG-binding protein 2 (MeCPE), La ribonucleoprotein domain family member 7 (LARP7), hexamethylene bis-acetamide inducible (HEXIM)-1, and the small nuclear ribonucleoprotein (snRNP) 7SK (Zhou et al., 2012), and the second state as an activated complex. The activated complexes of P-TEFb are formed with either BRD4 (Wu & Chiang, 2007) or the larger super elongation complex (SEC) (Lin et al., 2011; Smith, Lin, & Shilatifard, 2011). Of these, the BRD4·CDK9 complex is the most abundant form of activated P-TEFb and the most intensively studied. BRD4 plays multiple roles in the P-TEFb complex, including preventing HEXIM1/2·7SK snRNA binding; phosphorylating CDK9 and Ser 2 in the CTD of RNA Pol II; and serving as an adapter to interact with acetylated histones and transcription factors, including acetylated

NFκB (Brasier et al., 2011; Itzen, Greifenberg, Bosken, & Geyer, 2014; Jang et al., 2005; Zhou et al., 2009; Zou et al., 2014). This latter interaction is mediated through its repeated bromodomain (BD) motif, which has emerged as a therapeutic target (Shi & Vakoc, 2014). The distribution of active and inactive P-TEFb complexes is in dynamic equilibrium; in resting cells, P-TEFb complexes are in a 1:1 ratio of inactive:active complexes (Nguyen, Kiss, Michels, & Bensaude, 2001); by contrast, cells activated by viral molecular patterns show a dramatic shift into the activated form (Tian et al., 2013; Yang et al., 2015).

In unstimulated cells, BRD4's affinity for binding diacetylated histone H4 (with preference for Lys5/8, 8/12, and 12/16), triacetylated H4 (Lys12/16/20), and diacetylated H3 (on Lys9/14) results in its primary distribution throughout the genome associated with H3K27Ac-enriched enhancers of actively expressed genes (Jung et al., 2014; Kanno et al., 2014). Although a number of studies have implicated BRD4 in the control of housekeeping gene expression, BRD4 interaction is dynamically reconfigured through indirect transcription factor interaction and directly binding acetylated Lys side chains of regulatory histones. In innate signaling, BRD4 binds NFκB/RelA activated via Ser276 phosphorylation-dependent p300/CBP acetylation (Brasier et al., 2011; Huang, Yang, Zhou, Ozato, & Chen, 2009; Nowak et al., 2008). Detailed mutagenesis studies of the BRD4 BDs indicate that the upstream BD1 preferentially binds acetylated RelA (Huang et al., 2009). ChIP-seq studies have shown an important role of NFκB/RelA in the formation of dense BRD4 foci ("super-enhancers") upstream of inflammatory gene networks (Brown et al., 2014).

The SEC comprises elongation factors such as eleven-nineteen Lys-rich leukemia (ELL), eleven-nineteen leukemia (ENL), and AF4/FMR2 family member 4 (AFF4) (Lin et al., 2010; Luo, Lin, & Shilatifard, 2012; Smith et al., 2011) and can vary in composition depending on the cellular context and target gene specificity (Luo, Lin, Guest, et al., 2012). The SEC can interact with a subset of coactivators such as mediator, polymerase-associated factor 1 (PAF1), and integrator, the latter of which is a complex that interacts with the CTD of Pol II (Gardini et al., 2014; He et al., 2011; Kim, Guermah, & Roeder, 2010; Takahashi et al., 2011; Wier, Mayekar, Heroux, Arndt, & VanDemark, 2013). Moreover, the SEC can also colocalize with BRD4 on gene promoters (Flajollet et al., 2013). BRD4 and SEC predominantly mediate the recruitment P-TEFb, but their regulatory importance and composition with regard to paused Pol II release vary across different genes, cell types, and stimuli. Ultimately, recruitment of these cofactors depends on the transcription factors that bind to promoters or enhancers.

Although recruitment of P-TEFb to inducible genes may occur via multiple mechanisms, P-TEFb targeting in innate inflammation is mediated by interaction sequence-specific transcription factors including NFκB (Barboric, Nissen, Kanazawa, Jabrane-Ferrat, & Peterlin, 2001; Brasier et al., 2011; Diamant & Dikstein, 2013; Nowak et al., 2008) and STAT3 (Hou et al., 2007). As discussed earlier, activated NFκB/RelA produced by Ser276 phosphorylation-coupled Lys310 acetylation interacts with BD1 of BRD4 and tethers CDK9 into the complex (Brasier et al., 2011; Fang et al., 2014; Huang et al., 2009; Nowak et al., 2008; Zou et al., 2014). Adding to this complexity, CDK9 kinase activity is itself regulated by multiple activating and inhibitory posttranslational modifications. Phosphorylation of Thr186 in the T-loop of CDK9 induces conformational changes to promote substrate interaction and increase CDK9 kinase activity (Chen, Yang, & Zhou, 2004; Garber et al., 2000), whereas phosphorylation of Thr49 is inhibitory (Zhou et al., 2009). Similarly, acetylation of Lys44/48 by GCN5 inhibits kinase activity (Fu, Yoon, Qin, & Wong, 2007). Moreover, protein phosphatase 1G (PPM1G) may remove a phosphate group from the inactive form of P-TEFb to facilitate its release and activation from the inhibitory HEXIM–7SK complex after induction of the NFκB pathway through an interaction with RelA (McNamara, McCann, Gudipaty, & D'Orso, 2013). The interplay and requirements of these CDK9-regulating posttranslational modifications in the IIR are incompletely understood.

In addition to inflammatory and antiviral gene expression control, genes encoding proteins at important regulatory nodes of the NFκB pathway are controlled by transcriptional elongation (Gilchrist et al., 2012), indicating that both the direct response to NFκB activation and the modulation and fine-tuning of the components of the NFκB pathway are regulated at the level of promoter-proximal pausing. In addition to NFκB-dependent gene regulation, IFN-inducible genes are regulated by transcriptional elongation in response to type II IFN stimulation (Patel et al., 2013) and infection with respiratory syncytial virus (RSV) (Tian et al., 2013). In these examples, the mechanisms for inducible P-TEFb recruitment are not fully understood, but are dependent on the histone acetyl lysine-binding activity of the BRD4 protein.

5. MECHANISMS LINKING CHRONIC INFLAMMATION TO MESENCHYMAL TRANSITION AND AIRWAY FIBROSIS

Epidemiological studies have found that frequent upper respiratory viral infections are strongly associated with declines in pulmonary function, yet the mechanism underlying this relationship is not fully understood (Al-Muhsen et al., 2011; Bousquet, Jeffery, & Busse, 2000; Hogg et al., 2004; Johnston, 2007). In addition to their central role in triggering protective acute inflammation, chronic PRR activation produces coupled phenomena of epithelial injury, adaptive inflammation, and fibrosis. Mechanistically PRR activation produces epithelial injury by ROS production, resulting in defective epithelial barrier function and leakage of plasma proteins into the interstitial space. Epithelial injury stimulates the release of growth factors and cytokines to produce airway remodeling (Davies, 2009; Lambrecht & Hammad, 2012; Sisson et al., 2010), a process mediated by transforming growth factor (TGFβ) (Chakir et al., 2003; Hackett et al., 2009; Harris et al., 2013; Holgate et al., 2004). Additionally, PRR-activated epithelial cells release chemokine mediators that recruit CCR2+ macrophages and dendritic cells and activate these leukocyte populations via cell–cell interactions (Hammad & Lambrecht, 2008; Holt & Strickland, 2010). Dendritic cells are migratory antigen-presenting cells serving at the interface of adaptive arms of immunity, triggering activation of specific populations of T helper (Th) 2 and Th17 lymphocytes. Th2 lymphocytes synthesize interleukins (IL)-4 and IL-13 responsible for interstitial recruitment and retention eosinophils, leukocytes that secrete TGFβ. Th17 cells similarly are rich in TGFβ (Oukka, 2008).

TGFβ plays a major role in the progression of pulmonary disease by its pleiotrophic actions to promote mucosal repair, airway remodeling, and obstruction (Chakir et al., 2003; Hackett et al., 2009; Harris et al., 2013; Holgate et al., 2004). Airway remodeling is a term that refers to a structural alteration of the respiratory tree produced by epithelial–mesenchymal transition (EMT), subepithelial fibrosis, myofibroblast hyperplasia and smooth muscle hypertrophy, largely irreversible processes resulting in reduced airway compliance and gas exchange (Al-Muhsen et al., 2011). The actions of TGFβ to trigger EMT and induce transdifferentiation of fibroblasts to myofibroblasts are important in airway remodeling because both cell types secrete extracellular matrix (ECM) components and remodeling enzymes, such as the matrix metalloproteinases (MMPs) (Lamouille, Xu, & Derynck, 2014; Willis & Borok, 2007; Yang et al., 2012). Airway fibrosis and remodeling are largely irreversible processes that impair lung function and air exchange (Abiatari et al., 2009).

There are currently no therapies to reverse the process of airway remodeling, representing a major unmet need in treatment of asthma and chronic obstructive lung disease. Consequently, TGFβ signaling has been a focus of much investigation to modify fibrosis and airway remodeling.

6. THE RelA–BRD4 PATHWAY LINKS INFLAMMATION, FIBROSIS, AND AIRWAY REMODELING

EMT is transcriptional reprogramming event that dedifferentiates highly specialized epithelial cells to assume fibroblast- and stem cell-like characteristics (Kalluri & Weinberg, 2009). TGFβ signaling leads to repression of epithelial cadherin (CDH1), producing loss of apical–basal polarity and disrupting epithelial adherens junctions (Ijaz et al., 2014; Kalluri & Weinberg, 2009). In addition, mesenchymal transitioned epithelial cells express smooth muscle cell actin (α-SMA)-containing stress fibers and intermediate filament vimentin (VIM) and produce ECM remodeling through the secretion of collagen, fibronectin (FN1), and MMPs (Ijaz et al., 2014).

Next-generation sequencing studies have shown that TGFβ-induced EMT is a global transcriptional reprogramming event, altering expression of ∼3000 genes. Although the various forms of epigenetic regulation may play roles, the majority of type II EMT (that occurring in primary epithelial cells) is driven predominately by histone modifications (McDonald, Wu, Timp, Doi, & Feinberg, 2011; Tian et al., 2015). The EMT is maintained by the activity of a core set of transcription repressors/coactivators; the best known include snail family zinc finger 1 (SNAI1), twist family bHLH transcription factor 1 (TWIST1), and Zinc finger E-box homeodomain 1 (ZEB1). SNAI1 is a homeobox protein that upregulates TGFβ1 and VIM (Katsuno, Lamouille, & Derynck, 2013; Lamouille et al., 2014; Willis & Borok, 2007). TWIST is a zinc finger-containing transcription factor responsible for repressing epithelial CDH1 (Lamouille et al., 2014). Finally, an important focus has been on the role of ZEB1/TCF8/DeltaEF1, a basic helix-loop-helix transcription family member whose expression is required for inducing the EMT (Gonzalez & Medici, 2014) by binding bipartite E-box regions (Peinado, Olmeda, & Cano, 2007). Collectively, these core regulators produce coordinate activation and repression of genes important in the EMT.

ZEB1 plays a nonredundant role in epigenetic reprogramming of inflammation-linked fibrosis. Transcriptional activation or repression produced by ZEB1 occurs by recruiting cosuppressors or coactivators via several protein-binding domains, including the p300-P/CAF-binding domain (CBD), the Smad interaction domain (SID), and CtBP interaction domain. For instance, ZEB1 directly binds to the bipartite E-box regions of the CDH1 promoter (Eger et al., 2005) and represses CDH1 transcription by recruiting the CtBP corepressor (Shi et al., 2003) and/or the SWI/SNF chromatin-remodeling protein BRG1 (Sanchez-Tillo et al., 2010). This core repression complex also contains the histone deacetylases SIRT1 (Byles et al., 2012), HDAC1 and HDAC2 (Aghdassi et al., 2012; Meidhof et al., 2015; Wang et al., 2009), as well as the Enhancer of Zeste 2 (EZH2)/PRC2. EZH2 is a methyltransferase that produces repressive H3K27me3 chromatin marks and has been a focus of medical chemistry efforts discussed later (Zhang et al., 2011). It has been shown in prostate cancer cells that SIRT1 is recruited to the CDH1 proximal promoter by ZEB1 to deacetylate histone H3 and decrease the binding of RNA Pol II, ultimately suppressing CDH1 transcription. This indicates that ZEB1 acts as a transcriptional repressor and an epigenetic modulator of epithelial genes to induce EMT.

On the other hand, ZEB1 may also activate gene transcription by interacting with acetyltransferase p300-P/CAF and activated Smad proteins through its CBD and SID domains. This mechanism underlies the activation of target genes that contribute to mesenchymal differentiation, such as MMPs, important in matrix-remodeling processes associated with the EMT (Nishimura et al., 2006; Postigo, 2003; Postigo, Depp, Taylor, & Kroll, 2003). The mechanisms controlling the repressive or activating mechanisms of ZEB1 are determined in a gene-specific manner and are not well understood.

Our studies have recently demonstrated that NFκB/RelA is a master transcription factor of the EMT program in normal epithelial cells, functioning upstream of the EMT core regulators (Fig. 3). Initially, studies integrating the experimentally determined TGFβ transcriptional program with ENCODE ChIP-Seq studies identified three major transcription factor clusters associated with a stable EMT. Of these, the NFκB pathway was a prominent component of the mesenchymal signature. The central role of NFκB was further demonstrated by the inhibitory effects of IKK inhibition and RelA silencing on the EMT (Tian et al., 2015). Mechanistically, we observed that TGFβ signaling induces NFκB/RelA translocation via a paracrine mechanism that preferentially activates a subset of phospho-Ser276 RelA-dependent immediate-early genes, indicating that the central role of transcriptional elongation is also required for triggering the EMT. Our studies further demonstrated that phospho-Ser276 RelA is required for the EMT by complexing with BRD4, recruiting it to the promoters of the core EMT transcription factors SNAI1/ZEB1/Twist1, whose actions are required for initiation of the type II EMT and its fibrotic program (Tian, Zhao, et al., 2016).

To further our understanding of the relationship between innate inflammation and fibrosis, we have developed and studied a model of recurrent

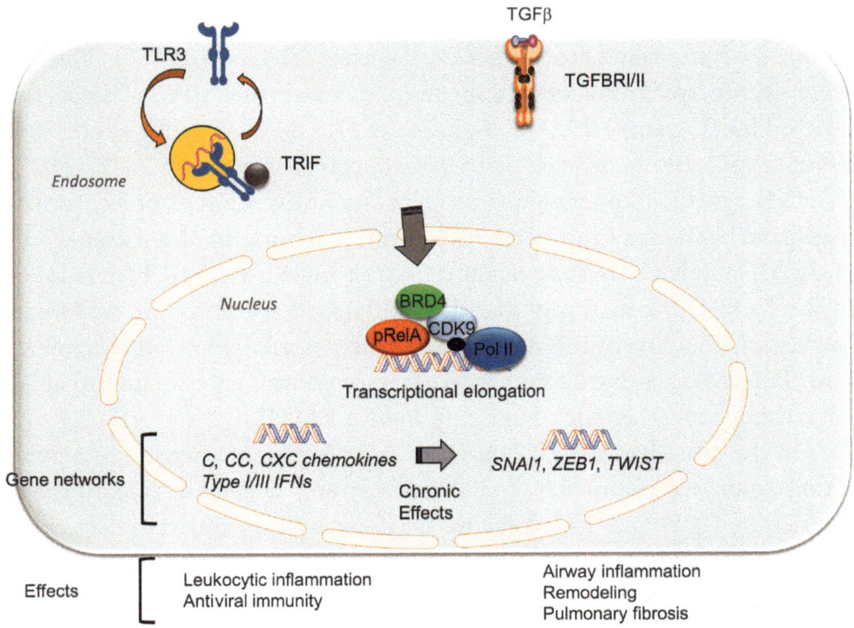

Fig. 3 Integration of the NFκB–BRD4 in inflammation and fibrosis. TGFβ or poly(I:C) stimulation activates an NFκB-dependent transcriptional elongation complex. Short-term stimulation produces the expression of the innate inflammatory response including chemokine and mucosal IFN production. Chronic or repetitive stimulation triggers the expression of the core EMT regulators SNAI1, TWIST, and ZEB through the transcriptional elongation process. In this way, the NFκB-dependent BRD4 recruitment is a major determinant of chronic airway inflammation, type II mesenchymal transition of airway epithelium, and pulmonary fibrosis.

viral infections by prolonged stimulation with the synthetic viral PAMP poly(I:C). Poly(I:C) activates TLR3 and stimulates ROS generation, producing a robust innate response. In normal human small airway epithelial cells, poly(I:C) induces the expression of the core EMT regulators SNAI1, ZEB1, as well as mesenchymal intermediate filaments (VIM), and ECM proteins FN1 and collagen 1A (Tian, Patrikeev, et al., 2016). This mesenchymal transition was prevented by silencing the expression of NFκB/RelA or treatment with a small-molecule IKK inhibitor. Acute poly(I:C) exposure in vivo induced profound neutrophilic airway inflammation and, when administered repetitively, resulted in enhanced fibrosis demonstrated by lung microCT, second harmonic imaging of optically cleared lung tissue, and immunohistochemical analyses. Epithelial flattening, expansion of the epithelial–mesenchymal trophic unit, as well as enhanced SNAI1 and FN1 expression in airway epithelium were also observed. Repetitive poly(I:C)-induced airway remodeling, fibrosis, and mesenchymal transition of the airway epithelium were inhibited by administration of BMS345541, an inhibitor of the IKK. These data suggest that persistent innate inflammation induces the fibrotic program in vitro and in vivo. Moreover, these studies support the conclusion that NFκB is a major controller of EMT and pulmonary fibrosis. These data have important mechanistic implications linking innate inflammation with the EMT, fibrosis, and remodeling in chronic inflammatory diseases.

7. EFFECTS OF THE MESENCHYMAL CELL-STATE CHANGE ON INNATE RESPONSES

7.1 Dysregulation of Innate Signaling Pathways

From the convergence of studies described earlier, persistent activation of the innate pathway is sufficient for triggering a TGFβ-induced fibrotic program and mesenchymal state, yet conversely, the presence of the mesenchymal state reprograms the innate response. In a systems-level study, we compared intracellular signaling in normal epithelial cells with that in their EMT-transformed counterparts in response to perturbations with 8 innate activating ligands and 5 inhibitors, measuring the temporal activation of 13 phosphoproteins by bead-based ELISA (Desai et al., 2015). The 2496 independent data points (192 measurements for each phosphoprotein) were normalized by mixed-effects modeling to remove random effects due to technical artifacts. In epithelial cells, the strong couplings observed between the ROS generators TNFα and TGFβ (and their known phosphosubstrates) were substantially attenuated in the mesenchymal state. Moreover, network topology maps showed significant differences in the linkages between the two cellular states, including a loss of p38 mitogen-activated protein kinase HSP27 signaling (Desai et al., 2015). These rewired connections of signaling kinases represent therapeutic targets that could potentially reverse the EMT and restore a normal phenotype to the respiratory mucosa, and have important implications for the development of therapeutics for chronic fibrotic diseases.

To extend these studies, we applied unbiased quantitative proteomics and phosphoprotein profiling of the TNF response to understand the impact of the EMT on the IIR. Differential expression of 169 kinases in all 7 major kinase superfamilies was observed. In particular, we observed that the TNFα pathway is partially activated in the mesenchymal state, yet in response to stimulation, the pathway's response is highly attenuated (Zhao, Tian, Sadygov, Zhang, & Brasier, 2016). Despite this, a subset of

NFκB-dependent inflammatory genes were highly inducible. Chromatin immunoprecipitation (ChIP) analyses showed that the EMT induced increased levels of activating H3K4Ac marks, as well as the RelA-associated transcriptional elongation complex on the promoters of hyperresponsive innate genes. (Tian, Zhao, et al., 2016).

7.2 Epigenetic Silencing of Antiviral Interferons

Adding to this complexity, we observed that the mesenchymal transition dramatically affects the expression of antiviral genes by inhibiting IFN-type III (IFNL) expression (Yang, Tian, Sun, Garofalo, & Brasier, 2017). IFNL is an epithelium-directed cytokine that plays a significant role in restricting respiratory viruses associated with asthma exacerbations and airway remodeling (Contoli et al., 2006). Investigating the mechanism for IFNL suppression, we discovered that the mesenchymal state suppressed the expression of the IRF1 transcription factor (Yang et al., 2017). Here, ZEB1 and CtB binding was observed in ChIP experiments. CtB is a ZEB1-binding adapter, which links ZEB1 to the PRC2 repressive complex. Within the PRC2, EZH2 is the catalytic subunit responsible for histone methyltransferase activity. ZEB1–CtB binding was associated with an increase in silencing H3K27me3 epigenetic marks on IRF1. Predictably, IRF1 expression was reinduced by treatment with a small-molecule EZH2 inhibitor. This study provides insights into mechanisms for epigenetic regulation of mucosal IFNs in the mesenchymal state, mediated by ZEB1·PRC2/EZH2 complex interaction with the IRF1 promoter. In mesenchymal-transitioned cells, the abundant ZEB1·PRC2/EZH2 prevents IRF1 expression via a mechanism involving the formation of repressive H3K27(me3) marks. Consequently, IFNL expression in response to viral PAMPs is blunted (Fig. 4). These studies indicate that chronic activation of the NFκB–BRD4 pathway induces mesenchymal cell-state change associated with global rewiring of the innate pathway through the changes in phosphoprotein signaling, functional hyperinduction of inflammatory genes, and suppression of antiviral cytokine expression.

8. THERAPEUTIC TARGETS IN THE RelA-TRIGGERED INNATE/FIBROTIC PATHWAY

8.1 NFκB/RelA

A robust and rapid activation of the innate response is critical for limiting pathogen spread, yet a prolonged inflammatory response is detrimental.

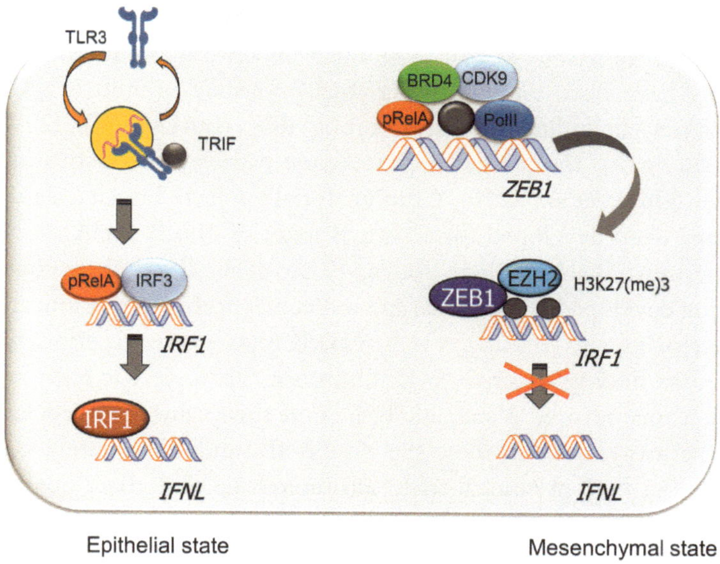

Fig. 4 Epigenetic modulation of mucosal antiviral immunity. *Left*: Schematic model of NFκB–IRF1 signaling in the epithelial state. Activation of TLR3 results in inducible IRF1 synthesis and downstream IFNL expression. *Right*: Epigenetic silencing of the IRF1 promoter in the mesenchymal transition. The EMT core regulator, ZEB1, is induced by RelA by a promoter pausing-release mechanism downstream of BRD4. Upregulated ZEB1–PRC2–EZH2 forms inactivating histone marks with H3K27(me3) mediated by methyltransferase activity of EZH2, functionally inactivating the IRF1 promoter, blocking downstream IFNL expression.

For this reason, the NFκB pathway is regulated by multiple negative regulatory loops triggered by RelA activation itself, including the NFκB–IκB autoregulatory loop that terminates RelA action in the nucleus; the NFκB–TNFAIP3/A20 negative feedback loop that terminates rate-limiting IKK activity; and the NFκB–p62 pathway that terminates inflammasome activity (Brasier, 2006; Zhong et al., 2016). Despite these levels of negative regulation, prolonged activation of RelA is associated with the pathogenesis of chronic inflammatory disease. This aspect is particularly true for severe lower respiratory tract infections with the paramyxovirus RSV (Shay, Holman, Roosevelt, Clarke, & Anderson, 2001). Inhibition of the NFκB pathway during RSV infection decreases disease manifestations, including weight loss, chemokine production, and airway inflammation in a mouse model, despite enhanced levels of viral replication (Haeberle et al., 2004).

Because of its central role in such inflammatory diseases such as asthma, atherosclerosis, diabetes, and cancer, over 700 compounds that block NFκB signaling have been evaluated in preclinical studies (Gupta, Sundaram, Reuter, & Aggarwal, 2010). A number of critical points in the NFκB pathway have been targeted using small-molecule inhibitors and natural compounds, including the rate-limiting IKK (Burke et al., 2003), IκB ubiquitin ligases, the ubiquitin–proteasome pathway, and activating phosphatases (Gupta et al., 2010). Combinatorially selected nucleic acid inhibitors have been developed for activated states of NFκB/RelA that bind its NH_2-terminal Rel homology domain (Zhao et al., 2011); these compounds are being developed as biological and molecular probes in preclinical studies. Because of its rate-limiting role in NFκB release, a large effort has been devoted to finding selective IKK inhibitors; this topic has been reviewed (Karin, Yamamoto, & Wang, 2004). Despite their value as molecular probes of the pathway, none of these specific NFκB inhibition strategies are currently being used in clinical trials. An interesting side effect observed was that IKK inhibitors paradoxically induce neutrophilia in vivo due to NFκB's antiinflammatory role of stimulating mitochondrial autophagy (Zhong et al., 2016). A tabulation of the inhibitors of the NFκB/RelA currently evaluated in clinical trials is shown in Table 1. For NFκB inhibitors to be clinically applicable, more selective inhibitors and strategies are needed.

One strategy might be to modulate NFκB function by inhibiting specific gene networks without globally interfering with its function. We currently understand that NFκB/RelA activation involves the formation of a transcriptional regulatory complex with chromatin-modifying cofactors (Bouwmeester et al., 2004; Brasier, 2008a, 2008b; Li et al., 2014). An important conceptual advance in understanding the complexity of the NFκB/RelA transcription factor function was generated by our combined analysis of the RelA interactome using unbiased protein profiling coupled with analyses of RelA activity inferred from gene expression microarray datasets (Li et al., 2014; Li, Zhu, Brasier, & Kudlicki, 2015). In this approach, a probabilistic model was developed for a three-way interaction of NFκB/RelA, its modulators (binding proteins), and its target genes (Yang et al., 2013). Over 562 RelA activity-modulating proteins were identified in 8349 triplets (Li et al., 2014). The important findings were the identification of 20 key modulators with the greatest number of target genes that fell into distinct biological functions. This validated approach was used to generate a modulatory network where groups of modulators control genes with similar genome ontology functions, including regulation of cell death, regulation of transcription, regulation of apoptosis, response to chemical stimulus, regulation of the TGFβ receptor signaling pathway, and others. This study

Table 1 Selected Inhibitors of NF-κB, CDK9, BRD4, and EZH2 in Clinical Trials

Target	Drug	Company	Status	Trial ID	Indications
NF-κB	Curcumin/Meriva	Thorne Research	Phase 2/3	NCT01740323; NCT02255370	Breast cancer; Crohn's disease
	CAT-1004	Catabasis	Phase 1/2	NCT02439216	Muscular dystrophy, Duchenne
	Dimethylfumarate	Universitätsmedizin Mannheim	Phase 2	NCT02546440	Cutaneous T-cell lymphoma
	Saxagliptin	Bristol-Myers Squibb	Phase 4	NCT02285985	Type 2 diabetes mellitus
	Probiotics		Phase 3	NCT01436448	Glucose intolerance
	RTA 408	Reata	Phase 1/2	NCT02259231	Melanoma
	Ixazomib	Millennium	Phase 1/2	NCT02447887; NCT02765854	Metastatic renal cell carcinoma; Recurrent plasma cell myeloma
CDKs/ CDK9	Roscovitine	Cyclacel	Phase 2	NCT02649751	Cystic fibrosis
	BAY1251152	Bayer	Phase 1	NCT02745743; NCT02635672	Hematologic neoplasms
	AT7519	Astex	Phase 1	NCT02503709	Adult solid neoplasm
	Flavopiridol	Sanofi-Aventis	Phase 2	NCT02520011	Acute myeloid leukemia
	SCH 727965	Schering-Plough	Phase 1	NCT01434316	Advanced malignant neoplasm
BRD4	GSK525762	GlaxoSmithKline	Phase 1	NCT01587703; NCT01943851; NCT02706535	Midline carcinoma; refractory hematologic malignancies
	CPI-0610	Constellation Pharmaceuticals	Phase 1	NCT01949883; NCT02157636; NCT02158858	Lymphoma; Multiple Myeloma; Myelodysplastic Syndrome (MDS); Myelofibrosis

Continued

Table 1 Selected Inhibitors of NF-κB, CDK9, BRD4, and EZH2 in Clinical Trials—cont'd

Target	Drug	Company	Status	Trial ID	Indications
	OTX015/MK-8628	OncoEthix/Merck	Phase 1	NCT02698176; NCT02698189	Acute lymphoblastic leukemia; multiple myeloma
	RVX-208	Resverlogix	Phase 3	NCT02586155	Diabetes mellitus; coronary artery disease; cardiovascular diseases
	TEN-010	Tensha	Phase 1	NCT02308761; NCT01987362	Myelodysplastic syndromes, acute myeloid leukemia; solid tumors
EZH2/PRC2	CPI-1205	Constellation	Phase 1	NCT02395601	B-cell lymphoma
	EPZ-6438	Epizyme	Phase 1/2	NCT01897571; NCT02601937; NCT02601950; NCT02875548; NCT02860286	Lymphomas; advanced solid tumors; rhabdoid tumors; synovial sarcoma; mesothelioma
	GSK126	GlaxoSmithKline	Phase 1	NCT02082977	Lymphomas; multiple myeloma

provides a blueprint for selectively affecting RelA-dependent biological functions through the disruption of modulator–RelA interactions. This provides an alternative approach that controls specific activities of RelA, avoiding the systemic toxicity of generalized RelA inhibition.

8.2 Cyclin-Dependent Kinase 9

CDK9 is one of the major Pol II-directed kinases of P-TEFb. A number of small-molecule inhibitors of CDK9 kinase activity directed to the ATP-binding pocket have been developed and tested (Krystof, Chamrad, Jorda, & Kohoutek, 2010). CDK9 is a member of a conserved cyclin kinase family first described in cell cycle control, where initial efforts were directed to develop anticancer compounds (Table 1). With a greater understanding of its biological roles in mediating inflammatory, fibrotic, and hypertrophic diseases, potential indications for CDK9 inhibitors expanded. The best-studied inhibitors include DRB, flavopiridol, and roscovitine. DRB and flavopiridol have global effects on mRNA expression and have inhibitory activity toward a number of CDK isoforms, particularly CDK-7 and -9, kinases that phosphorylate RNA Pol II. By contrast, roscovitine does not produce a global effect on mRNA expression. Derivatives of roscovitine have been produced with even greater CDK9 specificity and have entered clinical trials as anticancer compounds (Malumbres, Pevarello, Barbacid, & Bischoff, 2008). CAN508 is the most potent CDK9 inhibitor derived from 4-arylazo-3,5-diamino-1H-pyrazoles (Krystof et al., 2006); CAN508 and EXCEL-8647 are the only inhibitors that are CDK9 specific (Krystof et al., 2010).

Use of CDK9 inhibitors has been proposed for cancer and cardiac hypertrophy, and the state of the art has previously been reviewed (Krystof et al., 2010; Sonawane et al., 2016). Recent studies from our group have extended these indications for CDK9 inhibitors to the exuberant inflammatory response to virus-induced lung inflammation (Tian et al., 2013). CDK9 inhibitors are potent blockers of PRR-induced innate inflammation in lung airway cells (Tian et al., 2013). The CDK9 inhibitors currently undergoing clinical trial development are shown in Table 1.

Because of the high degree of conservation of the ATP-binding pocket of the CDK family, other strategies to disrupt CDK function may provide greater selectivity. In an approach similar to that used in our analysis of the NFκB/RelA interactome, we have applied unbiased proteomic studies to understand how innate signaling induces restructuring of the CDK9

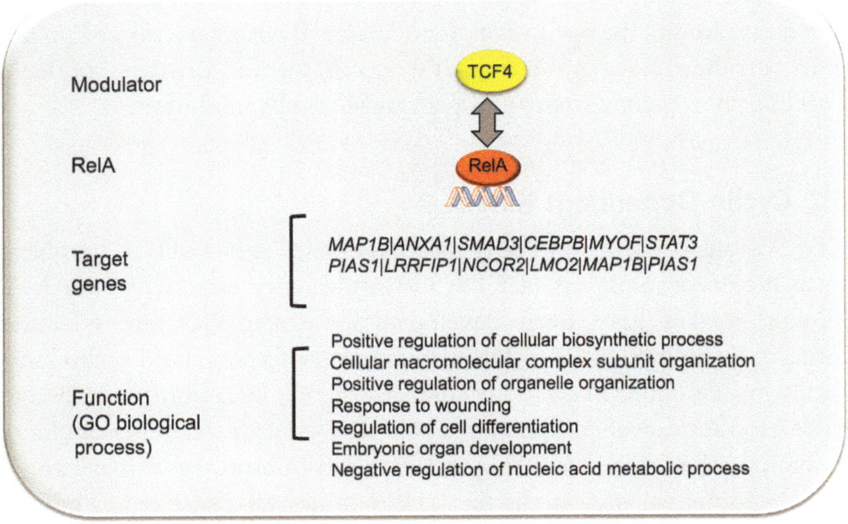

Fig. 5 RelA modulatory network. Understanding the RelA modulatory network provides actionable strategies for selectively modulating NFκB-dependent biological processes. NFκB modulators regulate RelA transcriptional activity on specific target genes; these target genes are grouped into distinct biological processes. Illustrated is TCF4, a top-level RelA modulator with its major target genes and biological processes predicted from our computational studies (Li et al., 2014).

complex (Fig. 5). Our findings showed that innate activation triggers reorganization of the CDK9 interactome to acquire new biological functions, including RNA/DNA helicase activity, ribosomal binding, translational elongation components, and unfolded protein response proteins (Yang et al., 2015). The innate activated complex consequently contains proteins involved in multiple steps in RNA processing—transcriptional elongation, mRNA splicing and transport, and initiation of translation, leading to the concept of CDK9's role in "cotranscriptionality" in innate signaling (Perales & Bentley, 2009; Yang et al., 2015). The global CDK9 modulatory network is predicted to be regulated by 257 modulators affecting 344 target genes. Under basal conditions the CDK9 complex is enriched in DEAD-box RNA helicase proteins; this diverse family of RNA-dependent ATPases is involved in gene expression as transcription factor coactivators, pre-mRNA processing via splicing, and termination through transcript release from chromatin (Fuller-Pace, 2013). By contrast, the innate pathway-activated CDK9 is associated with ribosomal proteins associated with EEF1A2 and the large 60S ribosomal subunit. Taken together, we speculate that the CDK9 complex may be functionally involved in export, ribosomal engagement, and translation of innate response genes. These data suggest that modulators of gene expression CDK9-binding proteins may be classified into two clusters that control two distinct biological pathways—one dominant in the basal state and the second induced by the innate response. Therapeutic targets that disrupt the CDK9 interactome may modify its activity in the same manner as described for RelA modulation, earlier.

8.3 Bromodomain-Containing Protein 4

BRD4 is a member of a bromodomain and extraterminal domain (BET) family of proteins that includes the ubiquitously expressed BRD2, BRD3, BRD4, and BRDT proteins. Among these, BRD4 is unique in that it interacts with CDK9 in the active P-TEFb complex through its C-terminal tail (Bisgrove, Mahmoudi, Henklein, & Verdin, 2007). As a component of the activated P-TEFb complex along with CDK9/RNA Pol II CTD kinase and histone acetyltransferase, BDR4 has also been a major focus of small-molecule inhibitor development (Yu et al., 2015). A variety of small molecules directed toward the acetyl Lys-binding BD have been developed for the treatment of cancers, inflammatory disease, and fibrosis, beginning with the prototypical JQ1 inhibitor (Filippakopoulos et al., 2010) (Table 1). BRD4 inhibition has shown promising results in cancer, by targeting c-MYC expression in MYC-driven oncogenesis (Delmore et al., 2011; Mertz et al., 2011).

As the biological roles of BRD4 in organ fibrosis (Ding et al., 2015) have been recognized, potential therapeutic applications for BRD4 inhibitors have been expanded. One example is the application of BRD4 inhibitors to the treatment of fibrosis associated with hypertrophic scar (HTS) formation. HTS produces significant morbidity through pruritus, compression, anatomic deformity, and decreased joint mobility, for which few effective treatments are available (Finnerty et al., 2016). In this pathology, transdifferentiation of quiescent dermal fibroblasts to secretory myofibroblasts plays a central role in wound healing and pathological scar formation. In dermal fibroblasts, TGFβ-induced myofibroblast formation is blocked by BRD4 inhibition, reversing hypertrophic skin fibroblast transdifferentiation. These data suggest that BRD4 inhibition may be a useful strategy to limit HTS formation after burn injury.

BRD4 inhibitors are likely to have important applications in inflammatory diseases. In vascular inflammation associated with atherosclerosis and aortic remodeling, inhibition of BRD4 produces global genomic effects by disruption of super-enhancers, thereby blocking transcription (Brown et al., 2014; Loven et al., 2013). The status of BRD4 inhibitors used in clinical trials is shown in Table 1. As discussed earlier, inhibition of the NFκB–BRD4 pathway shows promise to decrease disease manifestations of viral respiratory tract infections. Modulation of transcriptional elongation will almost certainly have a role in diseases beyond inflammation, fibrosis, and cancer, and these possibilities also need investigation.

New opportunities for affecting BRD4-mediated reprogramming include targeting the intrinsic kinase activity directed toward the RNA Pol II CTD, which is essential for transcriptional elongation (Devaiah et al., 2012; Jang et al., 2005). BRD4's kinase activity also regulates CDK9 activity (Zhou et al., 2009). Additionally, exciting studies have recently shown that BRD4 is a previously unrecognized atypical histone acetyltransferase whose activity is directed toward Lys residues on the periphery of the nucleosome, enabling RNA Pol II to more easily displace nucleosomes during the process of transcription (Devaiah et al., 2016). Therapeutics developed to target these activities may have very interesting actions.

8.4 EZH2 of the PRC2 Repressor Complex

The PRC2 complex is a dynamic complex that consists of four core subunits—the catalytic EZH2, embryonic ectoderm development (EED), suppressor of zeste 12 (SUZ12), and retinoblastoma (Rb)-associated proteins 46/48 (Kim & Roberts, 2016). EZH2 catalyzes trimethylation of histone H3 lysine 27 (to H3K27me3), a transcriptionally repressive epigenetic mark. Because of associations between the gain of activation and amplification of EZH2 in lymphomas, EZH2 has been the focus of small-molecule inhibitor development efforts (Table 1). For example, GSK126 is an S-adenosylmethionine-competitive small-molecule EZH2 inhibitor that decreases global H3K27me3 levels and reactivates silenced PRC2 target genes (McCabe et al., 2012). Similarly, EPZ005687 has been characterized as a potent EZH2 inhibitor with a K_i in the 20 nM range, with 50-fold selectivity against EZH1 (Knutson et al., 2012). Although EZH2 inhibitors induce apoptosis in lymphoma, this class of compounds induces a long-term memory response in normal lymphocytes, even at low doses, without an immediate cytotoxic effect (Baylin & Jones, 2011; Tsai et al., 2012). In mesenchymal-transitioned epithelial cells, we have found that EZH2 inhibitors reverse the inhibition of IRF1 and partially restore IFNL secretion (Yang et al., 2017). Since ZEB1 represses the transcription of epithelial genes via epigenetic mechanisms, the application of selected epigenetic drugs targeting HDAC or EZH2/PRC2 for the rational treatment of aberrant activation of an EMT or an associated disorder might be promising.

9. DISCUSSION AND OUTLOOK

Significant progress has been made in the understanding of the mechanism underlying the IIR and the role of transcriptional elongation in the production of latent-inducible genes that mediate protective mucosal immunity. Recent exciting work has provided a unifying mechanism linking persistent PRR activation with cell-state changes and fibrosis in the airway that will drive the further development of much-needed therapeutics that modify airway remodeling and dysregulation of the mucosal IFN response. Current therapeutic development has primarily focused on the regulatory kinases of NFκB (IKKs), the ATP-binding domain of CDK9, the acetylated lysine-binding domain of BRD4, and the histone methyltransferase activity of EZH2.

We contend that additional exploration into selective modulation of RelA- and CDK9-inducible protein–protein interactions with modulatory proteins that control specific biological functions, histone acetyltransferase, and kinase activity of BRD4 and modification of other enzymes within the PRC2 complex may produce more selective and efficacious therapeutics for airway remodeling, organ fibrosis, and cancer.

ACKNOWLEDGMENTS

Core laboratory support was provided by the UTMB Transgenic mouse facility, Next-Generation Sequencing Core, Sealy Center for Molecular Medicine Selected Reaction Monitoring Core, and Optical Imaging Core. This work was supported, in part, by NIH Grants NIAID AI062885 (A.R.B.), UL1TR001439 (A.R.B.), and NIEHS ES006676 (A.R.B., B.T., and J.Y.), and NSF Grant DMS-1361411/DMS-1361318 (A.R.B.). We thank Dr. David Konkel for critically editing the manuscript.

REFERENCES

Abiatari, I., DeOliveira, T., Kerkadze, V., Schwager, C., Esposito, I., Giese, N. A., … Kleeff, J. (2009). Consensus transcriptome signature of perineural invasion in pancreatic carcinoma. *Molecular Cancer Therapeutics*, *8*(6), 1494–1504.

Adelman, K., & Lis, J. T. (2012). Promoter-proximal pausing of RNA polymerase II: Emerging roles in metazoans. *Nature Reviews. Genetics*, *13*(10), 720–731. http://dx.doi.org/10.1038/nrg3293.

Aghdassi, A., Sendler, M., Guenther, A., Mayerle, J., Behn, C. O., Heidecke, C. D., … Weiss, F. U. (2012). Recruitment of histone deacetylases HDAC1 and HDAC2 by the transcriptional repressor ZEB1 downregulates E-cadherin expression in pancreatic cancer. *Gut*, *61*(3), 439–448. http://dx.doi.org/10.1136/gutjnl-2011-300060.

Akira, S., Uematsu, S., & Takeuchi, O. (2006). Pathogen recognition and innate immunity. *Cell*, *124*(4), 783–801.

Al-Muhsen, S., Johnson, J. R., & Hamid, Q. (2011). Remodeling in asthma. *The Journal of Allergy and Clinical Immunology*, *128*(3), 451–462. http://dx.doi.org/10.1016/j.jaci.2011.04.047. quiz 463–454.

Bals, R., & Hiemstra, P. S. (2004). Innate immunity in the lung: How epithelial cells fight against respiratory pathogens. *European Respiratory Journal*, *23*(2), 327–333.

Bao, X., Liu, T., Shan, Y., Li, K., Garofalo, R. P., & Casola, A. (2008). Human metapneumovirus glycoprotein G inhibits innate immune responses. *PLoS Pathogens*, *4*(5), e1000077.

Bao, X., Liu, T., Spetch, L., Kolli, D., Garofalo, R. P., & Casola, A. (2007). Airway epithelial cell response to human metapneumovirus infection. *Virology*, *368*(1), 91–101.

Barboric, M., Nissen, R. M., Kanazawa, S., Jabrane-Ferrat, N., & Peterlin, B. M. (2001). NF-kB binds P-TEFb to stimulate transcriptional elongation by RNA polymerase II. *Molecular Cell*, *3*, 327–337.

Baylin, S. B., & Jones, P. A. (2011). A decade of exploring the cancer epigenome—Biological and translational implications. *Nature Reviews. Cancer*, *11*(10), 726–734. http://dx.doi.org/10.1038/nrc3130.

Bertolusso, R., Tian, B., Zhao, Y., Vergara, L. A., Sabree, A., Iwanaszko, M., … Kimmel, M. (2014). Dynamic cross talk model of the epithelial innate immune response to double-stranded RNA stimulation: Coordinated dynamics emerging from cell-level noise. *PLoS One*, *9*(4), e93396. http://dx.doi.org/10.1371/journal.pone.0093396.

Bianchi, M. E. (2007). DAMPs, PAMPs and alarmins: All we need to know about danger. *Journal of Leukocyte Biology*, *81*(1), 1–5. http://dx.doi.org/10.1189/jlb.0306164.

Bisgrove, D. A., Mahmoudi, T., Henklein, P., & Verdin, E. (2007). Conserved P-TEFb-interacting domain of BRD4 inhibits HIV transcription. *Proceedings of the National Academy of Sciences of the United States of America*, *104*(34), 13690–13695. http://dx.doi.org/10.1073/pnas.0705053104.

Bousquet, J., Jeffery, P. K., & Busse, W. W. (2000). Asthma: From bronchoconstriction to airways inflammation and remodeling. *American Journal of Respiratory and Critical Care Medicine*, *161*, 1720–1745.

Bouwmeester, T., Bauch, A., Ruffner, H., Angrand, P. O., Bergamini, G., Croughton, K., … Superti-Furga, G. (2004). A physical and functional map of the human TNF-[alpha]/NF-[kappa]B signal transduction pathway. *Nature Cell Biology*, *6*(2), 97–105.

Brasier, A. R. (2006). The NF-kappaB regulatory network. *Cardiovascular Toxicology*, *6*(2), 111–130.

Brasier, A. R. (2008a). The NF-kB signaling network: Insights from systems approaches. In A. R. Brasier, S. M. Lemon, & A. Garcia-Sastre (Eds.), *Cellular signaling and innate immune responses to RNA virus infections* (pp. 119–135). Washington, DC: ASM Press.

Brasier, A. R. (2008b). Perspective: Expanding role of cyclin dependent kinases in cytokine inducible gene expression. *Cell Cycle*, 7(17), 2661–2666. http://dx.doi.org/10.4161/cc.7.17.6594.

Brasier, A. R., Garcia-Sastre, A., & Lemon, S. M. (2008). *Cellular signaling and innate immune response to RNA virus infections*. Washington, DC: ASM Press.

Brasier, A. R., Tian, B., Jamaluddin, M., Kalita, M. K., Garofalo, R. P., & Lu, M. (2011). RelA Ser276 phosphorylation-coupled Lys310 acetylation controls transcriptional elongation of inflammatory cytokines in respiratory syncytial virus infection. *Journal of Virology*, *85*(22), 11752–11769. http://dx.doi.org/10.1128/JVI.05360-11.

Brown, J. D., Lin, C. Y., Duan, Q., Griffin, G., Federation, A. J., Paranal, R. M., … Plutzky, J. (2014). NF-kappaB directs dynamic super enhancer formation in inflammation and atherogenesis. *Molecular Cell*, *56*(2), 219–231. http://dx.doi.org/10.1016/j.molcel.2014.08.024.

Burke, J. R., Pattoli, M. A., Gregor, K. R., Brassil, P. J., MacMaster, J. F., McIntyre, K. W., … Zusi, F. C. (2003). BMS-345541 is a highly selective inhibitor of IκB kinase that binds at an allosteric site of the enzyme and blocks NF-κB-dependent transcription in mice. *Journal of Biological Chemistry*, *278*(3), 1450–1456. http://dx.doi.org/10.1074/jbc.M209677200.

Byles, V., Zhu, L., Lovaas, J. D., Chmilewski, L. K., Wang, J., Faller, D. V., & Dai, Y. (2012). SIRT1 induces EMT by cooperating with EMT transcription factors and enhances prostate cancer cell migration and metastasis. *Oncogene*, *31*(43), 4619–4629. http://dx.doi.org/10.1038/onc.2011.612.

Chakir, J., Shannon, J., Molet, S., Fukakusa, M., Elias, J., Laviolette, M., … Hamid, Q. (2003). Airway remodeling-associated mediators in moderate to severe asthma: Effect of steroids on TGF-beta, IL-11, IL-17, and type I and type III collagen expression. *The Journal of Allergy and Clinical Immunology*, *111*(6), 1293–1298.

Chen, L. f., Williams, S. A., Mu, Y., Nakano, H., Duerr, J. M., Buckbinder, L., & Greene, W. C. (2005). NF-{kappa}B RelA phosphorylation regulates RelA acetylation. *Molecular and Cellular Biology*, *25*(18), 7966–7975.

Chen, R., Yang, Z., & Zhou, Q. (2004). Phosphorylated positive transcription elongation factor b (P-TEFb) is tagged for inhibition through association with 7SK snRNA. *The Journal of Biological Chemistry*, *279*(6), 4153–4160. http://dx.doi.org/10.1074/jbc.M310044200.

Chiba, K., Yamamoto, J., Yamaguchi, Y., & Handa, H. (2010). Promoter-proximal pausing and its release: Molecular mechanisms and physiological functions. *Experimental Cell Research*, *316*(17), 2723–2730. http://dx.doi.org/10.1016/j.yexcr.2010.05.036.

Choudhary, S., Boldogh, I., & Brasier, A. R. (2016). Inside-out signaling pathways from nuclear reactive oxygen species control pulmonary innate immunity. *Journal of Innate Immunity*, *8*(2), 143–155. http://dx.doi.org/10.1159/000442254.

Choudhary, S., Rosenblatt, K. P., Fang, L., Tian, B., Wu, Z. H., & Brasier, A. R. (2011). High throughput short interfering RNA (siRNA) screening of the human kinome identifies novel kinases controlling the canonical nuclear factor-kappaB (NF-kappaB) activation pathway. *The Journal of Biological Chemistry*, *286*(43), 37187–37195. http://dx.doi.org/10.1074/jbc.M111.224923.

Contoli, M., Message, S. D., Laza-Stanca, V., Edwards, M. R., Wark, P. A., Bartlett, N. W., … Johnston, S. L. (2006). Role of deficient type III interferon-lambda production in asthma exacerbations. *Nature Medicine*, *12*(9), 1023–1026. http://dx.doi.org/10.1038/nm1462.

Davies, D. E. (2009). The role of the epithelium in airway remodeling in asthma. *Proceedings of the American Thoracic Society*, *6*(8), 678–682. http://dx.doi.org/10.1513/pats.200907-067DP.

Delmore, J. E., Issa, G. C., Lemieux, M. E., Rahl, P. B., Shi, J., Jacobs, H. M., … Mitsiades, C. S. (2011). BET bromodomain inhibition as a therapeutic strategy to target c-Myc. *Cell*, *146*(6), 904–917. http://dx.doi.org/10.1016/j.cell.2011.08.017.

Desai, P., Yang, J., Tian, B., Sun, H., Kalita, M., Ju, H., … Sadygov, R. G. (2015). Mixed-effects model of epithelial-mesenchymal transition reveals rewiring of signaling networks. *Cellular Signalling*, *27*(7), 1413–1425. http://dx.doi.org/10.1016/j.cellsig.2015.03.024.

Devaiah, B. N., Case-Borden, C., Gegonne, A., Hsu, C. H., Chen, Q., Meerzaman, D., … Singer, D. S. (2016). BRD4 is a histone acetyltransferase that evicts nucleosomes from chromatin. *Nature Structural & Molecular Biology*, *23*(6), 540–548. http://dx.doi.org/10.1038/nsmb.3228.

Devaiah, B. N., Lewis, B. A., Cherman, N., Hewitt, M. C., Albrecht, B. K., Robey, P. G., … Singer, D. S. (2012). BRD4 is an atypical kinase that phosphorylates serine2 of the RNA polymerase II carboxy-terminal domain. *Proceedings of the National Academy of Sciences of the United States of America*, *109*(18), 6927–6932. http://dx.doi.org/10.1073/pnas.1120422109.

Diamant, G., & Dikstein, R. (2013). Transcriptional control by NF-kappaB: Elongation in focus. *Biochimica et Biophysica Acta*, *1829*(9), 937–945. http://dx.doi.org/10.1016/j.bbagrm.2013.04.007.

Ding, N., Hah, N., Yu, R. T., Sherman, M. H., Benner, C., Leblanc, M., … Evans, R. M. (2015). BRD4 is a novel therapeutic target for liver fibrosis. *Proceedings of the National Academy of Sciences of the United States of America*, *112*(51), 15713–15718. http://dx.doi.org/10.1073/pnas.1522163112.

Eger, A., Aigner, K., Sonderegger, S., Dampier, B., Oehler, S., Schreiber, M., … Foisner, R. (2005). DeltaEF1 is a transcriptional repressor of E-cadherin and regulates epithelial plasticity in breast cancer cells. *Oncogene*, *24*(14), 2375–2385. http://dx.doi.org/10.1038/sj.onc.1208429.

Fang, L., Choudhary, S., Tian, B., Boldogh, I., Yang, C., Ivanciuc, T., … Brasier, A. R. (2015). Ataxia telangiectasia mutated kinase mediates NF-kappaB serine 276 phosphorylation and interferon expression via the IRF7-RIG-I amplification loop in paramyxovirus infection. *Journal of Virology*, *89*(5), 2628–2642. http://dx.doi.org/10.1128/JVI.02458-14.

Fang, L., Choudhary, S., Zhao, Y., Edeh, C. B., Yang, C., Boldogh, I., & Brasier, A. R. (2014). ATM regulates NF-kappaB-dependent immediate-early genes via RelA Ser 276 phosphorylation coupled to CDK9 promoter recruitment. *Nucleic Acids Research*, *42*(13), 8416–8432. http://dx.doi.org/10.1093/nar/gku529.

Filippakopoulos, P., Qi, J., Picaud, S., Shen, Y., Smith, W. B., Fedorov, O., … Bradner, J. E. (2010). Selective inhibition of BET bromodomains. *Nature*, *468*(7327), 1067–1073. http://dx.doi.org/10.1038/nature09504.

Finnerty, C. C., Jeschke, M. G., Branski, L. K., Barret, J. P., Dziewsulski, P., & Herndon, D. N. (2016). Hypertrophic scarring: The greatest unmet challenge following burn injury. *The Lancet*, *388*(10052), 1427–1436.

Flajollet, S., Rachez, C., Ploton, M., Schulz, C., Gallais, R., Metivier, R., … Lefebvre, P. (2013). The elongation complex components BRD4 and MLLT3/AF9 are transcriptional coactivators of nuclear retinoid receptors. *PLoS One*, *8*(6), e64880. http://dx.doi.org/10.1371/journal.pone.0064880.

Fu, J., Yoon, H. G., Qin, J., & Wong, J. (2007). Regulation of P-TEFb elongation complex activity by CDK9 acetylation. *Molecular and Cellular Biology*, *27*(13), 4641–4651.

Fuller-Pace, F. V. (2013). The DEAD box proteins DDX5 (p68) and DDX17 (p72): Multi-tasking transcriptional regulators. *Biochimica et Biophysica Acta*, *1829*(8), 756–763. http://dx.doi.org/10.1016/j.bbagrm.2013.03.004.

Garber, M. E., Mayall, T. P., Suess, E. M., Meisenhelder, J., Thompson, N. E., & Jones, K. A. (2000). CDK9 autophosphorylation regulates high-affinity binding of the human immunodeficiency virus type 1 tat-P-TEFb complex to TAR RNA. *Molecular and Cellular Biology*, *20*(18), 6958–6969.

Gardini, A., Baillat, D., Cesaroni, M., Hu, D., Marinis, J. M., Wagner, E. J., … Shiekhattar, R. (2014). Integrator regulates transcriptional initiation and pause release following activation. *Molecular Cell*, *56*(1), 128–139. http://dx.doi.org/10.1016/j.molcel.2014.08.004.

Garofalo, R., Mei, F., Manganaro, M., Espejo, R., Ogra, P. L., & Reyes, V. E. (1994). Upregulation of class I major histocompatibility complex (MHC) molecules on respiratory syncytial virus (RSV)-infected airway epithelial cells. *American Journal of Respiratory and Critical Care Medicine, 149*, A987.

Gilchrist, D. A., Fromm, G., dos Santos, G., Pham, L. N., McDaniel, I. E., Burkholder, A., … Adelman, K. (2012). Regulating the regulators: The pervasive effects of Pol II pausing on stimulus-responsive gene networks. *Genes & Development, 26*(9), 933–944. http://dx.doi.org/10.1101/gad.187781.112.

Gonzalez, D. M., & Medici, D. (2014). Signaling mechanisms of the epithelial-mesenchymal transition. *Science Signaling*, 7(344), re8. http://dx.doi.org/10.1126/scisignal.2005189.

Gupta, S. C., Sundaram, C., Reuter, S., & Aggarwal, B. B. (2010). Inhibiting NF-κB activation by small molecules as a therapeutic strategy. *Biochimica et Biophysica Acta, 1799*(10–12), 775–787. http://dx.doi.org/10.1016/j.bbagrm.2010.05.004.

Hackett, T. L., Warner, S. M., Stefanowicz, D., Shaheen, F., Pechkovsky, D. V., Murray, L. A., … Knight, D. A. (2009). Induction of epithelial-mesenchymal transition in primary airway epithelial cells from patients with asthma by transforming growth factor-beta1. *American Journal of Respiratory and Critical Care Medicine, 180*(2), 122–133. http://dx.doi.org/10.1164/rccm.200811-1730OC.

Haeberle, H. A., Casola, A., Gatalica, Z., Petronella, S., Dieterich, H. J., Ernst, P. B., … Garofalo, R. P. (2004). IkappaB kinase is a critical regulator of chemokine expression and lung inflammation in respiratory syncytial virus infection. *Journal of Virology, 78*(5), 2232–2241.

Hammad, H., & Lambrecht, B. N. (2008). Dendritic cells and epithelial cells: Linking innate and adaptive immunity in asthma. *Nature Reviews. Immunology, 8*(3), 193–204. http://dx.doi.org/10.1038/nri2275.

Hargreaves, D. C., Horng, T., & Medzhitov, R. (2009). Control of inducible gene expression by signal-dependent transcriptional elongation. *Cell, 138*(1), 129–145. http://dx.doi.org/10.1016/j.cell.2009.05.047.

Harris, W. T., Kelly, D. R., Zhou, Y., Wang, D., Macewen, M., Hagood, J. S., … Sorscher, E. J. (2013). Myofibroblast differentiation and enhanced TGF-B signaling in cystic fibrosis lung disease. *PLoS One, 8*(8), e70196. http://dx.doi.org/10.1371/journal.pone.0070196.

He, N., Chan, C. K., Sobhian, B., Chou, S., Xue, Y., Liu, M., … Zhou, Q. (2011). Human polymerase-associated factor complex (PAFc) connects the super elongation complex (SEC) to RNA polymerase II on chromatin. *Proceedings of the National Academy of Sciences of the United States of America, 108*(36), E636–E645. http://dx.doi.org/10.1073/pnas.1107107108.

Hogg, J. C., Chu, F., Utokaparch, S., Woods, R., Elliott, W. M., Buzatu, L., … Pare, P. D. (2004). The nature of small-airway obstruction in chronic obstructive pulmonary disease. *The New England Journal of Medicine, 350*(26), 2645–2653.

Holgate, S. T., Holloway, J., Wilson, S., Bucchieri, F., Puddicombe, S., & Davies, D. E. (2004). Epithelial-mesenchymal communication in the pathogenesis of chronic asthma. *Proceedings of the American Thoracic Society, 1*(2), 93–98. http://dx.doi.org/10.1513/pats.2306034.

Holt, P. G., & Strickland, D. H. (2010). Interactions between innate and adaptive immunity in asthma pathogenesis: New perspectives from studies on acute exacerbations. *The Journal of Allergy and Clinical Immunology, 125*(5), 963–972.

Holt, P. G., Strickland, D. H., Wikstrom, M. E., & Jahnsen, F. L. (2008). Regulation of immunological homeostasis in the respiratory tract. *Nature Reviews. Immunology, 8*(2), 142–152. http://dx.doi.org/10.1038/nri2236.

Hopkins, P. A., & Sriskandan, S. (2005). Mammalian Toll-like receptors: To immunity and beyond. *Clinical & Experimental Immunology, 140*(3), 395–407.

Hosakote, Y. M., Liu, T., Castro, S. M., Garofalo, R. P., & Casola, A. (2009). Respiratory syncytial virus induces oxidative stress by modulating antioxidant enzymes. *American Journal of Respiratory Cell and Molecular Biology*, *41*(3), 348–357.

Hou, T., Ray, S., & Brasier, A. R. (2007). The functional role of an interleukin 6-inducible CDK9.STAT3 complex in human gamma-fibrinogen gene expression. *The Journal of Biological Chemistry*, *282*(51), 37091–37102. http://dx.doi.org/10.1074/jbc.M706458200.

Hou, F., Sun, L., Zheng, H., Skaug, B., Jiang, Q.-X., & Chen, Z. J. (2011). MAVS forms functional prion-like aggregates to activate and propagate antiviral innate immune response. *Cell*, *146*(3), 448–461. http://dx.doi.org/10.1016/j.cell.2011.06.041.

Huang, B., Yang, X. D., Zhou, M. M., Ozato, K., & Chen, L. f. (2009). Brd4 coactivates transcriptional activation of NF-{kappa}B via specific binding to acetylated RelA. *Molecular and Cellular Biology*, *29*(5), 1375–1387.

Ijaz, T., Pazdrak, K., Kalita, M., Konig, R., Choudhary, S., Tian, B., … Brasier, A. R. (2014). Systems biology approaches to understanding epithelial mesenchymal transition (EMT) in mucosal remodeling and signaling in asthma. *World Allergy Organization Journal*, 7(1), 13.

Itzen, F., Greifenberg, A. K., Bosken, C. A., & Geyer, M. (2014). Brd4 activates P-TEFb for RNA polymerase II CTD phosphorylation. *Nucleic Acids Research*, *42*(12), 7577–7590. http://dx.doi.org/10.1093/nar/gku449.

Jamaluddin, M., Tian, B., Boldogh, I., Garofalo, R. P., & Brasier, A. R. (2009). Respiratory syncytial virus infection induces a reactive oxygen species-MSK1-phospho-Ser-276 RelA pathway required for cytokine expression. *Journal of Virology*, *83*(20), 10605–10615. http://dx.doi.org/10.1128/JVI.01090-09.

Jamaluddin, M., Wang, S., Boldogh, I., Tian, B., & Brasier, A. R. (2007). TNF-alpha-induced NF-kappaB/RelA Ser(276) phosphorylation and enhanceosome formation is mediated by an ROS-dependent PKAc pathway. *Cellular Signalling*, *19*(7), 1419–1433. http://dx.doi.org/10.1016/j.cellsig.2007.01.020.

Jamaluddin, M., Wang, S., Garofalo, R. P., Elliott, T., Casola, A., Baron, S., & Brasier, A. R. (2001). IFN-beta mediates coordinate expression of antigen-processing genes in RSV-infected pulmonary epithelial cells. *American Journal of Physiology Lung Cellular and Molecular Physiology*, *280*(2), L248–L257.

Jang, M. K., Mochizuki, K., Zhou, M., Jeong, H. S., Brady, J. N., & Ozato, K. (2005). The bromodomain protein Brd4 is a positive regulatory component of P-TEFb and stimulates RNA polymerase II-dependent transcription. *Molecular Cell*, *19*(4), 523–534.

Johnston, S. L. (2007). Innate immunity in the pathogenesis of virus-induced asthma exacerbations. *Proceedings of the American Thoracic Society*, *4*(3), 267–270. http://dx.doi.org/10.1513/pats.200701-030AW.

Jung, M., Philpott, M., Muller, S., Schulze, J., Badock, V., Eberspacher, U., … Haendler, B. (2014). Affinity map of bromodomain protein 4 (BRD4) interactions with the histone H4 tail and the small molecule inhibitor JQ1. *The Journal of Biological Chemistry*, *289*(13), 9304–9319. http://dx.doi.org/10.1074/jbc.M113.523019.

Kalita, M. K., Sargsyan, K., Tian, B., Paulucci-Holthauzen, A., Najm, H. N., Debusschere, B. J., & Brasier, A. R. (2011). Sources of cell-to-cell variability in canonical nuclear factor-kappaB (NF-kappaB) signaling pathway inferred from single cell dynamic images. *The Journal of Biological Chemistry*, *286*(43), 37741–37757. http://dx.doi.org/10.1074/jbc.M111.280925.

Kalluri, R., & Weinberg, R. A. (2009). The basics of epithelial-mesenchymal transition. *The Journal of Clinical Investigation*, *119*(6), 1420–1428. http://dx.doi.org/10.1172/JCI39104.

Kanno, T., Kanno, Y., LeRoy, G., Campos, E., Sun, H. W., Brooks, S. R., … Ozato, K. (2014). BRD4 assists elongation of both coding and enhancer RNAs by interacting with

acetylated histones. *Nature Structural & Molecular Biology*, *21*(12), 1047–1057. http://dx.doi.org/10.1038/nsmb.2912.
Karin, M., Yamamoto, Y., & Wang, Q. M. (2004). The IKK NF-[kappa]B system: A treasure trove for drug development. *Nature Reviews. Drug Discovery*, *3*(1), 17–26.
Kato, H., Takeuchi, O., Mikamo-Satoh, E., Hirai, R., Kawai, T., Matsushita, K., … Akira, S. (2008). Length-dependent recognition of double-stranded ribonucleic acids by retinoic acid-inducible gene-I and melanoma differentiation associated gene 5. *The Journal of Experimental Medicine*, *205*(7), 1601–1610.
Kato, H., Takeuchi, O., Sato, S., Yoneyama, M., Yamamoto, M., Matsui, K., … Akira, S. (2006). Differential roles of MDA5 and RIG-I helicases in the recognition of RNA viruses. *Nature (London)*, *441*(7089), 101–105.
Katsuno, Y., Lamouille, S., & Derynck, R. (2013). TGF-beta signaling and epithelial-mesenchymal transition in cancer progression. *Current Opinion in Oncology*, *25*(1), 76–84. http://dx.doi.org/10.1097/CCO.0b013e32835b6371.
Kim, J., Guermah, M., & Roeder, R. G. (2010). The human PAF1 complex acts in chromatin transcription elongation both independently and cooperatively with SII/TFIIS. *Cell*, *140*(4), 491–503. http://dx.doi.org/10.1016/j.cell.2009.12.050.
Kim, K. H., & Roberts, C. W. M. (2016). Targeting EZH2 in cancer. *Nature Medicine*, *22*(2), 128–134. http://dx.doi.org/10.1038/nm.4036.
Knight, D. A., & Holgate, S. T. (2003). The airway epithelium: Structural and functional properties in health and disease. *Respirology*, *8*(4), 432–446.
Knutson, S. K., Wigle, T. J., Warholic, N. M., Sneeringer, C. J., Allain, C. J., Klaus, C. R., … Kuntz, K. W. (2012). A selective inhibitor of EZH2 blocks H3K27 methylation and kills mutant lymphoma cells. *Nature Chemical Biology*, *8*(11), 890–896. http://www.nature.com/nchembio/journal/v8/n11/abs/nchembio.1084.html#supplementary-information.
Krystof, V., Cankar, P., Frysova, I., Slouka, J., Kontopidis, G., Dzubak, P., … Strnad, M. (2006). 4-Arylazo-3,5-diamino-1H-pyrazole CDK inhibitors: SAR study, crystal structure in complex with CDK2, selectivity, and cellular effects. *Journal of Medicinal Chemistry*, *49*(22), 6500–6509. http://dx.doi.org/10.1021/jm0605740.
Krystof, V., Chamrad, I., Jorda, R., & Kohoutek, J. (2010). Pharmacological targeting of CDK9 in cardiac hypertrophy. *Medicinal Research Reviews*, *30*, 646–666.
Kwak, H., & Lis, J. T. (2013). Control of transcriptional elongation. *Annual Review of Genetics*, *47*, 483–508. http://dx.doi.org/10.1146/annurev-genet-110711-155440.
Lambrecht, B. N., & Hammad, H. (2012). The airway epithelium in asthma. *Nature Medicine*, *18*(5), 684–692. http://dx.doi.org/10.1038/nm.2737.
Lamouille, S., Xu, J., & Derynck, R. (2014). Molecular mechanisms of epithelial-mesenchymal transition. *Nature Reviews. Molecular Cell Biology*, *15*(3), 178–196. http://dx.doi.org/10.1038/nrm3758.
Li, J., & Gilmour, D. S. (2013). Distinct mechanisms of transcriptional pausing orchestrated by GAGA factor and M1BP, a novel transcription factor. *The EMBO Journal*, *32*(13), 1829–1841. http://dx.doi.org/10.1038/emboj.2013.111.
Li, X., Zhao, Y., Tian, B., Jamaluddin, M., Mitra, A., Yang, J., … Kudlicki, A. (2014). Modulation of gene expression regulated by the transcription factor NF-kappaB/RelA. *The Journal of Biological Chemistry*, *289*(17), 11927–11944. http://dx.doi.org/10.1074/jbc.M113.539965.
Li, X., Zhu, M., Brasier, A. R., & Kudlicki, A. S. (2015). Inferring genome-wide functional modulatory network: A case study on NF-kappaB/RelA transcription factor. *Journal of Computational Biology*, *22*(4), 300–312. http://dx.doi.org/10.1089/cmb.2014.0299.
Lin, C., Garrett, A. S., De Kumar, B., Smith, E. R., Gogol, M., Seidel, C., … Shilatifard, A. (2011). Dynamic transcriptional events in embryonic stem cells mediated by the super elongation complex (SEC). *Genes & Development*, *25*(14), 1486–1498. http://dx.doi.org/10.1101/gad.2059211.

Lin, C., Smith, E. R., Takahashi, H., Lai, K. C., Martin-Brown, S., Florens, L., … Shilatifard, A. (2010). AFF4, a component of the ELL/P-TEFb elongation complex and a shared subunit of MLL chimeras, can link transcription elongation to leukemia. *Molecular Cell, 37*(3), 429–437. http://dx.doi.org/10.1016/j.molcel.2010.01.026.

Lis, J. T., Mason, P., Peng, J., Price, D. H., & Werner, J. (2000). P-TEFb kinase recruitment and function at heat shock loci. *Genes & Development, 14*(7), 792–803.

Liu, T., Castro, S., Brasier, A. R., Jamaluddin, M., Garofalo, R. P., & Casola, A. (2004). Reactive oxygen species mediate virus-induced STAT activation: Role of tyrosine phosphastases. *Journal of Biological Chemistry, 279*(4), 2461–2469.

Liu, P., Jamaluddin, M., Li, K., Garofalo, R. P., Casola, A., & Brasier, A. R. (2007). Retinoic acid-inducible gene I mediates early antiviral response and toll-like receptor 3 expression in respiratory syncytial virus-infected airway epithelial cells. *Journal of Virology, 81*(3), 1401–1411.

Loo, Y.-M., & Gale, M., Jr. (2011). Immune signaling by RIG-I-like receptors. *Immunity, 34*(5), 680–692. http://dx.doi.org/10.1016/j.immuni.2011.05.003.

Loven, J., Hoke, H. A., Lin, C. Y., Lau, A., Orlando, D. A., Vakoc, C. R., … Young, R. A. (2013). Selective inhibition of tumor oncogenes by disruption of super-enhancers. *Cell, 153*(2), 320–334. http://dx.doi.org/10.1016/j.cell.2013.03.036.

Luo, Z., Lin, C., Guest, E., Garrett, A. S., Mohaghegh, N., Swanson, S., … Shilatifard, A. (2012a). The super elongation complex family of RNA polymerase II elongation factors: Gene target specificity and transcriptional output. *Molecular and Cellular Biology, 32*(13), 2608–2617. http://dx.doi.org/10.1128/MCB.00182-12.

Luo, Z., Lin, C., & Shilatifard, A. (2012b). The super elongation complex (SEC) family in transcriptional control. *Nature Reviews. Molecular Cell Biology, 13*(9), 543–547. http://dx.doi.org/10.1038/nrm3417.

Malumbres, M., Pevarello, P., Barbacid, M., & Bischoff, J. R. (2008). CDK inhibitors in cancer therapy: What is next? *Trends in Pharmacological Sciences, 29*(1), 16–21. http://dx.doi.org/10.1016/j.tips.2007.10.012.

Marie, I., Durbin, J. E., & Levy, D. E. (1998). Differential viral induction of distinct interferon-alpha genes by positive feedback through interferon regulatory factor-7. *The EMBO Journal, 17*(22), 6660–6669. http://dx.doi.org/10.1093/emboj/17.22.6660.

McCabe, M. T., Ott, H. M., Ganji, G., Korenchuk, S., Thompson, C., Van Aller, G. S., … Creasy, C. L. (2012). EZH2 inhibition as a therapeutic strategy for lymphoma with EZH2-activating mutations. *Nature, 492*(7427), 108–112. http://dx.doi.org/10.1038/nature11606.

McDonald, O. G., Wu, H., Timp, W., Doi, A., & Feinberg, A. P. (2011). Genome-scale epigenetic reprogramming during epithelial-to-mesenchymal transition. *Nature Structural & Molecular Biology, 18*(8), 867–874. http://dx.doi.org/10.1038/nsmb.2084.

McNamara, R. P., McCann, J. L., Gudipaty, S. A., & D'Orso, I. (2013). Transcription factors mediate the enzymatic disassembly of promoter-bound 7SK snRNP to locally recruit P-TEFb for transcription elongation. *Cell Reports, 5*(5), 1256–1268. http://dx.doi.org/10.1016/j.celrep.2013.11.003.

Medzhitov, R., & Horng, T. (2009). Transcriptional control of the inflammatory response. *Nature Reviews. Immunology, 9*(10), 692–703. http://dx.doi.org/10.1038/nri2634.

Meidhof, S., Brabletz, S., Lehmann, W., Preca, B. T., Mock, K., Ruh, M., … Brabletz, T. (2015). ZEB1-associated drug resistance in cancer cells is reversed by the class I HDAC inhibitor mocetinostat. *EMBO Molecular Medicine*, 7(6), 831–847. http://dx.doi.org/10.15252/emmm.201404396.

Mertz, J. A., Conery, A. R., Bryant, B. M., Sandy, P., Balasubramanian, S., Mele, D. A., … Sims, R. J., 3rd. (2011). Targeting MYC dependence in cancer by inhibiting BET bromodomains. *Proceedings of the National Academy of Sciences of the United States of America, 108*(40), 16669–16674. http://dx.doi.org/10.1073/pnas.1108190108.

Nguyen, V. T., Kiss, T., Michels, A. A., & Bensaude, O. (2001). 7SK small nuclear RNA binds to and inhibits the activity of CDK9/cyclin T complexes. *Nature, 414*(6861), 322–325.

Nishimura, G., Manabe, I., Tsushima, K., Fujiu, K., Oishi, Y., Imai, Y., … Nagai, R. (2006). DeltaEF1 mediates TGF-beta signaling in vascular smooth muscle cell differentiation. *Developmental Cell, 11*(1), 93–104. http://dx.doi.org/10.1016/j.devcel.2006.05.011.

Nowak, D. E., Tian, B., Jamaluddin, M., Boldogh, I., Vergara, L. A., Choudhary, S., & Brasier, A. R. (2008). RelA Ser276 phosphorylation is required for activation of a subset of NF-{kappa}B-dependent genes by recruiting cyclin-dependent kinase 9/cyclin T1 complexes. *Molecular and Cellular Biology, 28*(11), 3623–3638.

O'Neill, L. A. J., & Bowie, A. G. (2007). The family of five: TIR-domain-containing adaptors in Toll-like receptor signalling. *Nature Reviews. Immunology*, 7(5), 353–364.

Oshiumi, H., Matsumoto, M., & Seya, T. (2012). Ubiquitin-mediated modulation of the cytoplasmic viral RNA sensor RIG-I. *Journal of Biochemistry, 151*(1), 5–11. http://dx.doi.org/10.1093/jb/mvr111.

Oukka, M. (2008). Th17 cells in immunity and autoimmunity. *Annals of the Rheumatic Diseases, 67*(Suppl. 3), iii26–iii29. http://dx.doi.org/10.1136/ard.2008.098004.

Patel, M. C., Debrosse, M., Smith, M., Dey, A., Huynh, W., Sarai, N., … Ozato, K. (2013). BRD4 coordinates recruitment of pause release factor P-TEFb and the pausing complex NELF/DSIF to regulate transcription elongation of interferon-stimulated genes. *Molecular and Cellular Biology, 33*(12), 2497–2507. http://dx.doi.org/10.1128/MCB.01180-12.

Peinado, H., Olmeda, D., & Cano, A. (2007). Snail, Zeb and bHLH factors in tumour progression: An alliance against the epithelial phenotype? *Nature Reviews. Cancer*, 7(6), 415–428. http://dx.doi.org/10.1038/nrc2131.

Perales, R., & Bentley, D. (2009). "Cotranscriptionality": The transcription elongation complex as a nexus for nuclear transactions. *Molecular Cell, 36*(2), 178–191. http://dx.doi.org/10.1016/j.molcel.2009.09.018.

Peterlin, B. M., & Price, D. H. (2006). Controlling the elongation phase of transcription with P-TEFb. *Molecular Cell, 23*(3), 297–305. http://dx.doi.org/10.1016/j.molcel.2006.06.014.

Pichlmair, A., & Reis e Sousa, C. (2007). Innate recognition of viruses. *Immunity, 27*(3), 370–383. http://dx.doi.org/10.1016/j.immuni.2007.08.012.

Postigo, A. A. (2003). Opposing functions of ZEB proteins in the regulation of the TGFbeta/BMP signaling pathway. *The EMBO Journal, 22*(10), 2443–2452. http://dx.doi.org/10.1093/emboj/cdg225.

Postigo, A. A., Depp, J. L., Taylor, J. J., & Kroll, K. L. (2003). Regulation of Smad signaling through a differential recruitment of coactivators and corepressors by ZEB proteins. *The EMBO Journal, 22*(10), 2453–2462. http://dx.doi.org/10.1093/emboj/cdg226.

Sanchez-Tillo, E., Lazaro, A., Torrent, R., Cuatrecasas, M., Vaquero, E. C., Castells, A., … Postigo, A. (2010). ZEB1 represses E-cadherin and induces an EMT by recruiting the SWI/SNF chromatin-remodeling protein BRG1. *Oncogene, 29*(24), 3490–3500. http://dx.doi.org/10.1038/onc.2010.102.

Shay, D. K., Holman, R. C., Roosevelt, G. E., Clarke, M. J., & Anderson, L. J. (2001). Bronchiolitis-associated mortality and estimates of respiratory syncytial virus-associated deaths among US children, 1979-1997. *Journal of Infectious Diseases, 183*(1), 16–22.

Shi, Y., Sawada, J., Sui, G., Affar el, B., Whetstine, J. R., Lan, F., … Shi, Y. (2003). Coordinated histone modifications mediated by a CtBP co-repressor complex. *Nature, 422*(6933), 735–738. http://dx.doi.org/10.1038/nature01550.

Shi, J., & Vakoc, C. R. (2014). The mechanisms behind the therapeutic activity of BET bromodomain inhibition. *Molecular Cell, 54*(5), 728–736. http://dx.doi.org/10.1016/j.molcel.2014.05.016.

Sisson, T. H., Mendez, M., Choi, K., Subbotina, N., Courey, A., Cunningham, A., … Simon, R. H. (2010). Targeted injury of type II alveolar epithelial cells induces pulmonary fibrosis. *American Journal of Respiratory and Critical Care Medicine, 181*(3), 254–263. http://dx.doi.org/10.1164/rccm.200810-1615OC.

Skaug, B., Jiang, X., & Chen, Z. J. (2009). The role of ubiquitin in the NF-kB regulatory pathways. *Annual Review of Biochemistry, 78*(1), 769–796. http://dx.doi.org/10.1146/annurev.biochem.78.070907.102750.

Smale, S. T. (2012). Transcriptional regulation in the innate immune system. *Current Opinion in Immunology, 24*(1), 51–57. http://dx.doi.org/10.1016/j.coi.2011.12.008.

Smith, E., Lin, C., & Shilatifard, A. (2011). The super elongation complex (SEC) and MLL in development and disease. *Genes & Development, 25*(7), 661–672. http://dx.doi.org/10.1101/gad.2015411.

Sonawane, Y. A., Taylor, M. A., Napoleon, J. V., Rana, S., Contreras, J. I., & Natarajan, A. (2016). Cyclin dependent kinase 9 inhibitors for cancer therapy. *Journal of Medicinal Chemistry, 59*(19), 8667–8684. http://dx.doi.org/10.1021/acs.jmedchem.6b00150.

Takahashi, H., Parmely, T. J., Sato, S., Tomomori-Sato, C., Banks, C. A., Kong, S. E., … Conaway, J. W. (2011). Human mediator subunit MED26 functions as a docking site for transcription elongation factors. *Cell, 146*(1), 92–104. http://dx.doi.org/10.1016/j.cell.2011.06.005.

Tian, B., & Brasier, A. R. (2003). Identification of a nuclear factor kappa B-dependent gene network. *Recent Progress in Hormone Research, 58*, 95–130.

Tian, B., Li, X., Kalita, M., Widen, S. G., Yang, J., Bhavnani, S. K., … Brasier, A. R. (2015). Analysis of the TGFbeta-induced program in primary airway epithelial cells shows essential role of NF-kappaB/RelA signaling network in type II epithelial mesenchymal transition. *BMC Genomics, 16*(1), 529. http://dx.doi.org/10.1186/s12864-015-1707-x.

Tian, B., Patrikeev, I., Ochoa, L., Vargas, G., Bellanger, K. A., Litvino, J., … Brasier, A. R. (2016). NFkB Mediates mesenchymal transition, remodeling and pulmonary fibrosis in response to chronic inflammation by viral RNA patterns. *American Journal of Respiratory Cell & Molecular Biology*. in press.

Tian, B., Zhang, Y., Luxon, B. A., Garofalo, R. P., Casola, A., Sinha, M., & Brasier, A. R. (2002). Identification of NF-kappaB-dependent gene networks in respiratory syncytial virus-infected cells. *Journal of Virology, 76*(13), 6800–6814.

Tian, B., Zhao, Y., Kalita, M., Edeh, C. B., Paessler, S., Casola, A., … Brasier, A. R. (2013). CDK9-dependent transcriptional elongation in the innate interferon-stimulated gene response to respiratory syncytial virus infection in airway epithelial cells. *Journal of Virology, 87*(12), 7075–7092. http://dx.doi.org/10.1128/JVI.03399-12.

Tian, B., Zhao, Y., Sun, H., Zhang, Y., Yang, J., & Brasier, A. R. (2016). BRD4 Mediates NFkB-dependent epithelial-mesenchymal transition and pulmonary fibrosis via transcriptional elongation. *The American Journal of Physiology—Lung Cellular and Molecular Physiology*. http://dx.doi.org/10.1152/ajplung.00224.2016. [Epub ahead of print].

Tsai, H. C., Li, H., Van Neste, L., Cai, Y., Robert, C., Rassool, F. V., … Zahnow, C. A. (2012). Transient low doses of DNA-demethylating agents exert durable antitumor effects on hematological and epithelial tumor cells. *Cancer Cell, 21*(3), 430–446. http://dx.doi.org/10.1016/j.ccr.2011.12.029.

Wang, J., Lee, S., Teh, C. E., Bunting, K., Ma, L., & Shannon, M. F. (2009). The transcription repressor, ZEB1, cooperates with CtBP2 and HDAC1 to suppress IL-2 gene activation in T cells. *International Immunology, 21*(3), 227–235. http://dx.doi.org/10.1093/intimm/dxn143.

Weber, C. M., Ramachandran, S., & Henikoff, S. (2014). Nucleosomes are context-specific, H2A.Z-modulated barriers to RNA polymerase. *Molecular Cell, 53*(5), 819–830. http://dx.doi.org/10.1016/j.molcel.2014.02.014.

Whitsett, J. A., & Alenghat, T. (2015). Respiratory epithelial cells orchestrate pulmonary innate immunity. *Nature Immunology*, *16*(1), 27–35. http://dx.doi.org/10.1038/ni.3045.

Wier, A. D., Mayekar, M. K., Heroux, A., Arndt, K. M., & VanDemark, A. P. (2013). Structural basis for Spt5-mediated recruitment of the Paf1 complex to chromatin. *Proceedings of the National Academy of Sciences of the United States of America*, *110*(43), 17290–17295. http://dx.doi.org/10.1073/pnas.1314754110.

Willis, B. C., & Borok, Z. (2007). TGF-beta-induced EMT: Mechanisms and implications for fibrotic lung disease. *American Journal of Physiology. Lung Cellular and Molecular Physiology*, *293*(3), L525–L534. http://dx.doi.org/10.1152/ajplung.00163.2007.

Wu, S.-Y., & Chiang, C.-M. (2007). The double bromodomain-containing chromatin adaptor Brd4 and transcriptional regulation. *Journal of Biological Chemistry*, *282*(18), 13141–13145. http://dx.doi.org/10.1074/jbc.R700001200.

Yang, J., Mitra, A., Dojer, N., Fu, S., Rowicka, M., & Brasier, A. R. (2013). A probabilistic approach to learn chromatin architecture and accurate inference of the NF-kappaB/RelA regulatory network using ChIP-Seq. *Nucleic Acids Research*, *41*, 7240–7259. http://dx.doi.org/10.1093/nar/gkt493.

Yang, J., Tian, B., Sun, H., Garofalo, R., & Brasier, A. R. (2017). ZEB1 mediated epigenetic silencing of IRF1 dysregulates type III interferon responses to respiratory virus infection in epithelial to mesenchymal transition. *Nature Microbiology*. in press.

Yang, Y. C., Zhang, N., Van Crombruggen, K., Hu, G. H., Hong, S. L., & Bachert, C. (2012). Transforming growth factor-beta1 in inflammatory airway disease: A key for understanding inflammation and remodeling. *Allergy*, *67*(10), 1193–1202. http://dx.doi.org/10.1111/j.1398-9995.2012.02880.x.

Yang, J., Zhao, Y., Kalita, M., Li, X., Jamaluddin, M., Tian, B., … Brasier, A. R. (2015). Systematic determination of human cyclin dependent kinase (CDK)-9 interactome identifies novel functions in RNA splicing mediated by the DDX5/17 RNA helicases. *Molecular & Cellular Proteomics*, *14*(10), 2701–2721. http://dx.doi.org/10.1074/mcp.M115.049221.

Yu, L., Wang, Z., Zhang, Z., Ren, X., Lu, X., & Ding, K. (2015). Small-molecule BET inhibitors in clinical and preclinical development and their therapeutic potential. *Current Topics in Medicinal Chemistry*, *15*(8), 776–794.

Zhang, Y., Luxon, B. A., Casola, A., Garofalo, R. P., Jamaluddin, M., & Brasier, A. R. (2001). Expression of respiratory syncytial virus-induced chemokine gene networks in lower airway epithelial cells revealed by cDNA microarrays. *Journal of Virology*, *75*(19), 9044–9058. http://dx.doi.org/10.1128/JVI.75.19.9044-9058.2001.

Zhang, Y., Yang, X., Gui, B., Xie, G., Zhang, D., Shang, Y., & Liang, J. (2011). Corepressor protein CDYL functions as a molecular bridge between polycomb repressor complex 2 and repressive chromatin mark trimethylated histone lysine 27. *The Journal of Biological Chemistry*, *286*(49), 42414–42425. http://dx.doi.org/10.1074/jbc.M111.271064.

Zhao, Y., Jamaluddin, M., Zhang, Y., Sun, H., Ivanciuc, T., Garofalo, R. P., & Brasier, A. R. (2016). Systematic analysis of cell-type differences in the epithelial secretome reveals insights into pathogenesis of RSV-induced lower respiratory tract infections. *Journal of Immunology*. in press.

Zhao, Y., Tian, B., Sadygov, R. G., Zhang, Y., & Brasier, A. R. (2016). Integrative proteomic analysis reveals reprograming tumor necrosis factor signaling in epithelial mesenchymal transition. *Journal of Proteomics*, *148*, 126–138. http://dx.doi.org/10.1016/j.jprot.2016.07.014.

Zhao, Y. X., Widen, S. G., Jamaluddin, M., Tian, B., Wood, T. G., Edeh, C. B., & Brasier, A. R. (2011). Quantification of activated NF-kappa B/RelA complexes using ssDNA aptamer affinity—Stable isotope dilution-selected reaction monitoring-mass spectrometry. *Molecular & Cellular Proteomics*, *10*(6). http://dx.doi.org/10.1074/mcp.M111.008771.

Zhong, Z., Umemura, A., Sanchez-Lopez, E., Liang, S., Shalapour, S., Wong, J., ... Karin, M. (2016). NF-kB restricts inflammasome activation via elimination of damaged mitochondria. *Cell*, *164*(5), 896–910. http://dx.doi.org/10.1016/j.cell.2015.12.057.

Zhou, M., Huang, K., Jung, K.-J., Cho, W.-K., Klase, Z., Kashanchi, F., ... Brady, J. N. (2009). Bromodomain protein Brd4 regulates human immunodeficiency virus transcription through phosphorylation of CDK9 at threonine 29. *Journal of Virology*, *83*(2), 1036–1044. http://dx.doi.org/10.1128/jvi.01316-08.

Zhou, Q., Li, T., & Price, D. H. (2012). RNA polymerase II elongation control. *Annual Review of Biochemistry*, *81*, 119–143. http://dx.doi.org/10.1146/annurev-biochem-052610-095910.

Zou, Z., Huang, B., Wu, X., Zhang, H., Qi, J., Bradner, J., ... Chen, L. F. (2014). Brd4 maintains constitutively active NF-kappaB in cancer cells by binding to acetylated RelA. *Oncogene*, *33*(18), 2395–2404. http://dx.doi.org/10.1038/onc.2013.179.

CHAPTER TWO

HMGB1 Protein: A Therapeutic Target Inside and Outside the Cell

I. Ugrinova[1], E. Pasheva
"Roumen Tsanev" Institute of Molecular Biology, Bulgarian Academy of Sciences, Sofia, Bulgaria
[1]Corresponding author: e-mail address: ugryiva@gmail.com

Contents

Abstract

High-mobility group box 1 protein (HMGB1) is a nonhistone chromosomal protein discovered more than 30 years ago. It is an abundant nuclear protein that has a dual function—in the nucleus, it binds DNA and participates in practically all DNA-dependent processes serving as an architectural factor. Outside the cell, HMGB1 plays a different role—it acts as an alarmine that activates a large number of HMGB1-"competent" cells and mediates a

Advances in Protein Chemistry and Structural Biology, Volume 107
ISSN 1876-1623
http://dx.doi.org/10.1016/bs.apcsb.2016.10.001

broad range of physiological and pathological responses. This universality makes it an attractive target for innovative therapeutic strategies in the treatment of various diseases. Here we present an overview of the major nuclear and extracellular properties of HMGB1 and describe its interaction with different molecular partners as specific receptors or inhibitors, which are important for its role as a target in multiple diseases. We highlight its pivotal role as a target for cancer treatment at two aspects: first in terms of its substantial impact on the repair capacity of cancer cells, thus affecting the effectiveness of chemotherapy with the antitumor drug *cis*-platinum and, second, the possibility to be targeted by microRNAs influencing different pathways of human diseases, thus making it a promising candidate for a new strategy for therapeutic interventions against various pathological conditions but mainly cancer.

1. INTRODUCTION

High-mobility group box 1 protein (HMGB1) is an abundant nonhistone chromosomal protein. HMGB1 was isolated and biochemically characterized by Goodwin and Johns in the 1970s (Goodwin, Sanders, & Johns, 1973) who claimed: "The total high-mobility group proteins are only about 3%, by weight of the histone or DNA content, and they are therefore a small group of proteins compared to the histones. However, this quantity of proteins is still relatively large when compared to the quantity of protein one might expect if they were specific gene derepressors where only a few molecules per structural gene would be required." From today's point of view, this was remarkably foresighted statement; however, it did not inspire great enthusiasm for the next decade. There were two periods of HMGB1 resurrection: the first one was in the 1990s with the discovery of the HMG box motif—the DNA-binding domain of the protein (Jantzen, Admon, Bell, & Tjian, 1990) and the second one was almost 10 years later with the identification of a specific receptor for HMGB1—the receptor for advanced glycation end products (RAGEs) that put the emphasis on the extracellular functions of the protein (Degryse et al., 2001; He et al., 2011; Hori et al., 1995; Taguchi et al., 2000). Romani and colleagues calculated how abundant HMGB1 actually is—there are about 10^6 molecules of HMGB1 per cell, which is only an order of magnitude less than the core histones (Romani, Rodman, Vidali, & Bustin, 1979). However, HMGB1 is one of the most mobile proteins in the nucleus crossing this organelle in 1.5 s (Mollica et al., 2007; Phair et al., 2004; Scaffidi, Misteli, & Bianchi, 2002). Taken together, these observations lead to the conclusion that HMGB1 should not be tightly bound to the chromatin, but instead, it moves fast in a scanning manner searching for the places and processes where it is needed.

In this review, we provide novel insights into the nuclear and extracellular HMGB1 biological activities with respect to the potential strategies that target HMGB1 for the treatment of different pathologies.

2. HMGB1—GENERAL OVERVIEW

2.1 Structure

Human HMGB1 has 215 amino acid residues and forms 2 DNA-binding domains (HMG A box [aa 9–79], HMG B box [aa 89–162]) and one C-terminal acidic tail (aa 186–215) (Fig. 1A) (Bianchi, Falciola, Ferrari, & Lilley, 1992). The interest in this protein sharply increased in the early 1990s with the discovery of a new kind of structural domain involved in the interactions with DNA (Jantzen et al., 1990), which displayed similarity to two homologous repeats of an 80-amino acid sequence in HMGB1 (Fig. 1A).

This motif called HMG box was found in less abundant eukaryotic proteins, mainly general transcription factors and gene-specific transcriptional activators (Grosschedl, Giese, & Pagel, 1994). The DNA-binding motif confers them the peculiar feature to recognize bent DNA or to induce bending in linear duplex DNA. An interesting crystallographic work came up (Sanchez-Giraldo et al., 2015), showing how box A is able to bind linear unmodified DNA (Fig. 1B). The authors found out that two box A domains of HMGB1 collaborate in an unusual configuration in which the Phe37 residues of both domains stack together and intercalate the same CG base pair, generating highly kinked DNA. These results suggested that the simultaneous binding of the two domains could be indicative of a concerted action of two HMGB1 molecules to bend DNA in vivo. In addition to the two DNA-binding domains, HMGB1 contains an unstructured acidic tail comprising exclusively glutamic and aspartic acids (Stott, Watson, Howe, Grossmann, & Thomas, 2010; Stros, Stokrova, & Thomas, 1994; Weir et al., 1993).

2.2 Posttranslational Modifications

HMGB1 undergoes several extensive posttranslational modifications (PTMs) such as acetylation, phosphorylation, methylation, ADP–ribosylation, oxidation, and glycosylation (Ulloa & Messmer, 2006); here we will focus on the acetylation and oxidation as the modifications with the greatest impact onf the role of HMGB1 as a signal molecule.

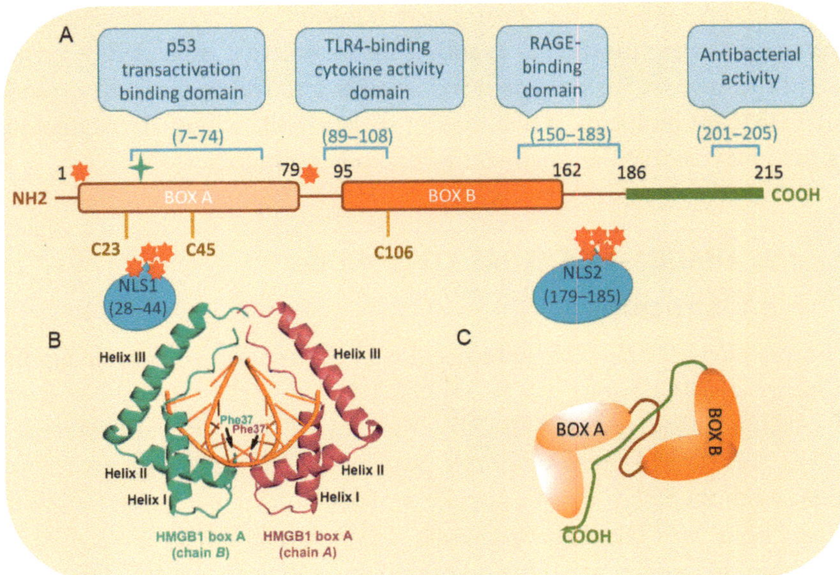

Fig. 1 Scheme of HMGB1 protein structure. HMGB1 contains two folded helical DNA-binding motifs, called A and B boxes, and an acidic tail that contains a string of glutamic and aspartic acids. HMGB1 has two NLSs located in the A box (aa 28–44) and in the B box (aa 179–185), respectively. Four conserved lysine residues are present in NLS1, and five are present in NLS2. They are susceptible to acetylation modification, resulting in nuclear exclusion and subsequent HMGB1 release (A). The two near-symmetric box A domains collaborate to bend DNA, Phe37 of both domains is indicated (B) (Sanchez-Giraldo et al., 2015). Scheme of HMGB1 protein, further used in this review (C). *Panel (B) of this figure is under Creative Commons Attribution License. This is an open-access article distributed under the terms of the Creative Commons Attribution License, which permits unrestricted use, distribution, and reproduction in any medium, provided the original authors and source are cited.*

2.2.1 *Posttranslational Acetylation*

Posttranslational acetylation of HMGB1 at lysine 2 is known for more than 35 years (Sterner, Vidali, & Allfrey, 1979); however, the properties of the acetylated protein were not studied until 2001 when our group showed that in vivo acetylation of HMGB1 at lysine 2 significantly enhanced its affinity to distorted DNA structures (Ugrinova, Pasheva, Armengaud, & Pashev, 2001). We also found that the protein is a substrate for CBP histone acetyltransferase and the removal of the acidic C-tail creates a novel site for acetylation at lysine 81 (Pasheva et al., 2004). By the mutation of the acetylation sites the importance of these modifications for the DNA-binding and -bending activities of HMGB1 was further analyzed

(Assenberg et al., 2008; Elenkov, Pelovsky, Ugrinova, Takahashi, & Pasheva, 2011). In our opinion, these modifications are mainly connected to the role of the protein as an architectural factor of the chromatin in the nucleus unlike the acetylation at the nuclear localization sites (NLSs), which is important for the translocation of the protein and its subsequent relocation and release by certain immune cells (Bonaldi et al., 2003; Venereau et al., 2016). Four conserved lysine residues are present in NLS1 (aa 28–44) and five in NLS2 (aa 179–185) (Fig. 1A). Therefore, it seems logical that an acetyltransferase different from CBP and maybe specific for those cells is responsible for the acetylation of the NLSs.

2.2.2 Oxidation

Another important modification of HMGB1 especially in terms of its biological activities outside the cell is the redox state of each of its three cysteines (C23, C45, and C106; Fig. 1A) (Yang et al., 2012). Fully reduced HMGB1, which expresses three cysteine thiol residues, exerts chemotactic activity. The cytokine-stimulating activity of HMGB1 requires C23 and C45 to form a disulfide bond—partially oxidized form and complete oxidation of cysteines to sulfonates by reactive oxygen species fully abolish its activities (Venereau et al., 2012). Thus, the PTMs of HMGB1 are responsible for its export (acetylation) and its extracellular activities (oxidation). Most likely, the PTMs of HMGB1 also control its binding to different receptors.

2.3 HMGB1-Receptor Families

HMGB1 signals via multiple receptors. To date, maybe more than 15 different HMGB1 receptors have been described: some of them interact directly with HMGB1 such as RAGE, TLR4/MD-2, TIM3, etc., and some require complex ligands where HMGB1 interacts with other molecules or complexes in order to switch on signaling. For instance, HMGB1 binds with CpG oligodeoxynucleotides (ODNs), or nucleosomes forming complexes first and then the complex fire on TLR9, or forming a heterocomplex with CXCL12, which binds to the CXCR4, etc. (Venereau et al., 2016; Yang, Wang, Chavan, & Andersson, 2015).

2.3.1 Receptor for Advanced Glycation End Products

RAGE is the first HMGB1 receptor that was identified as the amphoterin-binding receptor with a role in neurite outgrowth in the developing nervous system (Hori et al., 1995). As the name suggests, RAGE binds advanced glycation end products (AGEs), but more recently, it has been described to bind to a diverse array of structurally different molecules, including HMGB1, S100 family members, amyloid-β peptide, DNA, RNA, and other molecules to regulate multiple physiological and pathological processes. RAGE belongs to the immunoglobulin gene superfamily and is composed of an extracellular domain, a short transmembrane domain, and a 43-amino acid cytoplasmic tail. The HMGB1-induced RAGE activation contributes to various diseases including sepsis, cancer, rheumatoid arthritis, diabetes and diabetic nephropathy, atherosclerosis, and neurological disorders (Kang et al., 2014). Extracellular domain, containing one "V"-type domain and two "C"-type domains, is responsible for ligand binding, whereas a cytoplasmic tail is required for intracellular signaling transduction such as nuclear factor (NF)-kB signaling (Huttunen, Fages, & Rauvala, 1999). There are multiple spliced transcript variants of RAGEs, which encode different isoforms, as well as nonprotein-coding variants. In addition, the soluble form of RAGE (sRAGE) can be generated by ADAM10 or matrix metalloproteinase-mediated proteolysis (Raucci et al., 2008; Zhang et al., 2008). Interestingly, for proper ligand recognition, two RAGE molecules need to interact via their N-terminal V domains (Zong et al., 2010).

2.3.2 Toll-Like Receptors

The Toll-like receptors (TLRs) are type I transmembrane superfamily members able to recognize several danger signals, including pathogen-associated molecular patterns and damage-associated molecular patterns (DAMPs), and to activate the innate immunity response for defense against infection and injury (Akira & Takeda, 2004). HMGB1 can interact with TLRs (TLR2, TLR4, and TLR9) to activate the NF-kB pathway and produce cytokines and chemokines for the inflammation and immune response. HMGB1 also functions as the universal sentinel for nucleic acids as the absence of HMGBs severely impairs the activation of TLR3, TLR7, and TLR9 by their cognate nucleic acids (Yanai et al., 2009). As it was mentioned earlier the redox state of HMGB1 is important for its downstream signaling properties. The interaction of HMGB1 with TLRs leads to activation of different cell types through various signaling pathways, and HMGB1–TLR2 interactions for instance activate NK cells (Qiu et al., 2014) and stimulate self-renewal of stem cells by promoting NF-kB, STAT3, and Smad3 signaling (Conti et al., 2013). HMGB1–TLR4 signaling mediates anticancer immunity during radio- or chemotherapy, cell migration and adhesion (Apetoh et al., 2007; Bauer, Shapiro, Billiar, & Bauer, 2013; Wang, Sun, Wei, & Tian, 2013), and angiogenesis (Lin et al., 2011) and is involved in models of various disorders such as liver and lung damage, cancer, and epilepsy (Maroso et al., 2010; Mittal et al., 2010; Yang, Deng, et al., 2013).

2.3.3 Recently Discovered HMGB1 Receptors

Accumulating evidence demonstrates that HMGB1 has many other partners in providing various types of signaling. Among them CD24/Siglec-10 inhibits HMGB1-induced NF-kB activation and subsequent proinflammatory cytokine production. TIM-3 receptor expressed on different immune cells interacts with HMGB1 to interfere with the recruitment of nucleic acids into DC endosomes. As a consequence, it attenuates the therapeutic efficacy of DNA vaccination and chemotherapy by reducing immunogenicity of nucleic acids released from dying tumor cells (Chiba et al., 2012). Therefore, blockade of TIM-3 by neutralizing antibody could improve therapeutic responses in combination with DNA vaccines and chemotherapy (Kang et al., 2014). A recent study demonstrated that HMGB1 physiologically interacts with NMDAR via a specific region of HMGB1 localized in the B box upstream from the RAGE and downstream from the TLR4-binding sites. HMGB1 released by activated, stressed, and damaged or necrotic cells can facilitate NMDAR-mediated cell responses, both in the central nervous system and in peripheral tissues, independent of other known cell surface receptors for HMGB1 (Pedrazzi et al., 2012). Another studies found that the recently identified triggering receptor expressed on myeloid cell-1 is involved in mediating the HMGB1-induced cytokine production via the NF-kB pathway, which plays a critical role in sepsis (El Mezayen et al., 2007) and tumor growth (Wu et al., 2012). Months ago, it was discovered that haptoglobin, the acute phase protein that binds extracellular hemoglobin and targets cellular uptake through CD163 (a scavenger receptor), also binds HMGB1. These findings reveal a mechanism for haptoglobin modulation of the inflammatory action of HMGB1, with significant implications for developing experimental strategies, targeting HMGB1-dependent inflammatory diseases (Yang et al., 2016).

3. HMGB1 IN THE NUCLEUS—AN ARCHITECTURAL FACTOR

3.1 Nucleolar Functions

Nuclear HMGB1 acts as a DNA chaperone with DNA-binding and -bending activities. The protein interacts with different DNA structures

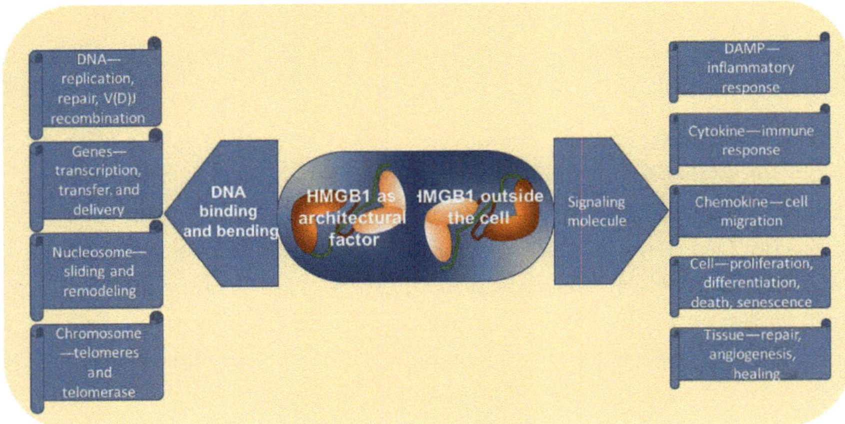

Fig. 2 Schematic presentation of HMGB1 functions. Nuclear HMGB1 acts as an architectural factor with DNA-binding and -bending activities. Extracellular HMGB1 acts as a signaling molecule with cytokine and chemokine activities.

without sequence specificity and regulates nuclear homeostasis and genome stability in several ways (Fig. 2).

3.1.1 DNA Replication, Repair, and V(D)J Recombination

Early studies indicated that HMGB1 isolated from rat liver acts as a DNA helix-destabilizing protein to stimulate the activity of DNA polymerases a and b in vitro (Bonne-Andrea, Harper, Sobczak, & De Recondo, 1984; Duguet, Soussi, Rossignol, Mechali, & De Recondo, 1977). The acetylated form of HMGB1 proteins can bind and stimulate the DNA polymerase activity in vitro (Alexandrova & Beltchev, 1988). A more recent study in our laboratory has shown that native, recombinant, and tailless HMGB1 variants act very differently in an in vitro replication assay. We found that native HMGB1 isolated from tumor cells and recombinant HMGB1 phosphorylated by PKC inhibited DNA replication. However, the inhibition effect could be reversed by HMGB1 acetylation and removal of the acidic tail (Topalova, Ugrinova, Pashev, & Pasheva, 2008). Despite all these findings the role of HMGB1 in DNA replication remains elusive. Loss of HMGB1 increases DNA damage and decreases DNA repair efficiency in response to chemotherapy, irradiation, and oxidative stress. HMGB1 is involved in practically all repair pathways, including nucleotide excision repair (NER), base excision repair, mismatch repair, and double-strand break repair via nonhomologous end-joining. Today, two general hypotheses have been proposed to explain the role of HMGB1 in repair—the "repair shielding" (a model in which HMGB1 interacts with the damaged DNA and masks it from the damage recognition proteins of NER) (Chu, 1994; Zamble & Lippard, 1995) and "repair enhancing" (the opposite model—HMGB1 enhances the interactions between damage recognition factors and the damage and promotes NER) (Lange & Vasquez, 2009). The elucidation of the role of HMGB1 in DNA repair is critical for the complete understanding of its intracellular functions, which is particularly relevant in the context of anti-HMGB1 therapies that are being developed to treat inflammatory diseases (Lange & Vasquez, 2009). V(D)J recombination is a mechanism of genetic recombination in the early stages of immunoglobulin and T-cell receptor production of the immune system. HMGB1 is required for efficient V(D)J recombination by enhancing recombination-activating gene binding to the recombination signal sequences (Kang, Zhang, Zeh, Lotze, & Tang, 2013).

3.1.2 Gene Transcription Transfer and Delivery

HMGB1 is able to change transcription rates and gene expression via several different mechanisms (Singh & Dixon, 1990): by sustaining nucleosome dynamics, or as it promotes interaction with the TBP (TATA-binding protein)/TATA-box complex (Das & Scovell, 2001) and enhanceosome assembly (Mitsouras, Wong, Arayata, Johnson, & Carey, 2002). HMGB1 regulates gene expression not only autonomously but also in complexes with several sequence-specific transcription factors bringing them to their cognate DNA such as p53, p73, the retinoblastoma protein (Rb), NF-kB, and the estrogen receptor. There is evidence that HMGB1 has the ability to enhance some DNA transposition systems like herpes simplex virus/sleeping beauty amplicon (de Silva, Lotta, Burris, & Bowers, 2010; Peterson, Mastrangelo, Federoff, & Bowers, 2007), and this could be used for improving gene transfer in gene therapy. HMGB1 significantly enhances transfection efficiency in several systems by its nuclear localization signals and DNA- binding ability (Siu, Li, Leung, Lee, & Li, 2012). Thus, HMGB1 may be useful as a nontoxic gene delivery carrier in gene therapy or at least for TAT-peptide- based vectors (Kim, Han, Kim, & Lee, 2008; Yi et al., 2012).

3.1.3 Nucleosome Structure and Dynamics

The nucleosome is the basic unit of chromatin, consisting of a short length of DNA wrapped around a core of histone proteins. HMGB1 binds to nucleosomes at the dyad axis, promotes nucleosome sliding, relaxes nucleosome structure, and makes chromatin more accessible by its ability to bend DNA

(Cato, Stott, Watson, & Thomas, 2008). The interaction between HMGB1 and linker histones (H1 and H5) occurs through their acidic and basic tails, respectively (Cato et al., 2008). Increased transient interaction between HMGB1 and nucleosomal linker DNA can activate ATP-utilizing chromatin assembly and the remodeling factor/chromatin accessibility complex (ACF/CHRAC) pathway, which in turn promotes nucleosome sliding, and the C-terminus of HMGB1 is required for activation of the ACF/CHRAC (Bonaldi, Langst, Strohner, Becker, & Bianchi, 2002). In addition, acetylated HMGB1 assists nucleosome mobilization induced by switch/sucrose nonfermentable, but does not affect its ATPase activity (Ugrinova, Pashev, & Pasheva, 2009). Surprisingly, HMGB1 regulates nucleosome biogenesis as well. Cells lacking HMGB1 contain 20–30% less histones and nucleosomes and more RNA transcripts (Celona et al., 2011), many of them promoting expression of inflammatory genes including chemokines. Exogenous HMGB1 promotes the assembly of chromatin in vitro by the virtue of its DNA chaperone activity (Celona et al., 2011). These findings indicate that HMGB1 contributes to genome chromatinization by sustaining the number, and possibly location, of nucleosomes.

3.1.4 Telomere Homeostasis

A telomere is a region of repetitive nucleotide sequences at the end of a chromosome. Loss of HMGB1 reduces the telomerase activity, decreases the telomere length, and increases the chromosomal instability. A recent study demonstrated that HMGB1 knockdown MEFs exhibited mild telomere shortening, but significantly decreased the telomerase activity and DNA damage. Interestingly, HMGB2 plays an opposing role in the regulation of telomerase activity.

3.2 HMGB1 as a Damage Recognition Protein Mediating the Antitumor Effect of the Chemotherapeutic Drug *cis*-Platin

HMGB1 was found to bind preferentially to distorted DNA such as UV-damaged DNA (Pasheva, Pashev, & Favre, 1998), synthetic cruciform DNA structures (Bianchi, Beltrame, & Paonessa, 1989), lesions formed on DNA by the anticancer drug *cis*-platin (Pil & Lippard, 1992), and semicatenated DNA (Gaillard & Strauss, 2000). The ability of HMG box-containing proteins to mediate bending in linear duplex DNA was demonstrated by several approaches (Paull, Haykinson, & Johnson, 1993; Pil, Chow, & Lippard, 1993).

When we speak about HMGB1 as an "architect" of chromatin, we must stress on the fact that the protein does not recognize any specific distorted DNA structure. The most striking example is the preference of HMGB1 to the DNA lesions induced by the widely used anticancer drug *cis*-diamminedichloroplatinum II (*cis*-DDP or *cis*-platin) but not to the geometric isomer *trans*-platin.

cis-Platin is a chemotherapeutic drug widely used for the treatment of several human cancers. It is generally accepted that the cytotoxic action of *cis*-platin involves the formation of covalent adducts with DNA capable of blocking DNA replication (Zamble & Lippard, 1995). Lesions produced in DNA are mainly 1,2-d(GpG) or d(ApG) and 1,3-d(GpXpG) intrastrand cross-links between neighboring purines. The therapeutic effect of *cis*-platin, however, cannot be attributed solely to its interaction with DNA because the geometric isomer *trans*-platin is clinically ineffective although it binds DNA and blocks replication and transcription (Ciccarelli, Solomon, Varshavsky, & Lippard, 1985). This is further complicated by studies with *cis*-platin: the immunochemical determination of *cis*-platin-DNA adducts in blood cell DNA from cancer patients revealed a positive correlation between adduct formation and favorable clinical outcome (Reed, Ozols, Tarone, Yuspa, & Poirier, 1987), while comparative studies with cultured cell lines reported differences in whether a relationship exists between the initial drug–DNA adduct level and cell survival (Terheggen et al., 1990). Important but yet unanswered questions concern the specificity of *cis*-platin for only certain types of cancer, the differential biological activity of the different platinum compounds, and the phenomenon of the intrinsic and acquired resistance to the drug. One possible reason for at least some of these issues is that cells differentially process and repair DNA adducts when bound by specific proteins referred to as damage recognition proteins (Pil & Lippard, 1992). HMGB1 is an excellent candidate for such a role. While testing new designed derivatives of *cis*-platinum, one should consider not only the effectiveness of DNA lesion formation but also the binding of HMGB1 to DNA adducts that also contributes to the cytotoxicity of *cis*-platin. Such a correlation between the antitumor activity of four *cis*-platin-based drugs—*cis*-diamminedichloroplatinum (II) (*cis*-platin); *cis*-diammine (1,1-cyclobutanedicarboxylate) platinum II (carboplatin); *cis*-ammine (cyclopentanecarboxylic acid hydrazide) dichloroplatinum II; and *cis*-ammine (cyclohexanecarboxylic acid hydrazide) dichloroplatinum II—and the binding affinity of HMGB1 to the respective DNA adducts was already reported (Pasheva, Ugrinova, Spassovska, & Pashev, 2002).

3.3 HMGB1 Protein—An Important Player in *cis*-platin-Based Therapy

3.3.1 HMGB1 Affects the Repair of DNA Damaged by the Antitumor Drug cis-Platin In Vitro

Elucidation of the cellular components responsive to chemotherapeutic agents such as *cis*-platin and its derivatives rationalizes the strategy for anticancer chemotherapy. The removal of the *cis*-platin/DNA lesions gives the chance to the cancer cells to survive and compromises the chemotherapy treatment. HMGB1 protein has the potential to affect cytotoxicity of the drug by interfering with lesion recognition and repair. The inhibition of repair of platinated DNA by HMGB1 is thought to result from shielding the lesion in DNA that hampers the access of the NER components of the repair machinery (Chu, 1994; Zamble & Lippard, 1995), sensitizing the cells to the antitumor agent. Logically the higher the affinity to the DNA lesion, the stronger the inhibitory effect of the damage recognition protein should be (Fig. 3). In our laboratory, we tested this hypothesis using naked platinated DNA (Mitkova, Ugrinova, Pashev, & Pasheva, 2005) and *cis*-platin-damaged DNA in nucleosomes, reconstituted on strongly positioned DNA fragments with a fixed length and strictly controlled level of DNA platination (Ugrinova, Zlateva, Pashev, & Pasheva, 2009). The experiments were carried out with three types of protein samples: full-length HMGB1, acetylated HMGB1 (acHMGB1), and truncated HMGB1, lacking the C-terminus negatively charged tail (HMGB1ΔC). The choice was

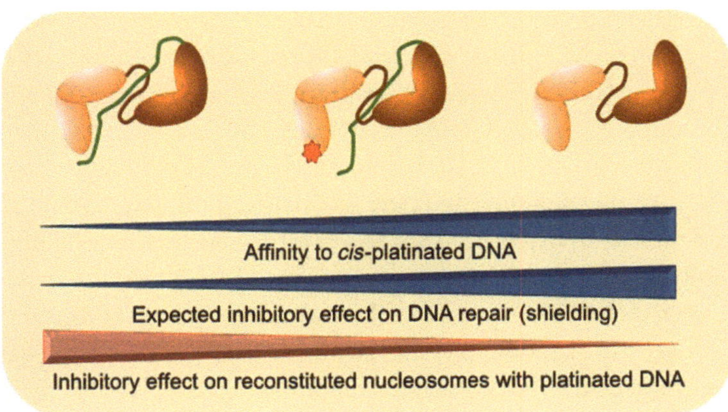

Fig. 3 Schematic presentation of the expected correlation between the affinity of HMGB1 protein samples to *cis*-platinated DNA and the inhibitory effect on DNA repair capacity based on the "shielding" hypothesis and registered inhibitory effect on reconstituted nucleosomes with controlled level of DNA platination.

motivated by the accumulating data that the posttranslational acetylation of HMGB1 at lysine 2 affects its chromatin interaction and binding affinity to DNA (Ugrinova, Mitkova, Moskalenko, Pashev, & Pasheva, 2007; Ugrinova et al., 2001). It was also found that lysine 2 physically interacted with the acidic tail of the protein (Ueda, Chou, Kawase, Shirakawa, & Yoshida, 2004) and its acetylation might disrupt this interaction. The acidic C-tail of HMGB1 also participates in protein–protein interactions (Cato et al., 2008; Ueda et al., 2004) and is necessary for nucleosome sliding (Bonaldi et al., 2002). Moreover, the identification in the nucleus of a specific protease that cleaves the C-terminal portion of HMG1 (Dyson & Walker, 1984) suggests a means for in vivo modulation of HMGB1–DNA interactions. The EMSA analysis revealed that the affinity of the truncated HMGB1 to *cis*-platinated DNA is at least an order of magnitude higher than the affinity of the full-length protein. Furthermore, we recently demonstrated that the affinity of HMGB1 to the same DNA was enhanced about six times upon in vivo acetylation (Ugrinova et al., 2001). The same way of behavior was demonstrated when platinated nucleosomes were used (Ugrinova, Zlateva, et al., 2009). Surprisingly, the expected increase of the inhibitory effect on the DNA repair in the presence of cell-free extract of both modified HMGB1 proteins as a result from eventual more efficient shielding was not observed. On the contrary, acetylated HMGB1 was a weaker inhibitor of adduct repair than the parental protein, while the truncated HMGB1 was without any effect, i.e., the inhibitory effect was completely alleviated. This striking finding was valid for naked platinated DNA as well as for damaged nucleosomes (Fig. 3). One may conclude that the C-terminus is responsible for the inhibition of DNA repair synthesis probably by protein–protein interactions with the components of the NER repair complex.

The main conclusion drawn is that the higher affinity to *cis*-platinated DNA does not result always in stronger inhibition of lesion removal. This fact should be considered for the design of recombinant proteins that are supposed to sensitized cancer cell to treatment with the antitumor drug *cis*-platinum.

3.3.2 The Impact of HMGB1 Protein for the Repair Potential of Human Cancer Cell Lines: A Significant Issue for Therapeutic Strategy

The effect of HMGB1 overexpression in vivo on *cis*-platin-based therapy is still unknown. The repair efficiency of the cancer cells is a substantial factor for the clinical outcome, and it is important to find molecular markers that could be used to discriminate between therapeutic failure and success. HMGB1 is found in most tumor types such as leukemia cells, hepatocellular carcinoma, and gastric and colorectal adenocarcinomas (Kang et al., 2014) and in a majority of human primary breast carcinomas where its level is higher than in the normal tissue (Brezniceanu et al., 2003; Kostova, Zlateva, Ugrinova, & Pasheva, 2010). The protein has a complex role in the cellular response to *cis*-platin lesions. Additionally, HMGB1 itself recognizes and binds such adducts and it interacts with proteins involved in the major repair pathway—NER—that removes these lesions, such as replication protein A (RPA) (Patrick & Turchi, 1999), xeroderma pigmentosum protein A (XPA) (Reddy, Christensen, & Vasquez, 2005), and XPC (xeroderma pigmentosum complementation group C) RAD23B (Lange, Reddy, & Vasquez, 2009). HMGB1 also binds DNA damage signaling proteins, i.e., the tumor suppressor protein p53 (Jayaraman et al., 1998). As mentioned earlier, HMGB1 protein may act at different steps of DNA damage response—from damage recognition to adduct removal and possibly affect each step differently. The effect of HMGB1 on DNA repair in living cells is still not well understood. Studies of its role in DNA repair could discover novel strategies for more effective chemotherapy.

To clarify the HMGB1 role for DNA repair of *cis*-platin damages in vivo, we studied the effect of HMGB1 overexpression and its truncated form on the repair capacity for *cis*-platinated DNA (Yusein-Myashkova, Stoykov, Gospodinov, Ugrinova, & Pasheva, 2016). The experiments were carried out with the two lung cancer cell lines H1299 (p53 null, invasive) and A549 (wt p53, noninvasive) as *cis*-platin is a standard therapeutic for first-line treatment of advanced NSCLC (non-small cell lung cancer) (Azzoli et al., 2009). Ectopically expressed HMGB1 inhibited DNA repair, and the effect was abolished when the truncated form was expressed in the cells, demonstrating the role of the C-tail in vivo. One possibility could be that the inhibition of adduct removal by HMGB1 results from an interaction of its acidic domain with a protein of the repair machinery. A candidate for such a role could be RPA, an eukaryotic single-stranded DNA-binding protein, which is indispensable in vivo for DNA replication, repair, and recombination. Studies with psoralen DNA intrastrand cross-linking demonstrated an interaction of HMGB1 with RPA at these lesions (Lange et al., 2009). The depletion of the cellular HMGB1 by transfection with esiRNA resulted in a well-demonstrated increase in DNA repair levels in H1299 cells, but no such effect was registered in A549-silenced cells. Overexpression of exogenous full-length HMGB1 variants in H1299-silenced cells resulted in a decrease of DNA repair levels. The DNA repair potential was not affected in H1299-silenced cells if a C-terminally truncated form was expressed. The same pattern of behavior was observed when HMGB1 variants were overexpressed in silenced A549 cells. These data provide additional evidence that the removal of the C-terminus eliminates the inhibitory effect of the protein in vivo. What should be pointed out is the difference in the cell repair potential of H1299- and A549-silenced cells, suggesting that unlike simple in vitro repair systems in the living cell, the cell type-dependent factors affect the outcome of the process. In H1299 (p53 null, invasive), obviously the repair efficiency of *cis*-platinated DNA strongly depends on the cellular HMGB1 level, while it was not such the case in A549 (wt p53, noninvasive). The substantial decrease of endogenous HMGB1 did not affect the repair synthesis. Both lines are lung cancer, but they differ in their p53 status. p53 participates directly in NER (Therrien, Drouin, Baril, & Drobetsky, 1999). The tumor suppressor factor p53 could control NER along with HMGB. Based on our findings a conclusion can be drawn that in H1299 lung cancer cell line considered to have higher invasive potential, the repair capacity for *cis*-platin-damaged DNA strongly depends on the expression level of HMGB1 protein. In A549—non-small lung cancer with better prognosis, the repair of *cis*-platinated DNA adducts is determined by the complex action of HMGB1 and p53 (Fig. 4). To further investigate the

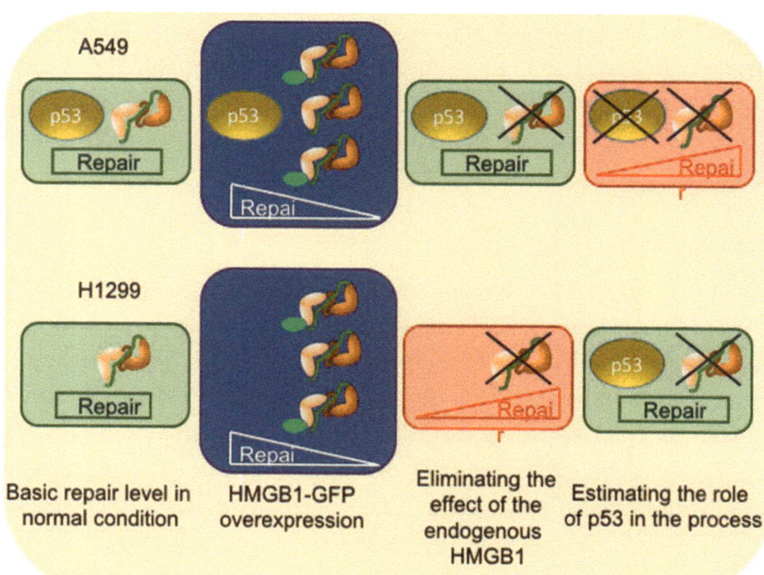

Fig. 4 Schematic presentation of the relationship between HMGB1 and p53 levels for the regulation of DNA repair in H1299 and A549 cell lines.

substantial role of HMGB1 in *cis*-platin repair, the cell viability of H1299 cell line overexpressing GFP-HMGB1 and GFP-HMGB1ΔC was tested (Yusein-Myashkova et al., 2016). The inhibition of the DNA repair by HMGB1 protein resulted in a significant decrease of cell viability and enhancement of the effect of *cis*-platin. The H1299 cells overexpressing the truncated protein were not sensitized to *cis*-platin treatment, and the viability was comparable to the control overexpressing GFP alone. Those findings demonstrated the close relation between the role of HMGB1 in repair of *cis*-platinated DNA and the efficiency of the anticancer drug. Those facts should be considered when chemotherapy is applied for cancer treatment.

4. HMGB1 OUTSIDE THE CELL

Matzinger (1994) discussed the idea that the innate immune system does not care about self and nonself, but acts in a way to protect against danger, by using host-derived mediators now known as DAMP molecules. DAMPs are, in general, nuclear and cytosolic endogenous proteins exerting well-defined intracellular roles in normal condition. However, under cellular stress caused by tissue damage or injury, these molecules promote innate and adaptive immune responses and do not maintain their previous intracellular activities. HMGB1 is one of the first identified members of the DAMP molecular family thanks to the remarkable work of Bianchi and coworkers (Scaffidi et al., 2002) in which they have shown that cells from HMGB1-deficient mice, in contrast to wild-type mice, have a greatly reduced ability to promote inflammation, which proves that the release of HMGB1 can signal the demise of a cell to its neighbors. This work marked the beginning of the new era of HMGB1 research.

4.1 HMGB1 as a Signaling Molecule

The story has began with the seminal work of Wang and colleagues who had discovered the extracellular role of HMGB1 in inflammation and infection by demonstrating that HMGB1 functions as a late mediator with cytokine activity in sepsis (Wang et al., 1999). Since then, growing evidence indicates that HMGB1 is a typical signaling molecule involved in a broad range of physiological and pathological processes including inflammatory and immune response, cell migration, differentiation, proliferation, death and senescence, tissue repair, and angiogenesis (Fig. 2). HMGB1 is released from cells via two mechanisms: the first one is by passive release from necrotic or damaged cells and the second one is by active secretion from the stimulated immune cells. Apoptotic cells on the other hand retain HMGB1 strongly bound to chromatin in membrane-encircled cellular fragments and do not induce any inflammatory response (Scaffidi et al., 2002). If cell death is a result of natural causes, then there is little threat to the organism, so there is "no danger" and the cell remains invisible to the immune system. However, if the cell dies as a consequence of injury or disease, then both defense and repair mechanisms are mobilized and DAMPs become "visible" (Kono & Rock, 2008).

HMGB1 signals via various receptors and can affect almost any type of cell (Pisetsky, Erlandsson-Harris, & Andersson, 2008) causing a huge variety of cellular responses—stimulation of proliferation and migration, induction of chemokine and cytokine responses, attraction and stimulation of immune cells, promotion of autophagy and impairment of apoptosis, and participation in cancerogenesis and metastasis. It is worth mentioning again that the extracellular activities of HMGB1 strongly depend on its redox state because the redox regulation of the three conserved Cys residues (C23, C45, and C106) modifies its receptor-binding ability and the biological consequences (Yang, Antoine, Andersson, & Tracey, 2013). For example, the cytokine-stimulating activity of HMGB1 requires C23 and C45 to be in a disulfide linkage, and at the same time the C106 must remain in its reduced form as a thiol and induce inflammation. Either reduction or oxidation impairs its cytokine activity (Venereau et al., 2012). Fully reduced HMGB1 functions as a chemoattractant for leukocyte migration, by interaction with chemokine CXCL12 (stromal cell-derived factor 1, SDF-1) and with CXCR4 in a synergistic manner. C106 expressing a thiol group in B box is indispensable for both TLR4/MD-2 binding and its subsequent cytokine role. Oxidation or selective mutation in C106 abrogates interaction and its cytokine effect (Yang et al., 2012), probably because of the adjacent TLR4-binding site. Focusing on the TLR4-cytokine axis in macrophages will help the design and development of therapeutics to inhibit HMGB1-mediated inflammation (Yang et al., 2010). The redox status of HMGB1 directly influences its extracellular activity fully oxidized HMGB1 is mainly released from apoptotic cells and it is known that apoptosis induces immunological tolerance and limits inflammation (Venereau et al., 2012). Binding of HMGB1 to TLR2 and TLR4 has been shown to activate MyD88 signaling pathway, thus resulting in the release of proinflammatory cytokines.

4.2 HMGB1 and Diseases

In 2014, Kang and colleagues published a remarkable review titled HMGB1 in health and disease (Kang et al., 2014) an article, which could be called a

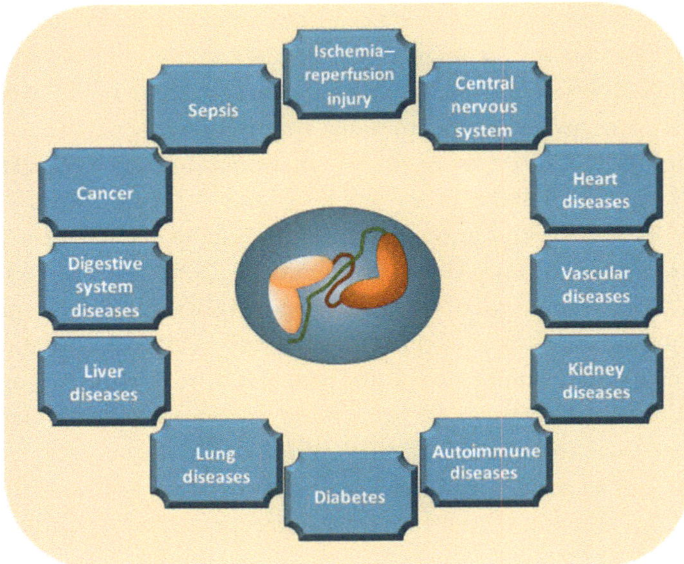

Fig. 5 HMGB1 and diseases.

textbook in the field. Here to avoid repetition and still pay attention to the broad contribution of HMGB1 in pathogenesis and pathophysiology of various diseases, we will do just a brief overview. HMGB1 is proposed to contribute to the pathogenesis of various diseases (Fig. 5), since Wang and colleagues have discovered that HMGB1 (Wang et al., 1999) is a late mediator of sepsis. Increasing body of evidence suggests that it could serve as a therapeutic target with a wider time window for clinical intervention compared to the early mediators such as TNF-α, IL-1β, IL-6, and INF-γ (Andersson & Tracey, 2003, 2011; Kang et al., 2014; Wang, Ward, & Sama, 2009). HMGB1 is an important mediator of ischemia/reperfusion (I/R) injury in the liver, heart, kidney, spinal cord, brain, and intestine and triggers innate immune response (Kang et al., 2014). HMGB1 levels were increased during mice liver I/R as early as 1 h after reperfusion and then increased in a time-dependent manner for up to 24 h (Tsung, Hoffman, et al., 2005). HMGB1 was initially recognized as a heparin-binding protein abundantly expressed in rat brain neurons promoting neurite outgrowth (Rauvala & Pihlaskari, 1987). Increasing number of publications revealed the role of HMGB1 in different pathological conditions of central nervous system including Huntington's (Min et al., 2013), Alzheimer's (Mazarati, Maroso, Iori, Vezzani, & Carli, 2011), and

Parkinson's (Song et al., 2014) diseases, multiple (Das, 2012) and amyotrophic sclerosis (Casula et al., 2011), and neuropathic pain (Wan et al., 2016). HMGB1 is implicated in vascular disorders, in particular systemic vasculitis and atherosclerosis (de Souza, Westra, Limburg, Bijl, & Kallenberg, 2012; Park, Yoon, Tae, & Shim, 2011). Evidence revealed that HMGB1 is involved in organ injury including ischemic, mechanical, allograft rejection and toxicity, and chronic diseases of the heart, kidneys, lungs, and brain (Asavarut, Zhao, Gu, & Ma, 2013). High serum levels of HMGB1 have been found in several inflammatory events, including rheumatoid arthritis (Andersson & Harris, 2010; Qin et al., 2014), chronic kidney disease (Chen, Guan, Zuo, Wang, & Yin, 2016), and systemic lupus erythematosus (Magna & Pisetsky, 2015; Schaper, Westra, & Bijl, 2014). HMGB1 is also implicated in diseases characterized by cell death and damage like diabetes (Koulis, de Haan, & Allen, 2012; Yu, Tang, & Kang, 2015). And last but not least, HMGB1 is demonstrated as one of the major players in cancer. The involvement of HMGB1 in cancer is complex, and intracellular/nuclear and extracellular forms of HMGB1 are involved in tumor formation, progression, and metastasis and in the response to chemotherapeutics. Elevated expression of HMGB1 occurs in several solid tumors, including melanoma, colon cancer, prostate cancer, pancreatic cancer, and breast cancer (Ellerman et al., 2007; Kang et al., 2013; Sims, Rowe, Rietdijk, Herbst, & Coyle, 2010; Tang, Kang, Zeh, & Lotze, 2010). HMGB1 has a dual role in tumorigenesis. The oncogenic functions are mainly connected with the extracellular HMGB1, which mediates and assists signalization and communication between tumor, normal, and immune cells in the tumor microenvironment that is important for tumor growth, invasiveness, metastasis, angiogenesis, and inhibition of antitumor immunity (Kang et al., 2013; Tang et al., 2010). The tumor suppression functions of HMGB1 are attributed to its role in sustaining the genome stability and autophagy regulation, as HMGB1 deficiency leads to an increase of genome instability and autophagy dysfunction and promotes tumorigenesis (Kang et al., 2014).

5. HMGB1 AS A THERAPEUTIC TARGET

The upregulation of HMGB1 is directly associated with the pathogenesis of various human diseases; therefore, it is considered as a promising therapeutic target. In fact, many diseases including sepsis, arthritis, cancer, and diabetes show the tissue damage by inflammatory effectors. It is believed that

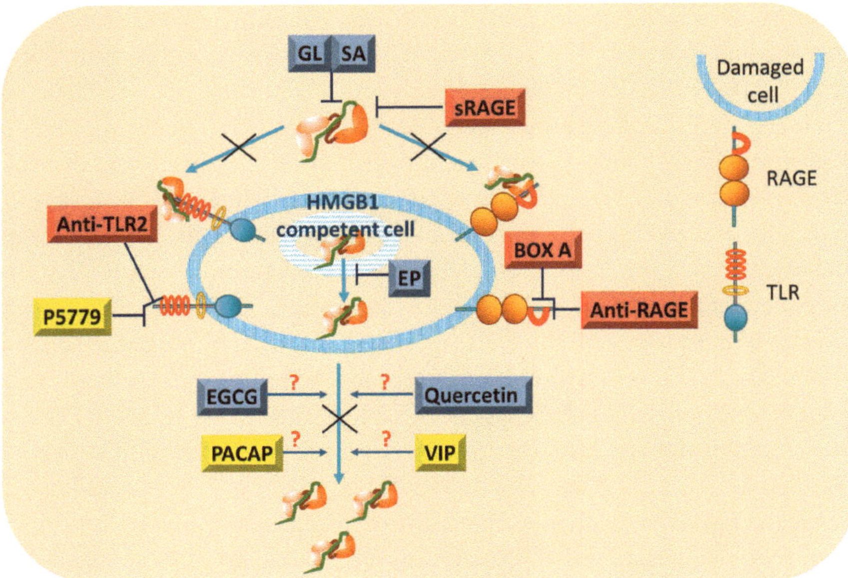

Fig. 6 HMGB1 as a therapeutic target. Multiple inhibitors block HMGB1 in various steps of its activities, revealing a broad range of possible therapeutic strategies.

the control of inflammation could downregulate the tissue damage inflicted by inflammatory mediators. Extracellular HMGB1 is a typical inflammatory mediator, which makes it a key molecular target in multiple diseases (Venereau et al., 2016). There are different approaches for targeting HMGB1 directly or toward HMGB1-receptor signaling (Fig. 6). Those include the use of anti-HMGB1 antibodies, peptide antagonists, soluble receptors, gene targeting, inhibitory molecules, and natural products. Here we will mention briefly the more important developments in this area as a wonderful review on the topic is accepted and soon will be published (Venereau et al., 2016). We will focus on one new but promising approach—HMGB1 targeting by microRNAs (miRs) as therapeutics.

5.1 Antibody-Based Targeting

Anti-HMGB1 polyclonal and monoclonal antibodies are available and are widely used to antagonize extracellular HMGB1 in many experimental models. When the extracellular role of HMGB1 as a late mediator of inflammation was discovered for the first time, it was demonstrated that the delayed administration of anti-HMGB1 antibodies attenuated endotoxin lethality in mice in a sepsis model (Wang et al., 1999). Since then, growing evidence has shown that appropriate administration of neutralizing anti-HMGB1 antibodies has a protective effect in experimental animal models of variety of pathologies including inflammatory disorders such as arthritis (Kokkola et al., 2003; Schierbeck et al., 2011), pancreatitis and pancreatic islet graft transplantation (Gao et al., 2010), pulmonary fibrosis (Hamada et al., 2008), liver diseases such as liver injury and hepatitis (Tsung, Sahai, et al., 2005), and attenuated neuropathic pain (Maeda, Ozaki, Kobayashi, Kiguchi, & Kishioka, 2013; Wan et al., 2016). Numerous studies attest to the therapeutic utility of anti-HMGB1 antibody for cancer treatment including inhibition of the prostate tumor progression (He et al., 2013), angiogenic process of colon cancer (van Beijnum et al., 2013), invasiveness of mesothelioma (Ohmori et al., 2010), and colon cancer development (Luo et al., 2013).

Generally, polyclonal antibodies are not suitable for clinical trials although they still could find application in some cases of digestive system diseases where the immunogenicity can be overcome by a per os application. Many efforts are directed to the development of new monoclonal anti-HMGB1 antibodies, and some of them have been demonstrated to be effective in preclinical models, but none has been tested in a clinical trial (Venereau et al., 2016).

5.2 Peptides and Protein Inhibitors of HMGB1

Another field of HMGB1 targeting is the search for and the development of HMGB1 inhibitors. Endogenous neuropeptides such as vasoactive intestinal peptide, pituitary adenylate cyclase-activating polypeptide, and urocortin were found to rescue mice with established endotoxemia from lethal sepsis by decreasing the level of secreted HMGB1 (Chorny & Delgado, 2008; Tang, Lv, Wang, Xiao, & Zuo, 2008). Thrombomodulin (TM) is a membrane protein, which consists of five domains (TMD1, TMD2, TMD3, TMD4, and TMD5) and serves as a receptor for thrombin inhibiting its procoagulant activity on the surface of vascular endothelial cells. Human-soluble thrombomodulin (ART-123) can physically bind to HMGB1, inhibit secretion of proinflammatory cytokines and HMGB1, and reduce the tissue damage in a rat model of LPS-induced inflammation (Hagiwara et al., 2010).

Screening of an HMGB1-derived peptide library revealed that a tetramer peptide (sequence FSSE, P5779) located within the HMGB1 box B domain (aa 105–108) selectively inhibits the interaction of disulfide HMGB1. P5779 binds MD-2 and disrupts HMGB1–TLR4/MD-2 ligation and subsequent cytokine induction.

Administration of P5779 ameliorated liver damage in sterile injury models induced by liver I/R or by acetaminophen overdose by reducing proinflammatory cytokine release and liver necrosis (Yang, Wang, Ju, et al., 2015).

5.3 Blockage of HMGB1-Receptor Signaling

A promising approach in pharmacology is the blockage of ligand–receptor interaction and downstream signaling. HMGB-A box acts as a competitive antagonist of HMGB1 extracellular function, inhibiting its proinflammatory action mainly by competing with the binding of the full-length protein to RAGE, but does not activate the receptor, as it lacks the proinflammatory cytokine activity localized in the B box (Li et al., 2003). Its administration in vivo protects from lethality in septic mice and ameliorates the symptoms of collagen-induced arthritis in rodents (Kokkola et al., 2003; Sitia, Iannacone, Muller, Bianchi, & Guidotti, 2007; Yang et al., 2004). In vitro studies showed that A box competitively inhibits HMGB1 cell surface binding and attenuates HMGB1-induced proinflammatory cytokine release in macrophage-like RAW 264.7 cells (Yang et al., 2004). HMGB-A box was described to interact with CXCL12 and thus competes with the full-length HMGB1 for signaling via CXCR4 (Schiraldi et al., 2012).

Currently, the HMGB1–RAGE signaling axis represents an important potential target for diseases such as diabetes (Manigrasso, Juranek, Ramasamy, & Schmidt, 2014), neurogeneration (Li, Sun, et al., 2014), cancer (Sims et al., 2010), and inflammatory and autoimmune diseases (Lamore, Cabello, & Wondrak, 2010). It was shown that anti-RAGE antibody inhibits, partially, HMGB1-mediated cytokine secretion (Susa, Masuda, Imaizumi, & Namiki, 2009) and attenuates diaphragm dysfunction in septic rats. Yu et al. showed that the use of anti-TLR2 antibodies decreased or prevented inflammation by attenuating HMGB1-induced IL-8 release in embryonic kidney cells (HEK283) and by reducing HMGB1 cell surface binding on murine macrophage-like RAW 264.7 cells (Yu et al., 2006). Treating animals with the RAGE antagonist FPS-ZM1 significantly reduced inflammatory cell infiltration and expression of IL-1β, and matrix metalloproteinase-9 in the perihematoma after collagenase-induced intracerebral hemorrhage (ICH) in the rat model; interestingly, treatment with

TLR2/4 antagonists did not significantly affect these post-ICH outcomes (Li, Lei, et al., 2015). The soluble form of RAGE (sRAGE) acts as a decoy receptor and has been used successfully to block HMGB1–RAGE signaling in a number of disease models. Even more likely is that sRAGE can sequester ligands, thereby preventing their interaction with other receptors in addition to RAGE. sRAGE has shown to decrease inflammation and the lesion area in a model of atherosclerosis in diabetic mice (Bucciarelli et al., 2002). Administration of recombinant sRAGE decreases the negative outcomes connected to intestinal barrier dysfunction following hemorrhagic shock in mice model (Raman et al., 2006). sRAGE has also been shown to block tumor growth and metastasis (Taguchi et al., 2000).

5.4 Small Molecules Inhibiting HMGB1

HMGB1 is involved in the development of many diseases and pathological conditions, and the interest toward compounds that can reduce its adverse effects is constantly growing. Interestingly but not surprisingly, small molecules from natural sources (mainly well-known medical herbs) proved an effective tool for inhibiting HMGB1 pathological activity. The first candidate in this group is glycyrrhizin (GL)—a natural triterpene found in roots and rhizomes of licorice (*Glycyrrhiza glabra*) and known for its antiinflammatory and antiviral properties. Sakamoto et al. first isolated HMGB1 and 2 proteins from a Gly-affinity column proving a direct interaction (Sakamoto, Okano, Takena, & Ohtsuki, 2001). In a seminal work of Mollica and colleagues, it was shown that GL acts as an effective and direct inhibitor of HMGB1 chemoattractant and mitogenic activities. The authors showed that GL binds directly to the shallow concave surface at the junction of the two arms of both HMGB boxes, near the DNA-binding site (Mollica et al., 2007). Since its discovery as the direct inhibitor of HMGB1, GL has been routinely used in several in vitro and in vivo models. It has been found that GL inhibited the expression of HMGB1 by Kupffer cells in rats after liver injury caused by I/R (Ogiku, Kono, Hara, Tsuchiya, & Fujii, 2011). GL was found to attenuate the transient spinal cord ischemic injury in rats via reducing inflammatory cytokines and inhibiting the release of HMGB1 (Gong et al., 2012), to alleviate rat myocardial I/R-induced injury through direct inhibition of extracellular HMGB1 cytokine activity (Zhai et al., 2012), to inhibit HMGB1–TLR-4 signaling that contributes to the attenuation of cold ischemic injury of liver in a rat model (Su et al., 2016), and to reduce myocardial inflammation in mice

(Bangert et al., 2016). Another interesting feature is the significant improvement of the effectiveness of anticancer therapies based on the combination of GL with proper drug-like peptide CAMEL, which induces the necrotic type of tumor cell death (Smolarczyk et al., 2012).

Another famous and widely used natural antiinflammatory drug is salicylic acid (SA), and its acetyl ester known as aspirin, which is a well-described irreversible inhibitor of the cyclooxygenases 1 and 2. Recently, it was discovered that HMGB1 is an SA-binding protein and even more important that SA binds to both Boxes A and B to the same binding sites as GL (Choi et al., 2015). Quercetin is another natural compound reported to reduce circulating levels of HMGB1 in animals with established endotoxemia. Quercetin is a plant-derived flavonoid found in fruits, vegetables, leaves, and grains. In macrophage cultures, quercetin inhibited the release and the cytokine activities of HMGB1 and limited the activation of mitogen-activated protein kinase (Tang et al., 2009). The major component of the green tea (the plant *Camellia sinensis*)—(−)-epigallocatechin-3-gallate—was found to protect mice against lethal endotoxemia and rescued them from lethal sepsis by lowering HMGB1-mediated inflammatory responses (Li et al., 2007). A folic acid antagonist methotrexate MTX conventionally used as the chemotherapeutic agent and the antiinflammatory drug was found to bind to two independent sites of HMGB1, one of which includes the RAGE-binding region. In mouse macrophage-like cells, MTX blocks tumor necrosis factor (TNF)-α release as well as the mitogenic activity induced by specific RAGE stimulation (Kuroiwa et al., 2013). Ethyl pyruvate (EP) is a stable aliphatic ester of pyruvic acid. Several studies have confirmed that EP is a potent inhibitor of HMGB1 nuclear-to-cytoplasmic translocation and secretion (Dave et al., 2009; Ulloa et al., 2002). EP exerts antitumor effects on the growth of gastric (Dave et al., 2009) and gallbladder cancer (GBC) cells via downregulation of HMGB1–RAGE axis. EP administration decreased the expression of HMGB1, RAGE, PCNA, and matrix metallopeptidase-9 (MMP-9), reduced human GBC cell proliferation, inhibited the invasive potential, and induced apoptosis and cycle arrest in S-phase GBC cells in vitro (Li et al., 2012).

5.5 Oligonucleotide (ODN)-Based HMGB1 Inhibitors

This strategy is based on the role of HMGB1 in the nucleus where it acts as an architectural factor. More specifically on its ability to recognize with high affinity a wide range of distorted DNA structures as it was described earlier. Musumeci and coworkers have shown that suitable DNA-based ligands containing kinked DNA duplexes were recognized by HMGB1 with high affinity ($K_D = 20$ nM) and were able to inhibit HMGB1-induced cell proliferation and migration in a concentration-dependent manner (Bucci et al., 2007; Musumeci et al., 2011, 2007). After optimization and chemical modifications the authors invented a kinked hairpin-loop DNA able to inhibit the protein-induced cellular migration in a concentration-dependent manner, with IC_{50} three orders of magnitude lower than the known HMGB1-inhibitor GLg (Musumeci, Roviello, & Montesarchio, 2014).

6. HMGB1 AS A TARGET OF miRs

miRs are small, noncoding RNA sequences about 22 nt nucleotides long, which bind to complementary sequences generally located within the 3′-untranslated region (UTR) of target transcript RNAs (Fig. 7A). They act as intracellular mediators that are necessary for various biological processes. miRs targeting pathways of human disease provide a new and potential powerful candidate for therapeutic intervention against various pathological conditions. HMGB1 is involved in the pathogenesis of multiple diseases, because of its regulatory role in multiple signaling pathways, including inflammation, immunity, genome stability, proliferation, metastasis, metabolism, apoptosis, and autophagy. Now we know that all of these processes are more or less controlled by the effects of different miRs. Not surprisingly, growing number of studies have conclusively demonstrated that several miRNAs directly bind to the 3′-UTR of HMGB1 gene and are involved in the

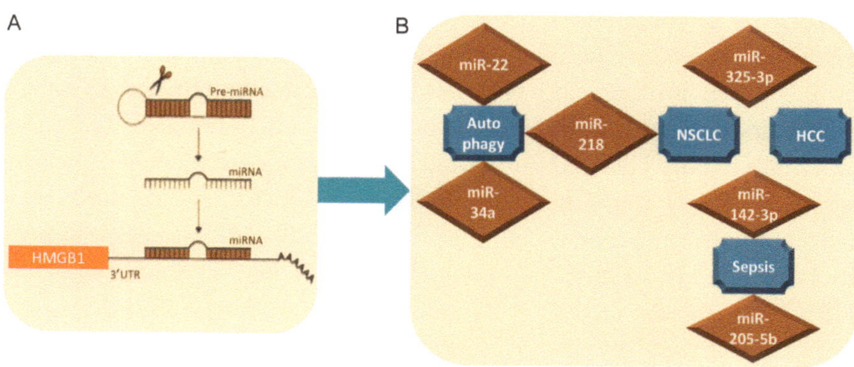

Fig. 7 microRNAs targeting HMGB1. 3′-UTR of HMGB1 is a target of many miRs (A). The inhibition of HMGB1 expression by specific miRs (indicated with yellow diamonds) causes positive outcome in different pathologies (indicated with blue rectangles) (B).

regulation of different pathologies (Fig. 7B). However in some reports a more complex mechanism is pointed out where HMGB1 through certain signaling pathways is able to regulate the expression of some miRs.

6.1 HMGB1 as a Direct Target of miRs

Accumulating evidences have shown that miRs may be involved in NSCLC pathogenesis, providing new insights into disease biology. By now at least three of them, miR-142-3p (Xiao & Liu, 2015), miR-325-3p (Yao, Zhao, & Jin, 2015), and miR-218 (Zhang, Ge, Hu, Yang, & Zhang, 2013), have been shown to function as potential tumor suppressors directly targeting and negatively regulating HMGB1 in NSCLC. Interestingly, miR-142-3p was also found to regulate directly and negatively RAC1 in HCC cells (Wu et al., 2011). Overexpression of these miRs inhibits NSCLC cell proliferation, migration and invasion, and induced cell apoptosis. Downregulation of these miRs in NSCLC might be used as biomarker for bad prognosis.

Another study has shown that the miR-325 inhibits hepatocellular carcinoma progression by targeting HMGB1 (Li, Huang, & Luo, 2015). The authors showed that miR-325 is decreased and HMGB1 is increased in 99 patients with hepatocellular carcinoma. miR-325 inhibition promoted cell invasion and proliferation, while miR-325 upregulation negatively regulates HMGB1 and inhibits cell invasion and proliferation. Here again downregulation of miR-325 predicts poor prognosis for HCC patients.

miR-22 was found to inhibit osteosarcoma cell (OSC) proliferation, migration, and invasion by blockage of autophagy. It was demonstrated that miR-22 well paired with the 3′-UTR of HMGB1 downregulated the HMGB1 expression and blocked the HMGB1-facilitated autophagy (Guo et al., 2014). miR-22 and HMGB1 were found upregulated in OSCs, postchemotherapy. However, the overexpressed miR-22 targeted HMGB1 and inhibited the HMGB1-promoted autophagy and so increased the drug sensitivity (Li, Wang, Chen, Liu, & Yang, 2014). Other two miRs were reported with similar activities. The first one miR-218 was demonstrated to bind directly to the 3′-UTR of HMGB1 gene and to inhibit the HMGB1-mediated autophagy in endometrial carcinoma cells during paclitaxel chemotherapy in vitro (Ran et al., 2015). The other one miR-34a was shown to downregulate HMGB1, which leads to inhibition of autophagy and enhances chemotherapy-induced apoptosis and tumor cell death in the retinoblastoma cell (Liu et al., 2014).

The role of miRs as HMGB1 translational regulators is not connected only with cancer pathology. For instance, Wang and coworkers (2016) have shown that miR-142-3p was significantly reduced in the articular cartilage tissues from experimental osteoarthritis OA mice. Moreover, the overexpression of miR-142-3p significantly inhibited cell apoptosis, NF-kB, and the production of proinflammatory cytokines, including interleukin (IL)-1, IL-6, and TNF-α induced by LPS. As it was mentioned, HMGB1 is a target of miR-142-3p. In OA mice, the overexpression of miR-142-3p significantly inhibited HMGB1 expression, NF-kB signaling, and proinflammatory cytokines. In conclusion the authors suggested that miR-142-3p inhibits chondrocyte apoptosis and inflammation in OA by inhibiting the HMGB1-mediated NF-kB signaling pathway. This study indicates that miR-142-3p might be a potential molecular target for OA treatment.

miR-205-5b was found to inhibit HMGB1 expression in LPS-induced sepsis (Zhou, Wang, Li, Li, & Sang, 2016). Animal experiments with miR-205-5b have shown its high degree of expression in the serum, colon, spleen, liver, and lungs following the downregulation of HMGB1 in the tissues, suggesting translational regulation. HMGB1 expression negatively correlated with miR-205-5b expression in LPS-induced sepsis. So the upregulation of miR-205-5b targeting HMGB1 could be a novel therapeutic strategy against late sepsis. HMGB1 was reported to play an important role in myocardial I/R injury. In a model of neonatal rat ventricular cardiomyocytes, it was shown that miR-451 targets HMGB1 and upregulation of miR'451 could protect against I/R-induced cardiomyocyte injury by inhibiting HMGB1 expression (Xie et al., 2016).

6.2 HMGB1 as a Regulator of miR Expression

In addition to "being regulated," HMGB1 was also found as a regulator of miR functions. miR-221 and miR-222 have been found to be deregulated in human papillary thyroid carcinomas. They are involved in cell proliferation through the inhibition of the cell cycle regulator, p27kip1, in human papillary carcinomas. It was shown that HMGB1 increases the expression of miR-221 and miR-222 in primary cultures of excised papillary lesions and in an established papillary cancer cell line (BC PAP) (Mardente et al., 2012). The overexpression of oncogenic miR-221 and miR-222 caused by HMGB1 was associated with an increase in malignancy scores, namely cell growth and motility. Moreover, it was shown that the extracellular HMGB1 interaction with RAGE enhances the expression of oncogenic cluster miR-221/222 that in turn inhibits tumor suppressor gene *PTEN* in two cell lines derived from human thyroid anaplastic and papillary cancers (Mardente et al., 2015). The proposed novel pathway including HMGB1/RAGE/miR-221/222 may represent an effective way of tumor escape from immune surveillance that could be further developed in new therapeutic strategies against anaplastic tumors.

An original work of Chen and coworkers has shown that HMGB1 promotes hepatocellular carcinoma progression through miR-21-mediated matrix metalloproteinase activity (Chen et al., 2015). During liver inflammation, HMGB1 is released and provides a favorable environment for HCC growth. It was demonstrated that HMGB1 signaling increases miR-21 expression to mediate the enhanced activity of MMPs through RECK and TIMP3. Thus a novel mechanism for HMGB1-mediated HCC progression through the IL6/Stat3-miR-21 axis has been provided.

7. CONCLUDING REMARKS

HMGB1 is an important effector molecule and plays a significant role in a large number of pathologies including inflammation, injury, cancer, and autoimmune diseases. HMGB1 overexpression is associated with poorer prognosis in various types of cancer, suggesting that it is a proper prognostic biomarker. The abundance of the protein is of big importance for its signaling properties; for instance, one dying necrotic cell provides approximately 10^6 HMGB1 molecules, which are potential signals for danger. Numerous strategies are being employed to inhibit HMGB1 release and extracellular activity most of them with good potential. The results are promising but we should keep in mind that all attempts are performed on animal models. A lot of work has to be done to apply these strategies in preclinical and clinical trials. No doubt, however, HMGB1 is unambiguously proved as a promising therapeutic target for multiple diseases. It is time for this unique molecule to become a subject of intensive pharmacological studies and drug design.

REFERENCES

Akira, S., & Takeda, K. (2004). Toll-like receptor signalling. *Nature Reviews. Immunology*, *4*(7), 499–511. http://dx.doi.org/10.1038/nri1391.

Alexandrova, E. A., & Beltchev, B. G. (1988). Acetylated HMG1 protein interacts specifically with homologous DNA polymerase alpha in vitro. *Biochemical and Biophysical Research Communications*, *154*(3), 918–927.

Andersson, U., & Harris, H. E. (2010). The role of HMGB1 in the pathogenesis of rheumatic disease. *Biochimica et Biophysica Acta, 1799*(1–2), 141–148. http://dx.doi.org/10.1016/j.bbagrm.2009.11.003.

Andersson, U., & Tracey, K. J. (2003). HMGB1 in sepsis. *Scandinavian Journal of Infectious Diseases, 35*(9), 577–584.

Andersson, U., & Tracey, K. J. (2011). HMGB1 is a therapeutic target for sterile inflammation and infection. *Annual Review of Immunology, 29*, 139–162. http://dx.doi.org/10.1146/annurev-immunol-030409-101323.

Apetoh, L., Ghiringhelli, F., Tesniere, A., Obeid, M., Ortiz, C., Criollo, A., … Zitvogel, L. (2007). Toll-like receptor 4-dependent contribution of the immune system to anticancer chemotherapy and radiotherapy. *Nature Medicine, 13*(9), 1050–1059. http://dx.doi.org/10.1038/nm1622.

Asavarut, P., Zhao, H., Gu, J., & Ma, D. (2013). The role of HMGB1 in inflammation-mediated organ injury. *Acta Anaesthesiologica Taiwanica, 51*(1), 28–33. http://dx.doi.org/10.1016/j.aat.2013.03.007.

Assenberg, R., Webb, M., Connolly, E., Stott, K., Watson, M., Hobbs, J., & Thomas, J. O. (2008). A critical role in structure-specific DNA binding for the acetylatable lysine residues in HMGB1. *The Biochemical Journal, 411*(3), 553–561. http://dx.doi.org/10.1042/BJ20071613.

Azzoli, C. G., Baker, S., Jr., Temin, S., Pao, W., Aliff, T., Brahmer, J., … American Society of Clinical Oncology (2009). American society of clinical oncology clinical practice guideline update on chemotherapy for stage IV non-small-cell lung cancer. *Journal of Clinical Oncology, 27*(36), 6251–6266. http://dx.doi.org/10.1200/JCO.2009.23.5622.

Bangert, A., Andrassy, M., Muller, A. M., Bockstahler, M., Fischer, A., Volz, C. H., … Kaya, Z. (2016). Critical role of RAGE and HMGB1 in inflammatory heart disease. *Proceedings of the National Academy of Sciences of the United States of America, 113*(2), E155–E164. http://dx.doi.org/10.1073/pnas.1522288113.

Bauer, E. M., Shapiro, R., Billiar, T. R., & Bauer, P. M. (2013). High mobility group Box 1 inhibits human pulmonary artery endothelial cell migration via a Toll-like receptor 4- and interferon response factor 3-dependent mechanism(s). *The Journal of Biological Chemistry, 288*(2), 1365–1373. http://dx.doi.org/10.1074/jbc.M112.434142.

Bianchi, M. E., Beltrame, M., & Paonessa, G. (1989). Specific recognition of cruciform DNA by nuclear protein HMG1. *Science, 243*(4894 Pt. 1), 1056–1059.

Bianchi, M. E., Falciola, L., Ferrari, S., & Lilley, D. M. (1992). The DNA binding site of HMG1 protein is composed of two similar segments (HMG boxes), both of which have counterparts in other eukaryotic regulatory proteins. *EMBO Journal, 11*(3), 1055–1063.

Bonaldi, T., Langst, G., Strohner, R., Becker, P. B., & Bianchi, M. E. (2002). The DNA chaperone HMGB1 facilitates ACF/CHRAC-dependent nucleosome sliding. *The EMBO Journal, 21*(24), 6865–6873.

Bonaldi, T., Talamo, F., Scaffidi, P., Ferrera, D., Porto, A., Bachi, A., … Bianchi, M. E. (2003). Monocytic cells hyperacetylate chromatin protein HMGB1 to redirect it towards secretion. *The EMBO Journal, 22*(20), 5551–5560. http://dx.doi.org/10.1093/emboj/cdg516.

Bonne-Andrea, C., Harper, F., Sobczak, J., & De Recondo, A. M. (1984). The role of HMG1 protein in nucleosome assembly and in chromatin replication. *Adv. Exp. Med. Biol., 179*, 479–488.

Brezniceanu, M. L., Volp, K., Bosser, S., Solbach, C., Lichter, P., Joos, S., & Zornig, M. (2003). HMGB1 inhibits cell death in yeast and mammalian cells and is abundantly expressed in human breast carcinoma. *The FASEB Journal, 17*(10), 1295–1297. http://dx.doi.org/10.1096/fj.02-0621fje.

Bucci, E. M., Valente, M., Musumeci, D., Sapio, R., Anro, V., Roviello, G. N., & Moccia, M. (2007). ODN-based drugs for targeting of extracellular proteins. *Nucleosides, Nucleotides & Nucleic Acids, 26*(8–9), 1047–1050. http://dx.doi.org/10.1080/15257770701509602.

Bucciarelli, L. G., Wendt, T., Qu, W., Lu, Y., Lalla, E., Rong, L. L., ... Schmidt, A. M. (2002). RAGE blockade stabilizes established atherosclerosis in diabetic apolipoprotein E-null mice. *Circulation, 106*(22), 2827–2835.

Casula, M., Iyer, A. M., Spliet, W. G., Anink, J. J., Steentjes, K., Sta, M., ... Aronica, E. (2011). Toll-like receptor signaling in amyotrophic lateral sclerosis spinal cord tissue. *Neuroscience, 179*, 233–243. http://dx.doi.org/10.1016/j.neuroscience.2011.02.001.

Cato, L., Stott, K., Watson, M., & Thomas, J. O. (2008). The interaction of HMGB1 and linker histones occurs through their acidic and basic tails. *Journal of Molecular Biology, 384*(5), 1262–1272. http://dx.doi.org/10.1016/j.jmb.2008.10.001.

Celona, B., Weiner, A., Di Felice, F., Mancuso, F. M., Cesarini, E., Rossi, R. L., ... Agresti, A. (2011). Substantial histone reduction modulates genomewide nucleosomal occupancy and global transcriptional output. *PLoS Biology*. *9*(6). e1001086. http://dx.doi.org/10.1371/journal.pbio.1001086.

Chen, Q., Guan, X., Zuo, X., Wang, J., & Yin, W. (2016). The role of high mobility group box 1 (HMGB1) in the pathogenesis of kidney diseases. *Acta Pharmaceutica Sinica B, 6*(3), 183–188. http://dx.doi.org/10.1016/j.apsb.2016.02.004.

Chen, M., Liu, Y., Varley, P., Chang, Y., He, X. X., Huang, H., ... Tsung, A. (2015). High-mobility group box 1 promotes hepatocellular carcinoma progression through miR-21-mediated matrix metalloproteinase activity. *Cancer Research, 75*(8), 1645–1656. http://dx.doi.org/10.1158/0008-5472.CAN-14-2147.

Chiba, S., Baghdadi, M., Akiba, H., Yoshiyama, H., Kinoshita, I., Dosaka-Akita, H., ... Jinushi, M. (2012). Tumor-infiltrating DCs suppress nucleic acid-mediated innate immune responses through interactions between the receptor TIM-3 and the alarmin HMGB1. *Nature Immunology, 13*(9), 832–842. http://dx.doi.org/10.1038/ni.2376.

Choi, H. W., Tian, M., Song, F., Venereau, E., Preti, A., Park, S. W., ... Klessig, D. F. (2015). Aspirin's active metabolite salicylic acid targets high mobility group box 1 to modulate inflammatory responses. *Molecular Medicine, 21*, 526–535. http://dx.doi.org/10.2119/molmed.2015.00148.

Chorny, A., & Delgado, M. (2008). Neuropeptides rescue mice from lethal sepsis by down-regulating secretion of the late-acting inflammatory mediator high mobility group box 1. *The American Journal of Pathology, 172*(5), 1297–1307. http://dx.doi.org/10.2353/ajpath.2008.070969.

Chu, G. (1994). Cellular responses to cisplatin. The roles of DNA-binding proteins and DNA repair. *The Journal of Biological Chemistry, 269*(2), 787–790.

Ciccarelli, R. B., Solomon, M. J., Varshavsky, A., & Lippard, S. J. (1985). In vivo effects of cis- and trans-diamminedichloroplatinum(II) on SV40 chromosomes: Differential repair, DNA-protein cross-linking, and inhibition of replication. *Biochemistry, 24*(26), 7533–7540.

Conti, L., Lanzardo, S., Arigoni, M., Antonazzo, R., Radaelli, E., Cantarella, D., ... Cavallo, F. (2013). The noninflammatory role of high mobility group box 1/Toll-like receptor 2 axis in the self-renewal of mammary cancer stem cells. *The FASEB Journal, 27*(12), 4731–4744. http://dx.doi.org/10.1096/fj.13-230201.

Das, U. N. (2012). Is multiple sclerosis a proresolution deficiency disorder? *Nutrition, 28*(10), 951–958. http://dx.doi.org/10.1016/j.nut.2011.12.016.

Das, D., & Scovell, W. M. (2001). The binding interaction of HMG-1 with the TATA-binding protein/TATA complex. *The Journal of Biological Chemistry, 276*(35), 32597–32605. http://dx.doi.org/10.1074/jbc.M011792200.

Dave, S. H., Tilstra, J. S., Matsuoka, K., Li, F., DeMarco, R. A., Beer-Stolz, D., ... Plevy, S. E. (2009). Ethyl pyruvate decreases HMGB1 release and ameliorates murine colitis. *Journal of Leukocyte Biology, 86*(3), 633–643. http://dx.doi.org/10.1189/jlb.1008662.

de Silva, S., Lotta, L. T., Jr., Burris, C. A., & Bowers, W. J. (2010). Virion-associated cofactor high-mobility group DNA-binding protein-1 facilitates transposition from the herpes simplex virus/Sleeping Beauty amplicon vector platform. *Human Gene Therapy, 21*(11), 1615–1622.

de Souza, A. W., Westra, J., Limburg, P. C., Bijl, M., & Kallenberg, C. G. (2012). HMGB1 in vascular diseases: Its role in vascular inflammation and atherosclerosis. *Autoimmunity Reviews, 11*(12), 909–917. http://dx.doi.org/10.1016/j.autrev.2012.03.007.

Degryse, B., Bonaldi, T., Scaffidi, P., Muller, S., Resnati, M., Sanvito, F., ... Bianchi, M. E. (2001). The high mobility group (HMG) boxes of the nuclear protein HMG1 induce chemotaxis and cytoskeleton reorganization in rat smooth muscle cells. *The Journal of Cell Biology, 152*(6), 1197–1206.

Duguet, M., Soussi, T., Rossignol, J. M., Mechali, M., & De Recondo, A. M. (1977). Stimulation of rat liver alpha- and beta-type DNA polymerases by an homologous DNA-unwinding protein. *FEBS Letters, 79*(1), 160–164.

Dyson, M., & Walker, J. M. (1984). Chromatin associated protease from calf thymus. *International Journal of Peptide and Protein Research, 24*(3), 201–207.

El Mezayen, R., El Gazzar, M., Seeds, M. C., McCall, C. E., Dreskin, S. C., & Nicolls, M. R. (2007). Endogenous signals released from necrotic cells augment inflammatory responses to bacterial endotoxin. *Immunology Letters, 111*(1), 36–44. http://dx.doi.org/10.1016/j.imlet.2007.04.011.

Elenkov, I., Pelovsky, P., Ugrinova, I., Takahashi, M., & Pasheva, E. (2011). The DNA binding and bending activities of truncated tail-less HMGB1 protein are differentially affected by Lys-2 and Lys-81 residues and their acetylation. *International Journal of Biological Sciences*, 7(6), 691–699.

Ellerman, J. E., Brown, C. K., de Vera, M., Zeh, H. J., Billiar, T., Rubartelli, A., & Lotze, M. T. (2007). Masquerader: High mobility group box-1 and cancer. *Clinical Cancer Research, 13*(10), 2836–2848. http://dx.doi.org/10.1158/1078-0432.CCR-06-1953.

Gaillard, C., & Strauss, F. (2000). DNA loops and semicatenated DNA junctions. *BMC Biochemistry, 1*, 1. http://dx.doi.org/10.1186/1471-2091-1-1.

Gao, Q., Ma, L. L., Gao, X., Yan, W., Williams, P., & Yin, D. P. (2010). TLR4 mediates early graft failure after intraportal islet transplantation. *American Journal of Transplantation, 10*(7), 1588–1596. http://dx.doi.org/10.1111/j.1600-6143.2010.03151.x.

Gong, G., Yuan, L. B., Hu, L., Wu, W., Yin, L., Hou, J. L., ... Zhou, L. S. (2012). Glycyrrhizin attenuates rat ischemic spinal cord injury by suppressing inflammatory cytokines and HMGB1. *Acta Pharmacologica Sinica, 33*(1), 11–18. http://dx.doi.org/10.1038/aps.2011.151.

Goodwin, G. H., Sanders, C., & Johns, E. W. (1973). A new group of chromatin-associated proteins with a high content of acidic and basic amino acids. *European Journal of Biochemistry, 38*(1), 14–19.

Grosschedl, R., Giese, K., & Pagel, J. (1994). HMG domain proteins: Architectural elements in the assembly of nucleoprotein structures. *Trends in Genetics, 10*(3), 94–100.

Guo, S., Bai, R., Liu, W., Zhao, A., Zhao, Z., Wang, Y., ... Wang, W. (2014). miR-22 inhibits osteosarcoma cell proliferation and migration by targeting HMGB1 and inhibiting HMGB1-mediated autophagy. *Tumour Biology, 35*(7), 7025–7034. http://dx.doi.org/10.1007/s13277-014-1965-2.

Hagiwara, S., Iwasaka, H., Matsumoto, S., Hasegawa, A., Yasuda, N., & Noguchi, T. (2010). In vivo and in vitro effects of the anticoagulant, thrombomodulin, on the inflammatory response in rodent models. *Shock, 33*(3), 282–288. http://dx.doi.org/10.1097/SHK.0b013e3181b0ef7b.

Hamada, N., Maeyama, T., Kawaguchi, T., Yoshimi, M., Fukumoto, J., Yamada, M., ... Nakanishi, Y. (2008). The role of high mobility group box1 in pulmonary fibrosis.

American Journal of Respiratory Cell and Molecular Biology, *39*(4), 440–447. http://dx.doi.org/10.1165/rcmb.2007-0330OC.

He, M., Kubo, H., Morimoto, K., Fujino, N., Suzuki, T., Takahasi, T., … Yamamoto, H. (2011). Receptor for advanced glycation end products binds to phosphatidylserine and assists in the clearance of apoptotic cells. *EMBO Reports*, *12*(4), 358–364. http://dx.doi.org/10.1038/embor.2011.28.

He, Y., Zha, J., Wang, Y., Liu, W., Yang, X., & Yu, P. (2013). Tissue damage-associated "danger signals" influence T-cell responses that promote the progression of preneoplasia to cancer. *Cancer Research*, *73*(2), 629–639. http://dx.doi.org/10.1158/0008-5472.CAN-12-2704.

Hori, O., Brett, J., Slattery, T., Cao, R., Zhang, J., Chen, J. X., et al. (1995). The receptor for advanced glycation end products (RAGE) is a cellular binding site for amphoterin. Mediation of neurite outgrowth and co-expression of rage and amphoterin in the developing nervous system. *The Journal of Biological Chemistry*, *270*(43), 25752–25761.

Huttunen, H. J., Fages, C., & Rauvala, H. (1999). Receptor for advanced glycation end products (RAGE)-mediated neurite outgrowth and activation of NF-kappaB require the cytoplasmic domain of the receptor but different downstream signaling pathways. *The Journal of Biological Chemistry*, *274*(28), 19919–19924.

Jantzen, H. M., Admon, A., Bell, S. P., & Tjian, R. (1990). Nucleolar transcription factor hUBF contains a DNA-binding motif with homology to HMG proteins. *Nature*, *344*(6269), 830–836. http://dx.doi.org/10.1038/344830a0.

Jayaraman, L., Moorthy, N. C., Murthy, K. G., Manley, J. L., Bustin, M., & Prives, C. (1998). High mobility group protein-1 (HMG-1) is a unique activator of p53. *Genes & Development*, *12*(4), 462–472.

Kang, R., Chen, R., Zhang, Q., Hou, W., Wu, S., Cao, L., … Tang, D. (2014). HMGB1 in health and disease. *Molecular Aspects of Medicine*, *40*, 1–116. http://dx.doi.org/10.1016/j.mam.2014.05.001.

Kang, R., Zhang, Q., Zeh, H. J., 3rd, Lotze, M. T., & Tang, D. (2013). HMGB1 in cancer: Good, bad, or both? *Clinical Cancer Research*, *19*(15), 4046–4057. http://dx.doi.org/10.1158/1078-0432.CCR-13-0495.

Kim, K., Han, J. S., Kim, H. A., & Lee, M. (2008). Expression, purification and characterization of TAT-high mobility group box-1A peptide as a carrier of nucleic acids. *Biotechnology Letters*, *30*(8), 1331–1337. http://dx.doi.org/10.1007/s10529-008-9695-4.

Kokkola, R., Li, J., Sundberg, E., Aveberger, A. C., Palmblad, K., Yang, H., … Harris, H. E. (2003). Successful treatment of collagen-induced arthritis in mice and rats by targeting extracellular high mobility group box chromosomal protein 1 activity. *Arthritis and Rheumatism*, *48*(7), 2052–2058. http://dx.doi.org/10.1002/art.11161.

Kono, H., & Rock, K. L. (2008). How dying cells alert the immune system to danger. *Nature Reviews. Immunology*, *8*(4), 279–289. http://dx.doi.org/10.1038/nri2215.

Kostova, N., Zlateva, S., Ugrinova, I., & Pasheva, E. (2010). The expression of HMGB1 protein and its receptor RAGE in human malignant tumors. *Molecular and Cellular Biochemistry*, *337*(1–2), 251–258. http://dx.doi.org/10.1007/s11010-009-0305-0.

Koulis, C., de Haan, J. B., & Allen, T. J. (2012). Novel pathways and therapies in experimental diabetic atherosclerosis. *Expert Review of Cardiovascular Therapy*, *10*(3), 323–335. http://dx.doi.org/10.1586/erc.12.13.

Kuroiwa, Y., Takakusagi, Y., Kusayanagi, T., Kuramochi, K., Imai, T., Hirayama, T., … Sugawara, F. (2013). Identification and characterization of the direct interaction between methotrexate (MTX) and high-mobility group box 1 (HMGB1) protein. *PLoS One*, *8*(5). e63073. http://dx.doi.org/10.1371/journal.pone.0063073.

Lamore, S. D., Cabello, C. M., & Wondrak, G. T. (2010). HMGB1-directed drug discovery targeting cutaneous inflammatory dysregulation. *Current Drug Metabolism*, *11*(3), 250–265.

Lange, S. S., Reddy, M. C., & Vasquez, K. M. (2009). Human HMGB1 directly facilitates interactions between nucleotide excision repair proteins on triplex-directed psoralen interstrand crosslinks. *DNA Repair (Amst)*, *8*(7), 865–872. http://dx.doi.org/10.1016/j.dnarep.2009.04.001.

Lange, S. S., & Vasquez, K. M. (2009). HMGB1: The jack-of-all-trades protein is a master DNA repair mechanic. *Molecular Carcinogenesis*, *48*(7), 571–580. http://dx.doi.org/10.1002/mc.20544.

Li, W., Ashok, M., Li, J., Yang, H., Sama, A. E., & Wang, H. (2007). A major ingredient of green tea rescues mice from lethal sepsis partly by inhibiting HMGB1. *PLoS One*, *2*(11). e1153. http://dx.doi.org/10.1371/journal.pone.0001153.

Li, H., Huang, W., & Luo, R. (2015). The microRNA-325 inhibits hepatocellular carcinoma progression by targeting high mobility group box 1. *Diagnostic Pathology*, *10*, 117. http://dx.doi.org/10.1186/s13000-015-0323-z.

Li, J., Kokkola, R., Tabibzadeh, S., Yang, R., Ochani, M., Qiang, X., ... Yang, H. (2003). Structural basis for the proinflammatory cytokine activity of high mobility group box 1. *Molecular Medicine*, *9*(1–2), 37–45.

Li, D., Lei, C., Zhang, S., Zhang, S., Liu, M., & Wu, B. (2015). Blockade of high mobility group box-1 signaling via the receptor for advanced glycation end-products ameliorates inflammatory damage after acute intracerebral hemorrhage. *Neuroscience Letters*, *609*, 109–119. http://dx.doi.org/10.1016/j.neulet.2015.10.035.

Li, M., Sun, L., Luo, Y., Xie, C., Pang, Y., & Li, Y. (2014). High-mobility group box 1 released from astrocytes promotes the proliferation of cultured neural stem/progenitor cells. *International Journal of Molecular Medicine*, *34*(3), 705–714. http://dx.doi.org/10.3892/ijmm.2014.1820.

Li, X., Wang, S., Chen, Y., Liu, G., & Yang, X. (2014). miR-22 targets the 3′ UTR of HMGB1 and inhibits the HMGB1-associated autophagy in osteosarcoma cells during chemotherapy. *Tumour Biology*, *35*(6), 6021–6028. http://dx.doi.org/10.1007/s13277-014-1797-0.

Li, M. L., Wang, X. F., Tan, Z. J., Dong, P., Gu, J., Lu, J. H., ... Liu, Y. B. (2012). Ethyl pyruvate administration suppresses growth and invasion of gallbladder cancer cells via downregulation of HMGB1-RAGE axis. *International Journal of Immunopathology and Pharmacology*, *25*(4), 955–965.

Lin, Q., Yang, X. P., Fang, D., Ren, X., Zhou, H., Fang, J., ... Su, S. B. (2011). High-mobility group box-1 mediates toll-like receptor 4-dependent angiogenesis. *Arteriosclerosis, Thrombosis, and Vascular Biology*, *31*(5), 1024–1032. http://dx.doi.org/10.1161/ATVBAHA.111.224048.

Liu, K., Huang, J., Xie, M., Yu, Y., Zhu, S., Kang, R., ... Duan, X. (2014). MIR34A regulates autophagy and apoptosis by targeting HMGB1 in the retinoblastoma cell. *Autophagy*, *10*(3), 442–452. http://dx.doi.org/10.4161/auto.27418.

Luo, Y., Chihara, Y., Fujimoto, K., Sasahira, T., Kuwada, M., Fujiwara, R., ... Kuniyasu, H. (2013). High mobility group box 1 released from necrotic cells enhances regrowth and metastasis of cancer cells that have survived chemotherapy. *European Journal of Cancer*, *49*(3), 741–751. http://dx.doi.org/10.1016/j.ejca.2012.09.016.

Maeda, T., Ozaki, M., Kobayashi, Y., Kiguchi, N., & Kishioka, S. (2013). HMGB1 as a potential therapeutic target for neuropathic pain. *Journal of Pharmacological Sciences*, *123*(4), 301–305.

Magna, M., & Pisetsky, D. S. (2015). The role of cell death in the pathogenesis of SLE: Is pyroptosis the missing link? *Scandinavian Journal of Immunology*, *82*(3), 218–224. http://dx.doi.org/10.1111/sji.12335.

Manigrasso, M. B., Juranek, J., Ramasamy, R., & Schmidt, A. M. (2014). Unlocking the biology of RAGE in diabetic microvascular complications. *Trends in Endocrinology and Metabolism*, *25*(1), 15–22. http://dx.doi.org/10.1016/j.tem.2013.08.002.

Mardente, S., Mari, E., Consorti, F., Di Gioia, C., Negri, R., Etna, M., … Antonaci, A. (2012). HMGB1 induces the overexpression of miR-222 and miR-221 and increases growth and motility in papillary thyroid cancer cells. *Oncology Reports*, *28*(6), 2285–2289. http://dx.doi.org/10.3892/or.2012.2058.

Mardente, S., Mari, E., Massimi, I., Fico, F., Faggioni, A., Pulcinelli, F., … Zicari, A. (2015). HMGB1-induced cross talk between PTEN and miRs 221/222 in thyroid cancer. *BioMed Research International*, *2015*, 512027. http://dx.doi.org/10.1155/2015/512027.

Maroso, M., Balosso, S., Ravizza, T., Liu, J., Aronica, E., Iyer, A. M., … Vezzani, A. (2010). Toll-like receptor 4 and high-mobility group box-1 are involved in ictogenesis and can be targeted to reduce seizures. *Nature Medicine*, *16*(4), 413–419. http://dx.doi.org/10.1038/nm.2127.

Matzinger, P. (1994). Tolerance, danger, and the extended family. *Annual Review of Immunology*, *12*, 991–1045. http://dx.doi.org/10.1146/annurev.iy.12.040194.005015.

Mazarati, A., Maroso, M., Iori, V., Vezzani, A., & Carli, M. (2011). High-mobility group box-1 impairs memory in mice through both toll-like receptor 4 and receptor for advanced glycation end products. *Experimental Neurology*, *232*(2), 143–148. http://dx.doi.org/10.1016/j.expneurol.2011.08.012.

Min, H. J., Ko, E. A., Wu, J., Kim, E. S., Kwon, M. K., Kwak, M. S., … Shin, J. S. (2013). Chaperone-like activity of high-mobility group box 1 protein and its role in reducing the formation of polyglutamine aggregates. *Journal of Immunology*, *190*(4), 1797–1806. http://dx.doi.org/10.4049/jimmunol.1202472.

Mitkova, E., Ugrinova, I., Pashev, I. G., & Pasheva, E. A. (2005). The inhibitory effect of HMGB-1 protein on the repair of cisplatin-damaged DNA is accomplished through the acidic domain. *Biochemistry*, *44*(15), 5893–5898. http://dx.doi.org/10.1021/bi047712c.

Mitsouras, K., Wong, B., Arayata, C., Johnson, R. C., & Carey, M. (2002). The DNA architectural protein HMGB1 displays two distinct modes of action that promote enhanceosome assembly. *Molecular and Cellular Biology*, *22*(12), 4390–4401.

Mittal, D., Saccheri, F., Venereau, E., Pusterla, T., Bianchi, M. E., & Rescigno, M. (2010). TLR4-mediated skin carcinogenesis is dependent on immune and radioresistant cells. *The EMBO Journal*, *29*(13), 2242–2252. http://dx.doi.org/10.1038/emboj.2010.94.

Mollica, L., De Marchis, F., Spitaleri, A., Dallacosta, C., Pennacchini, D., Zamai, M., … Bianchi, M. E. (2007). Glycyrrhizin binds to high-mobility group box 1 protein and inhibits its cytokine activities. *Chemistry & Biology*, *14*(4), 431–441. http://dx.doi.org/10.1016/j.chembiol.2007.03.007.

Musumeci, D., Bucci, E. M., Roviello, G. N., Sapio, R., Valente, M., Moccia, M., … Pedone, C. (2011). DNA-based strategies for blocking HMGB1 cytokine activity: Design, synthesis and preliminary in vitro/in vivo assays of DNA and DNA-like duplexes. *Molecular BioSystems*, 7(5), 1742–1752. http://dx.doi.org/10.1039/c1mb05009e.

Musumeci, D., Roviello, G. N., Moccia, M., Pedone, C., Bucci, E. M., Sapio, R., … Fumero, S. (2007). Bent oligonucleotide duplexes as HMGB1 inhibitors: A comparative study. *Nucleosides, Nucleotides & Nucleic Acids*, *26*(10–12), 1447–1450. http://dx.doi.org/10.1080/15257770701542330.

Musumeci, D., Roviello, G. N., & Montesarchio, D. (2014). An overview on HMGB1 inhibitors as potential therapeutic agents in HMGB1-related pathologies. *Pharmacology & Therapeutics*, *141*(3), 347–357. http://dx.doi.org/10.1016/j.pharmthera.2013.11.001.

Ogiku, M., Kono, H., Hara, M., Tsuchiya, M., & Fujii, H. (2011). Glycyrrhizin prevents liver injury by inhibition of high-mobility group box 1 production by Kupffer cells after ischemia-reperfusion in rats. *The Journal of Pharmacology and Experimental Therapeutics*, *339*(1), 93–98. http://dx.doi.org/10.1124/jpet.111.182592.

Ohmori, H., Luo, Y., Fujii, K., Sasahira, T., Shimomoto, T., Denda, A., & Kuniyasu, H. (2010). Dietary linoleic acid and glucose enhances azoxymethane-induced colon cancer and metastases via the expression of high-mobility group box 1. *Pathobiology*, *77*(4), 210–217. http://dx.doi.org/10.1159/000296305.

Park, S., Yoon, S. J., Tae, H. J., & Shim, C. Y. (2011). RAGE and cardiovascular disease. *Frontiers in Bioscience (Landmark edition)*, *16*, 486–497.

Pasheva, E. A., Pashev, I. G., & Favre, A. (1998). Preferential binding of high mobility group 1 protein to UV-damaged DNA. Role of the COOH-terminal domain. *The Journal of Biological Chemistry*, *273*(38), 24730–24736.

Pasheva, E., Sarov, M., Bidjekov, K., Ugrinova, I., Sarg, B., Lindner, H., & Pashev, I. G. (2004). In vitro acetylation of HMGB-1 and -2 proteins by CBP: The role of the acidic tail. *Biochemistry*, *43*(10), 2935–2940. http://dx.doi.org/10.1021/bi035615y.

Pasheva, E. A., Ugrinova, I., Spassovska, N. C., & Pashev, I. G. (2002). The binding affinity of HMG1 protein to DNA modified by cis-platin and its analogs correlates with their antitumor activity. *The International Journal of Biochemistry & Cell Biology*, *34*(1), 87–92.

Patrick, S. M., & Turchi, J. J. (1999). Replication protein A (RPA) binding to duplex cisplatin-damaged DNA is mediated through the generation of single-stranded DNA. *The Journal of Biological Chemistry*, *274*(21), 14972–14978.

Paull, T. T., Haykinson, M. J., & Johnson, R. C. (1993). The nonspecific DNA-binding and -bending proteins HMG1 and HMG2 promote the assembly of complex nucleoprotein structures. *Genes & Development*, 7(8), 1521–1534.

Pedrazzi, M., Averna, M., Sparatore, B., Patrone, M., Salamino, F., Marcoli, M., … Melloni, E. (2012). Potentiation of NMDA receptor-dependent cell responses by extracellular high mobility group box 1 protein. *PLoS One*, 7(8). e44518. http://dx.doi.org/10.1371/journal.pone.0044518.

Peterson, E. B., Mastrangelo, M. A., Federoff, H. J., & Bowers, W. J. (2007). Neuronal specificity of HSV/sleeping beauty amplicon transduction in utero is driven primarily by tropism and cell type composition. *Molecular Therapy*, *15*(10), 1848–1855.

Phair, R. D., Scaffidi, P., Elbi, C., Vecerova, J., Dey, A., Ozato, K., … Misteli, T. (2004). Global nature of dynamic protein-chromatin interactions in vivo: Three-dimensional genome scanning and dynamic interaction networks of chromatin proteins. *Molecular and Cellular Biology*, *24*(14), 6393–6402. http://dx.doi.org/10.1128/MCB.24.14.6393-6402.2004.

Pil, P. M., Chow, C. S., & Lippard, S. J. (1993). High-mobility-group 1 protein mediates DNA bending as determined by ring closures. *Proceedings of the National Academy of Sciences of the United States of America*, *90*(20), 9465–9469.

Pil, P. M., & Lippard, S. J. (1992). Specific binding of chromosomal protein HMG1 to DNA damaged by the anticancer drug cisplatin. *Science*, *256*(5054), 234–237.

Pisetsky, D. S., Erlandsson-Harris, H., & Andersson, U. (2008). High-mobility group box protein 1 (HMGB1): An alarmin mediating the pathogenesis of rheumatic disease. *Arthritis Research & Therapy*, *10*(3), 209. http://dx.doi.org/10.1186/ar2440.

Qin, Y., Chen, Y., Wang, W., Wang, Z., Tang, G., Zhang, P., … Shen, Q. (2014). HMGB1-LPS complex promotes transformation of osteoarthritis synovial fibroblasts to a rheumatoid arthritis synovial fibroblast-like phenotype. *Cell Death & Disease*, *5*. e1077. http://dx.doi.org/10.1038/cddis.2014.48.

Qiu, Y., Yang, J., Wang, W., Zhao, W., Peng, F., Xiang, Y., … Feng, J. (2014). HMGB1-promoted and TLR2/4-dependent NK cell maturation and activation take part in rotavirus-induced murine biliary atresia. *PLoS Pathogens*, *10*(3). e1004011. http://dx.doi.org/10.1371/journal.ppat.1004011.

Raman, K. G., Sappington, P. L., Yang, R., Levy, R. M., Prince, J. M., Liu, S., … Fink, M. P. (2006). The role of RAGE in the pathogenesis of intestinal barrier dysfunction after hemorrhagic shock. *American Journal of Physiology. Gastrointestinal and Liver Physiology*, *291*(4), G556–G565. http://dx.doi.org/10.1152/ajpgi.00055.2006.

Ran, X., Yang, J., Liu, C., Zhou, P., Xiao, L., & Zhang, K. (2015). MiR-218 inhibits HMGB1-mediated autophagy in endometrial carcinoma cells during chemotherapy. *International Journal of Clinical and Experimental Pathology*, *8*(6), 6617–6626.

Raucci, A., Cugusi, S., Antonelli, A., Barabino, S. M., Monti, L., Bierhaus, A., … Bianchi, M. E. (2008). A soluble form of the receptor for advanced glycation end-products (RAGE) is produced by proteolytic cleavage of the membrane-bound form by the sheddase a disintegrin and metalloprotease 10 (ADAM10). *The FASEB Journal*, *22*(10), 3716–3727. http://dx.doi.org/10.1096/fj.08-109033.

Rauvala, H., & Pihlaskari, R. (1987). Isolation and some characteristics of an adhesive factor of brain that enhances neurite outgrowth in central neurons. *The Journal of Biological Chemistry*, *262*(34), 16625–16635.

Reddy, M. C., Christensen, J., & Vasquez, K. M. (2005). Interplay between human high mobility group protein 1 and replication protein A on psoralen-cross-linked DNA. *Biochemistry*, *44*(11), 4188–4195. http://dx.doi.org/10.1021/bi047902n.

Reed, E., Ozols, R. F., Tarone, R., Yuspa, S. H., & Poirier, M. C. (1987). Platinum-DNA adducts in leukocyte DNA correlate with disease response in ovarian cancer patients receiving platinum-based chemotherapy. *Proceedings of the National Academy of Sciences of the United States of America*, *84*(14), 5024–5028.

Romani, M., Rodman, T. C., Vidali, G., & Bustin, M. (1979). Serological analysis of species specificity in the high mobility group chromosomal proteins. *The Journal of Biological Chemistry*, *254*(8), 2918–2922.

Sakamoto, R., Okano, M., Takena, H., & Ohtsuki, K. (2001). Inhibitory effect of glycyrrhizin on the phosphorylation and DNA-binding abilities of high mobility group proteins 1 and 2 in vitro. *Biological & Pharmaceutical Bulletin*, *24*(8), 906–911.

Sanchez-Giraldo, R., Acosta-Reyes, F. J., Malarkey, C. S., Saperas, N., Churchill, M. E., & Campos, J. L. (2015). Two high-mobility group box domains act together to underwind and kink DNA. *Acta Crystallographica. Section D, Biological Crystallography*, *71*(Pt. 7), 1423–1432. http://dx.doi.org/10.1107/S1399004715007452.

Scaffidi, P., Misteli, T., & Bianchi, M. E. (2002). Release of chromatin protein HMGB1 by necrotic cells triggers inflammation. *Nature*, *418*(6894), 191–195. http://dx.doi.org/10.1038/nature00858.

Schaper, F., Westra, J., & Bijl, M. (2014). Recent developments in the role of high-mobility group box 1 in systemic lupus erythematosus. *Molecular Medicine*, *20*, 72–79. http://dx.doi.org/10.2119/molmed.2014.00019.

Schierbeck, H., Lundback, P., Palmblad, K., Klevenvall, L., Erlandsson-Harris, H., Andersson, U., & Ottosson, L. (2011). Monoclonal anti-HMGB1 (high mobility group box chromosomal protein 1) antibody protection in two experimental arthritis models. *Molecular Medicine*, *17*(9–10), 1039–1044. http://dx.doi.org/10.2119/molmed.2010.00264.

Schiraldi, M., Raucci, A., Munoz, L. M., Livoti, E., Celona, B., Venereau, E., … Uguccioni, M. (2012). HMGB1 promotes recruitment of inflammatory cells to damaged tissues by forming a complex with CXCL12 and signaling via CXCR4. *The Journal of Experimental Medicine*, *209*(3), 551–563. http://dx.doi.org/10.1084/jem.20111739.

Sims, G. P., Rowe, D. C., Rietdijk, S. T., Herbst, R., & Coyle, A. J. (2010). HMGB1 and RAGE in inflammation and cancer. *Annual Review of Immunology*, *28*, 367–388. http://dx.doi.org/10.1146/annurev.immunol.021908.132603.

Singh, J., & Dixon, G. H. (1990). High mobility group proteins 1 and 2 function as general class II transcription factors. *Biochemistry*, *29*(26), 6295–6302.

Sitia, G., Iannacone, M., Muller, S., Bianchi, M. E., & Guidotti, L. G. (2007). Treatment with HMGB1 inhibitors diminishes CTL-induced liver disease in HBV transgenic mice. *Journal of Leukocyte Biology*, *81*(1), 100–107. http://dx.doi.org/10.1189/jlb.0306173.

Siu, Y. S., Li, L., Leung, M. F., Lee, K. L., & Li, P. (2012). Polyethylenimine-based amphiphilic core-shell nanoparticles: Study of gene delivery and intracellular trafficking. *Biointerphases*, 7(1–4), 16. http://dx.doi.org/10.1007/s13758-011-0016-4.

Smolarczyk, R., Cichon, T., Matuszczak, S., Mitrus, I., Lesiak, M., Kobusinska, M., … Szala, S. (2012). The role of Glycyrrhizin, an inhibitor of HMGB1 protein, in anticancer therapy. *Archivum Immunologiae et Therapiae Experimentalis*, *60*(5), 391–399. http://dx.doi.org/10.1007/s00005-012-0183-0.

Song, J. X., Lu, J. H., Liu, L. F., Chen, L. L., Durairajan, S. S., Yue, Z., … Li, M. (2014). HMGB1 is involved in autophagy inhibition caused by SNCA/alpha-synuclein overexpression: A process modulated by the natural autophagy inducer corynoxine B. *Autophagy*, *10*(1), 144–154. http://dx.doi.org/10.4161/auto.26751.

Sterner, R., Vidali, G., & Allfrey, V. G. (1979). Studies of acetylation and deacetylation in high mobility group proteins. Identification of the sites of acetylation in HMG-1. *The Journal of Biological Chemistry*, *254*(22), 11577–11583.

Stott, K., Watson, M., Howe, F. S., Grossmann, J. G., & Thomas, J. O. (2010). Tail-mediated collapse of HMGB1 is dynamic and occurs via differential binding of the acidic tail to the A and B domains. *Journal of Molecular Biology*, *403*(5), 706–722. http://dx.doi.org/10.1016/j.jmb.2010.07.045.

Stros, M., Stokrova, J., & Thomas, J. O. (1994). DNA looping by the HMG-box domains of HMG1 and modulation of DNA binding by the acidic C-terminal domain. *Nucleic Acids Research*, *22*(6), 1044–1051.

Su, S., Wu, J., Gong, T., He, K., Feng, C., Zhang, M., … Xia, X. (2016). Inhibition of high mobility group box 1-toll-like receptor-4 signaling by glycyrrhizin contributes to the attenuation of cold ischemic injury of liver in a rat model. *Transplantation Proceedings*, *48*(1), 191–198. http://dx.doi.org/10.1016/j.transproceed.2015.12.014.

Susa, Y., Masuda, Y., Imaizumi, H., & Namiki, A. (2009). Neutralization of receptor for advanced glycation end-products and high mobility group box-1 attenuates septic diaphragm dysfunction in rats with peritonitis. *Critical Care Medicine*, *37*(9), 2619–2624. http://dx.doi.org/10.1097/CCM.0b013e3181a930f7.

Taguchi, A., Blood, D. C., del Toro, G., Canet, A., Lee, D. C., Qu, W., … Schmidt, A. M. (2000). Blockade of RAGE-amphoterin signalling suppresses tumour growth and metastases. *Nature*, *405*(6784), 354–360. http://dx.doi.org/10.1038/35012626.

Tang, D., Kang, R., Xiao, W., Zhang, H., Lotze, M. T., Wang, H., & Xiao, X. (2009). Quercetin prevents LPS-induced high-mobility group box 1 release and proinflammatory function. *American Journal of Respiratory Cell and Molecular Biology*, *41*(6), 651–660. http://dx.doi.org/10.1165/rcmb.2008-0119OC.

Tang, D., Kang, R., Zeh, H. J., 3rd, & Lotze, M. T. (2010). High-mobility group box 1 and cancer. *Biochimica et Biophysica Acta*, *1799*(1–2), 131–140. http://dx.doi.org/10.1016/j.bbagrm.2009.11.014.

Tang, Y., Lv, B., Wang, H., Xiao, X., & Zuo, X. (2008). PACAP inhibit the release and cytokine activity of HMGB1 and improve the survival during lethal endotoxemia. *International Immunopharmacology*, *8*(12), 1646–1651. http://dx.doi.org/10.1016/j.intimp.2008.07.014.

Terheggen, P. M., Emondt, J. Y., Floot, B. G., Dijkman, R., Schrier, P. I., den Engelse, L., & Begg, A. C. (1990). Correlation between cell killing by cis-diamminedichloroplatinum(-II) in six mammalian cell lines and binding of a cis-diamminedichloroplatinum(II)-DNA antiserum. *Cancer Research*, *50*(12), 3556–3561.

Therrien, J. P., Drouin, R., Baril, C., & Drobetsky, E. A. (1999). Human cells compromised for p53 function exhibit defective global and transcription-coupled nucleotide excision repair, whereas cells compromised for pRb function are defective only in global repair. *Proceedings of the National Academy of Sciences of the United States of America*, *96*(26), 15038–15043.

Topalova, D., Ugrinova, I., Pashev, I. G., & Pasheva, E. A. (2008). HMGB1 protein inhibits DNA replication in vitro: A role of the acetylation and the acidic tail. *The International Journal of Biochemistry & Cell Biology*, *40*(8), 1536–1542. http://dx.doi.org/10.1016/j.biocel.2007.11.014.

Tsung, A., Hoffman, R. A., Izuishi, K., Critchlow, N. D., Nakao, A., Chan, M. H., … Billiar, T. R. (2005). Hepatic ischemia/reperfusion injury involves functional TLR4 signaling in nonparenchymal cells. *Journal of Immunology*, *175*(11), 7661–7668.

Tsung, A., Sahai, R., Tanaka, H., Nakao, A., Fink, M. P., Lotze, M. T., … Billiar, T. R. (2005). The nuclear factor HMGB1 mediates hepatic injury after murine liver ischemia-reperfusion. *The Journal of Experimental Medicine*, *201*(7), 1135–1143. http://dx.doi.org/10.1084/jem.20042614.

Ueda, T., Chou, H., Kawase, T., Shirakawa, H., & Yoshida, M. (2004). Acidic C-tail of HMGB1 is required for its target binding to nucleosome linker DNA and transcription stimulation. *Biochemistry*, *43*(30), 9901–9908. http://dx.doi.org/10.1021/bi035975l.

Ugrinova, I., Mitkova, E., Moskalenko, C., Pashev, I., & Pasheva, E. (2007). DNA bending versus DNA end joining activity of HMGB1 protein is modulated in vitro by acetylation. *Biochemistry*, *46*(8), 2111–2117. http://dx.doi.org/10.1021/bi0614479.

Ugrinova, I., Pashev, I. G., & Pasheva, E. A. (2009). Nucleosome binding properties and Co-remodeling activities of native and in vivo acetylated HMGB-1 and HMGB-2 proteins. *Biochemistry*, *48*(27), 6502–6507. http://dx.doi.org/10.1021/bi9004304.

Ugrinova, I., Pasheva, E. A., Armengaud, J., & Pashev, I. G. (2001). In vivo acetylation of HMG1 protein enhances its binding affinity to distorted DNA structures. *Biochemistry*, *40*(48), 14655–14660.

Ugrinova, I., Zlateva, S., Pashev, I. G., & Pasheva, E. A. (2009). Native HMGB1 protein inhibits repair of cisplatin-damaged nucleosomes in vitro. *The International Journal of Biochemistry & Cell Biology*, *41*(7), 1556–1562. http://dx.doi.org/10.1016/j.biocel.2009.01.010.

Ulloa, L., & Messmer, D. (2006). High-mobility group box 1 (HMGB1) protein: Friend and foe. *Cytokine & Growth Factor Reviews*, *17*(3), 189–201. http://dx.doi.org/10.1016/j.cytogfr.2006.01.003.

Ulloa, L., Ochani, M., Yang, H., Tanovic, M., Halperin, D., Yang, R., … Tracey, K. J. (2002). Ethyl pyruvate prevents lethality in mice with established lethal sepsis and systemic inflammation. *Proceedings of the National Academy of Sciences of the United States of America*, *99*(19), 12351–12356. http://dx.doi.org/10.1073/pnas.192222999.

van Beijnum, J. R., Nowak-Sliwinska, P., van den Boezem, E., Hautvast, P., Buurman, W. A., & Griffioen, A. W. (2013). Tumor angiogenesis is enforced by autocrine regulation of high-mobility group box 1. *Oncogene*, *32*(3), 363–374. http://dx.doi.org/10.1038/onc.2012.49.

Venereau, E., Casalgrandi, M., Schiraldi, M., Antoine, D. J., Cattaneo, A., De Marchis, F., … Bianchi, M. E. (2012). Mutually exclusive redox forms of HMGB1 promote cell recruitment or proinflammatory cytokine release. *The Journal of Experimental Medicine*, *209*(9), 1519–1528. http://dx.doi.org/10.1084/jem.20120189.

Venereau, E., De Leo, F., Mezzapelle, R., Careccia, G., Musco, G., & Bianchi, M. E. (2016). HMGB1 as biomarker and drug target. *Pharmacological Research*, *111*, 534–544. http://dx.doi.org/10.1016/j.phrs.2016.06.031.

Wan, W., Cao, L., Khanabdali, R., Kalionis, B., Tai, X., & Xia, S. (2016). The emerging role of HMGB1 in neuropathic pain: A potential therapeutic target for neuroinflammation. *Journal of Immunology Research*, *2016*, 6430423. http://dx.doi.org/10.1155/2016/6430423.

Wang, H., Bloom, O., Zhang, M., Vishnubhakat, J. M., Ombrellino, M., Che, J., … Tracey, K. J. (1999). HMG-1 as a late mediator of endotoxin lethality in mice. *Science*, *285*(5425), 248–251.

Wang, X., Guo, Y., Wang, C., Yu, H., Yu, X., & Yu, H. (2016). MicroRNA-142-3p inhibits chondrocyte apoptosis and inflammation in osteoarthritis by targeting HMGB1. *Inflammation*, *39*(5), 1718–1728. http://dx.doi.org/10.1007/s10753-016-0406-3.

Wang, X., Sun, R., Wei, H., & Tian, Z. (2013). High-mobility group box 1 (HMGB1)-Toll-like receptor (TLR)4-interleukin (IL)-23-IL-17A axis in drug-induced damage-associated lethal hepatitis: Interaction of gammadelta T cells with macrophages. *Hepatology*, *57*(1), 373–384. http://dx.doi.org/10.1002/hep.25982.

Wang, H., Ward, M. F., & Sama, A. E. (2009). Novel HMGB1-inhibiting therapeutic agents for experimental sepsis. *Shock*, *32*(4), 348–357. http://dx.doi.org/10.1097/SHK.0b013e3181a551bd.

Weir, H. M., Kraulis, P. J., Hill, C. S., Raine, A. R., Laue, E. D., & Thomas, J. O. (1993). Structure of the HMG box motif in the B-domain of HMG1. *The EMBO Journal*, *12*(4), 1311–1319.

Wu, L., Cai, C., Wang, X., Liu, M., Li, X., & Tang, H. (2011). MicroRNA-142-3p, a new regulator of RAC1, suppresses the migration and invasion of hepatocellular carcinoma cells. *FEBS Letters*, *585*(9), 1322–1330. http://dx.doi.org/10.1016/j.febslet.2011.03.067.

Wu, J., Li, J., Salcedo, R., Mivechi, N. F., Trinchieri, G., & Horuzsko, A. (2012). The proinflammatory myeloid cell receptor TREM-1 controls Kupffer cell activation and development of hepatocellular carcinoma. *Cancer Research*, *72*(16), 3977–3986. http://dx.doi.org/10.1158/0008-5472.CAN-12-0938.

Xiao, P., & Liu, W. L. (2015). MiR-142-3p functions as a potential tumor suppressor directly targeting HMGB1 in non-small-cell lung carcinoma. *International Journal of Clinical and Experimental Pathology*, *8*(9), 10800–10807.

Xie, J., Hu, X., Yi, C., Hu, G., Zhou, X., & Jiang, H. (2016). MicroRNA451 protects against cardiomyocyte anoxia/reoxygenation injury by inhibiting high mobility group box 1 expression. *Molecular Medicine Reports*, *13*(6), 5335–5341. http://dx.doi.org/10.3892/mmr.2016.5192.

Yanai, H., Ban, T., Wang, Z., Choi, M. K., Kawamura, T., Negishi, H., … Taniguchi, T. (2009). HMGB proteins function as universal sentinels for nucleic-acid-mediated innate immune responses. *Nature*, *462*(7269), 99–103. http://dx.doi.org/10.1038/nature08512.

Yang, H., Antoine, D. J., Andersson, U., & Tracey, K. J. (2013). The many faces of HMGB1: Molecular structure-functional activity in inflammation, apoptosis, and chemotaxis. *Journal of Leukocyte Biology*, *93*(6), 865–873. http://dx.doi.org/10.1189/jlb.1212662.

Yang, Z., Deng, Y., Su, D., Tian, J., Gao, Y., He, Z., & Wang, X. (2013). TLR4 as receptor for HMGB1-mediated acute lung injury after liver ischemia/reperfusion injury. *Laboratory Investigation*, *93*(7), 792–800. http://dx.doi.org/10.1038/labinvest.2013.66.

Yang, H., Hreggvidsdottir, H. S., Palmblad, K., Wang, H., Ochani, M., Li, J., … Tracey, K. J. (2010). A critical cysteine is required for HMGB1 binding to Toll-like receptor 4 and activation of macrophage cytokine release. *Proceedings of the National Academy of Sciences of the United States of America*, *107*(26), 11942–11947. http://dx.doi.org/10.1073/pnas.1003893107.

Yang, H., Lundback, P., Ottosson, L., Erlandsson-Harris, H., Venereau, E., Bianchi, M. E., … Antoine, D. J. (2012). Redox modification of cysteine residues regulates the cytokine activity of high mobility group box-1 (HMGB1). *Molecular Medicine*, *18*, 250–259. http://dx.doi.org/10.2119/molmed.2011.00389.

Yang, H., Ochani, M., Li, J., Qiang, X., Tanovic, M., Harris, H. E., … Tracey, K. J. (2004). Reversing established sepsis with antagonists of endogenous high-mobility group box 1. *Proceedings of the National Academy of Sciences of the United States of America*, *101*(1), 296–301. http://dx.doi.org/10.1073/pnas.2434651100.

Yang, H., Wang, H., Chavan, S. S., & Andersson, U. (2015a). High mobility group box protein 1 (HMGB1): The prototypical endogenous danger molecule. *Molecular Medicine*, *21*(Suppl. 1), S6–S12. http://dx.doi.org/10.2119/molmed.2015.00087.

Yang, H., Wang, H., Ju, Z., Ragab, A. A., Lundback, P., Long, W., … Al-Abed, Y. (2015b). MD-2 is required for disulfide HMGB1-dependent TLR4 signaling. *The Journal of Experimental Medicine*, *212*(1), 5–14. http://dx.doi.org/10.1084/jem.20141318.

Yang, H., Wang, H., Levine, Y. A., Gunasekaran, M. K., Wang, Y., Addorisio, M., … Tracey, K. J. (2016). Identification of CD163 as an antiinflammatory receptor for HMGB1-haptoglobin complexes. *JCI Insight*, *1*(7), e85375. http://dx.doi.org/10.1172/jci.insight.85375.

Yao, S., Zhao, T., & Jin, H. (2015). Expression of MicroRNA-325-3p and its potential functions by targeting HMGB1 in non-small cell lung cancer. *Biomedicine & Pharmacotherapy*, *70*, 72–79. http://dx.doi.org/10.1016/j.biopha.2015.01.013.

Yi, W. J., Yang, J., Li, C., Wang, H. Y., Liu, C. W., Tao, L., … Zhang, X. Z. (2012). Enhanced nuclear import and transfection efficiency of TAT peptide-based gene delivery systems modified by additional nuclear localization signals. *Bioconjugate Chemistry*, *23*(1), 125–134. http://dx.doi.org/10.1021/bc2005472.

Yu, Y., Tang, D., & Kang, R. (2015). Oxidative stress-mediated HMGB1 biology. *Frontiers in Physiology*, *6*, 93. http://dx.doi.org/10.3389/fphys.2015.00093.

Yu, M., Wang, H., Ding, A., Golenbock, D. T., Latz, E., Czura, C. J., … Yang, H. (2006). HMGB1 signals through toll-like receptor (TLR) 4 and TLR2. *Shock*, *26*(2), 174–179. http://dx.doi.org/10.1097/01.shk.0000225404.51320.82.

Yusein-Myashkova, S., Stoykov, I., Gospodinov, A., Ugrinova, I., & Pasheva, E. (2016). The repair capacity of lung cancer cell lines A549 and H1299 depends on HMGB1 expression level and the p53 status. *Journal of Biochemistry*, *160*(1), 37–47. http://dx.doi.org/10.1093/jb/mvw012.

Zamble, D. B., & Lippard, S. J. (1995). Cisplatin and DNA repair in cancer chemotherapy. *Trends in Biochemical Sciences*, *20*(10), 435–439.

Zhai, C. L., Zhang, M. Q., Zhang, Y., Xu, H. X., Wang, J. M., An, G. P., … Li, L. (2012). Glycyrrhizin protects rat heart against ischemia-reperfusion injury through blockade of HMGB1-dependent phospho-JNK/Bax pathway. *Acta Pharmacologica Sinica*, *33*(12), 1477–1487. http://dx.doi.org/10.1038/aps.2012.112.

Zhang, L., Bukulin, M., Kojro, E., Roth, A., Metz, V. V., Fahrenholz, F., … Postina, R. (2008). Receptor for advanced glycation end products is subjected to protein ectodomain shedding by metalloproteinases. *The Journal of Biological Chemistry*, *283*(51), 35507–35516. http://dx.doi.org/10.1074/jbc.M806948200.

Zhang, C., Ge, S., Hu, C., Yang, N., & Zhang, J. (2013). MiRNA-218, a new regulator of HMGB1, suppresses cell migration and invasion in non-small cell lung cancer. *Acta Biochimica et Biophysica Sinica Shanghai*, *45*(12), 1055–1061. http://dx.doi.org/10.1093/abbs/gmt109.

Zhou, W., Wang, J., Li, Z., Li, J., & Sang, M. (2016). MicroRNA-2055b inhibits HMGB1 expression in LPS-induced sepsis. *International Journal of Molecular Medicine*, *38*(1), 312–318. http://dx.doi.org/10.3892/ijmm.2016.2613.

Zong, H., Madden, A., Ward, M., Mooney, M. H., Elliott, C. T., & Stitt, A. W. (2010). Homodimerization is essential for the receptor for advanced glycation end products (RAGE)-mediated signal transduction. *The Journal of Biological Chemistry*, *285*(30), 23137–23146. http://dx.doi.org/10.1074/jbc.M110.133827.

CHAPTER THREE

Targeting IKK and NF-κB for Therapy

J.K. Durand*,†, A.S. Baldwin†,1

*Curriculum in Genetics and Molecular Biology, University of North Carolina, Chapel Hill, NC, United States
†Lineberger Comprehensive Cancer Center, University of North Carolina, Chapel Hill, NC, United States
[1]Corresponding author: e-mail address: albert_baldwin@med.unc.edu

Contents

Abstract

In addition to regulating immune responses, the NF-κB family of transcription factors also promotes cellular proliferation and survival. NF-κB and its activating kinase, IKK, have become appealing therapeutic targets because of their critical roles in the progression of many diseases including chronic inflammation and cancer. Here, we discuss the conditions that lead to pathway activation, the effects of constitutive activation, and some of the strategies used to inhibit NF-κB signaling.

1. IKK AND NF-κB SIGNALING

Nuclear factor of κB (NF-κB) plays a crucial and evolutionarily conserved role in immunity (Ghosh, May, & Kopp, 1998). Initially thought

Advances in Protein Chemistry and Structural Biology, Volume 107
ISSN 1876-1623
http://dx.doi.org/10.1016/bs.apcsb.2016.11.006

to be a B-cell-specific transcription factor upon its discovery (Sen & Baltimore, 1986), NF-κB expression has since been found in virtually every cell type. In the absence of inducing stimuli, it is inactivated through interaction with inhibitor proteins of the IκB family. Upon activation, NF-κB is involved in the regulation of genes that impact not only immune responses but also cell proliferation, survival, and differentiation.

The NF-κB family of transcription factors consists of five structurally related protein subunits that share affinity for the κB DNA sequence motif (Ghosh et al., 1998; Kunsch, Ruben, & Rosen, 1992; Pierce, Lenardo, & Baltimore, 1988). This family of proteins can be divided into two classes: the NF-κB proteins (p105/p50 or NFKB1 and p100/p52 or NFKB2) and the Rel proteins (c-Rel, RelB, and RelA/p65) (Gilmore, 1999b). Through proteasome-mediated proteolysis, p105 and p100 are processed into the shorter DNA-binding subunits called p50 and p52, respectively (Betts & Nabel, 1996; Fan & Maniatis, 1991). The Rel proteins are distinguished by C-terminal transcriptional activation domains (TAD) (Schmitz & Baeuerle, 1991), while all five NF-κB/Rel subunits contain the N-terminal Rel homology domain (RHD), essential for DNA binding, dimerization, inhibitor of κB (IκB) interaction, and nuclear localization (Chen & Ghosh, 1999; Urban & Baeuerle, 1990; Fig. 1).

The canonical NF-κB transcription factor is a dimer composed of a p50 and RelA/p65 subunit, which largely resides in the cytoplasm as part of a latent complex with IκBα under basal conditions (Baeuerle & Baltimore, 1988a, 1988b). This highly inducible pathway is activated by intracellular or extracellular signals such as tumor necrosis factor alpha (TNF-α), interleukin-1 (IL-1), lipopolysaccharide (LPS), viral double-stranded RNA, and ionizing radiation (Brach et al., 1991; Chow, Young, Golenbock, Christ, & Gusovsky, 1999; Gil, Alcamí, & Esteban, 2000; Hoshino et al., 1999; Osborn, Kunkel, & Nabel, 1989; Fig. 2). Pathway activation leads to induction of the IκB kinase (IKK) complex comprised of the two catalytic subunits IKKα and IKKβ, and one scaffold subunit called nuclear factor κB essential modulator (NEMO) or IKKγ (Chen, Parent, & Maniatis, 1996; DiDonato, Hayakawa, Rothwarf, Zandi, & Karin, 1997; Mercurio et al., 1997; Rothwarf, Zandi, Natoli, & Karin, 1998; Yamaoka et al., 1998). IκBα is phosphorylated at serine 32 and 36 by the IKK complex (Brown, Gerstberger, Carlson, Franzoso, & Siebenlist, 1995; Traenckner et al., 1995; Whiteside et al., 1995), polyubiquitinated (Chen et al., 1995; DiDonato et al., 1996; Scherer, Brockman, Chen, Maniatis, & Ballard, 1995), and degraded by the proteasome (Miyamoto, Chiao, & Verma, 1994; Whiteside et al., 1995).

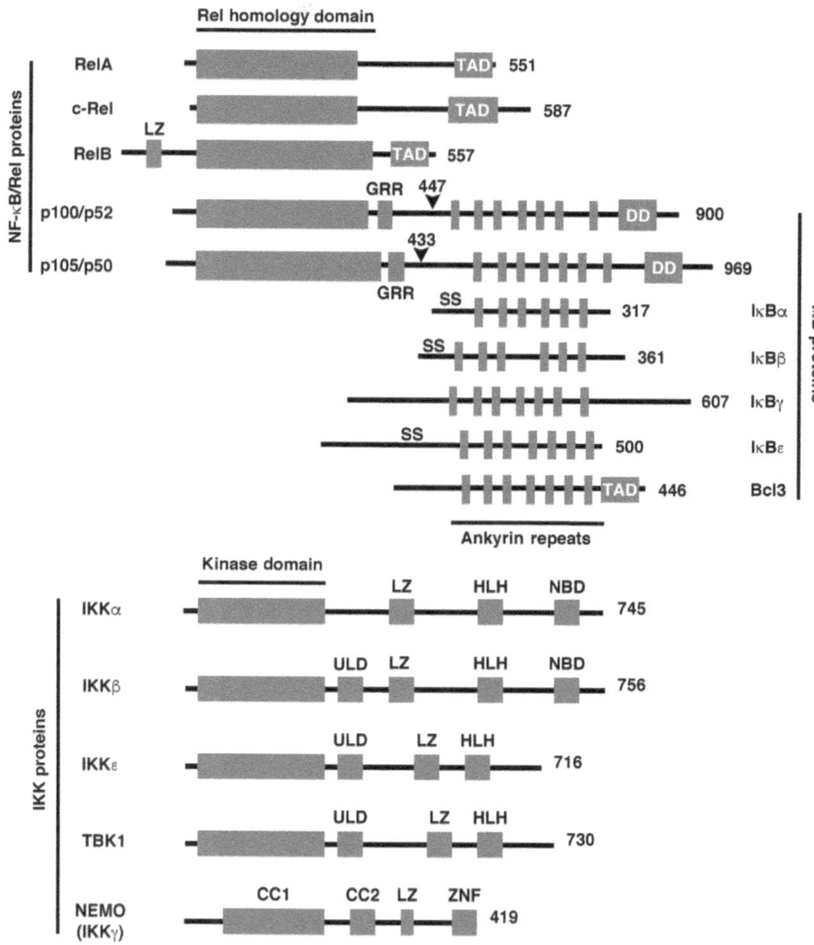

Fig. 1 Structure of the human NF-κB/Rel, IκB, and IKK proteins. The NF-κB/Rel family of proteins consists of five members containing the characteristic Rel homology domain (RHD). These protein subunits are able to stably interact with each other via the RHD to form various context-dependent dimers (Perkins, 2007). The RHD is also essential for DNA binding, IκB interaction, and nuclear localization (Chen & Ghosh, 1999; Urban & Baeuerle, 1990). RelA/p65, c-Rel, and RelB contain a transcriptional activation domain (TAD), and so NF-κB transcription factors containing one of these Rel subunits usually function to activate gene transcription (Oeckinghaus & Ghosh, 2009). The p100 and p105 precursor proteins contain a C-terminal ankyrin repeat domain (ARD) that is proteolytically cleaved away to produce the active subunits p52 and p50, respectively. The presumed cleavage sites are denoted by the *arrows* (Hayden & Ghosh, 2004; Lin & Kobayashi, 2003). Processing of p100 is regulated by the NF-κB-inducing kinase (NIK) and IKKα, which induce site-specific phosphorylation and ubiquitination (Xiao, Fong, & Sun, 2004). The C-terminus of the p105 precursor is selectively degraded by the 20S proteasome leading to p50 generation in an ubiquitin-independent manner.

(Continued)

The degradation of IκBα consequently frees the canonical NF-κB dimer, p50-RelA/p65, to translocate into the nucleus and activate gene transcription (Antonaki et al., 2011; Miyamoto et al., 1994).

Although the canonical NF-κB pathway has been more extensively studied, the noncanonical pathway is important within certain contexts (Dejardin, 2006). Activation of the "alternative" pathway begins with the induction of TNF receptor (TNFR) family members such as lymphotoxin-beta receptor or LTβR (Dejardin et al., 2002), receptor of B-cell-activating factor belonging to the TNF family or BAFF-R (Claudio, Brown, Park, Wang, & Siebenlist, 2002; Kayagaki et al., 2002), receptor activator of NF-κB (RANK) (Novack et al., 2003), and cluster of differentiation 40 (CD40) (Coope et al., 2002). Upon receptor activation, TNF receptor-associated factor (TRAF) proteins mediate the activity of NF-κB-inducing kinase (NIK) (Liao, Zhang, Harhaj, & Sun, 2004), which activates an IKKα homodimer (Ling, Cao, & Goeddel, 1998; Senftleben et al., 2001), leading to p100 processing into

Fig. 1—Cont'd The glycine-rich region (GRR) prevents the complete degradation of the precursors (Moorthy et al., 2006), while a processing-inhibitory domain (PID) (not shown) functions to suppress processing altogether (Cohen, Ciechanover, Kravtsova-Ivantsiv, Lapid, & Lahav-Baratz, 2009; Xiao, Harhaj, & Sun, 2001). Both precursor proteins harbor a death domain (DD) that is required for signal-induced proteolysis (Beinke, Belich, & Ley, 2002; Xiao et al., 2001). The ARD is the characteristic domain of the IκB proteins including IκBα, IκBβ, IκBγ, IκBε, and Bcl3. In fact, the p100 and p105 molecules themselves may act as inhibitors of NF-κB before their ARD is cleaved away. The ARD mediates interaction with the NF-κB/Rel proteins obstructing their nuclear localization signal (NLS) within the RHD of the Rel proteins and restricting nuclear translocation. Phosphorylation of two N-terminal serine residues (SS) of IκB triggers polyubiquitination and proteasome-mediated degradation. Bcl3 is considered an unusual IκB, which is inducible, restricted to the nucleus and is not degraded upon pathway activation. It also contains ankyrin repeats, but does not always act as a repressor. Due to its TAD, when Bcl3 assembles with p50 or p52, these new complexes can activate transcription. Two other unusual IκBs are IκBζ and IκBNS (not shown), which can activate IL-6 and IL-2, respectively (Siebenlist, Brown, & Claudio, 2005). The IκB kinase (IKK) complex consists of the catalytic subunits IKKα and IKKβ, and the regulatory subunit called the NF-κB essential modifier (NEMO, also called IKKγ). NEMO contains several domains that are crucial for its function as a regulatory subunit of the canonical IKK complex. The N-terminal coiled-coil domain (CC1) interacts with IKKα and IKKβ (Marienfeld, Palkowitsch, & Ghosh, 2006). The other coiled-coil region (CC2) together with the leucine zipper (LZ) forms the ubiquitin-binding domain. Finally, the C-terminal zinc finger (ZNF) region that may be involved in substrate interactions (Schröfelbauer, Polley, Behar, Ghosh, & Hoffmann, 2012). The kinase domain of IKKε shares 27% identity with IKKα and 24% with IKKβ. The IKK-related kinase, TBK1, shares 65% similarity to IKKε. *Bcl3*, B-cell lymphoma 3; *HLH*, helix-loop-helix; *IκB*, inhibitor of NF-κB; *NBD*, NEMO-binding domain; *SS*, conserved serines in IκB; *ULD*, ubiquitin-like domain.

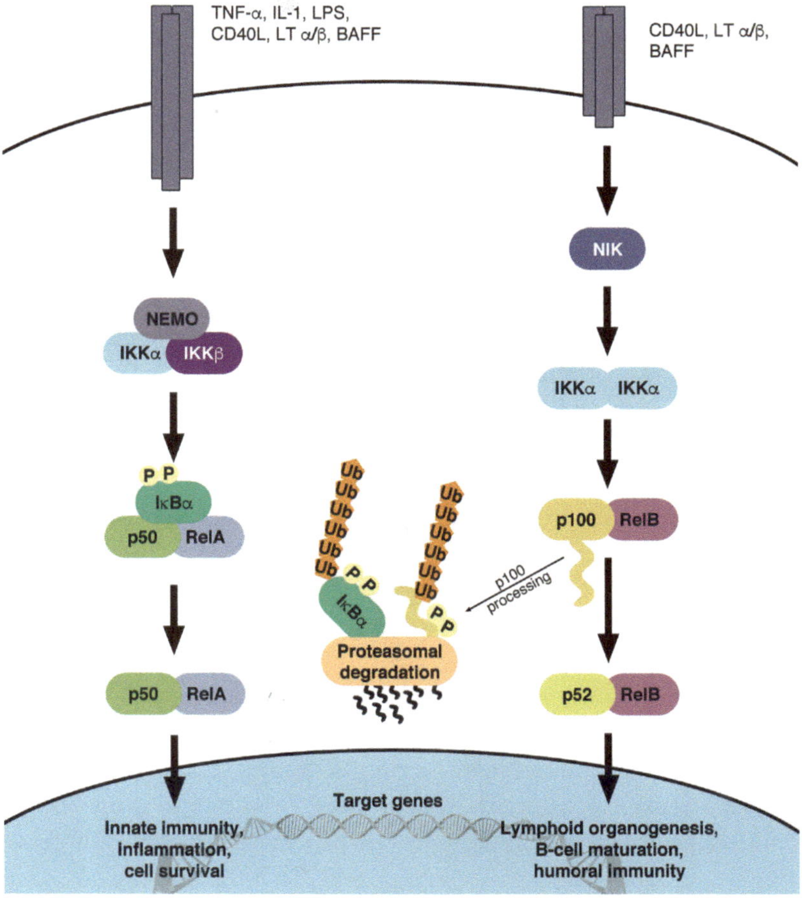

Fig. 2 The canonical and noncanonical NF-κB pathways. (*Left*) The canonical pathway is activated by TNF-α, IL-1, LPS, CD40 ligand (CD40L), lymphotoxin α/β (LT α/β), and BAFF. The IKK complex phosphorylates IκBα and induces its degradation, leading to activation of innate immunity, inflammation, inhibition of apoptosis, and/or proliferation. (*Right*) The noncanonical pathway is activated by LT α/β, CD40L, and BAFF and depends on NIK and IKKα homodimers to induce processing of p100 and nuclear translocation of p52-RelB dimers. This pathway is crucial for secondary lymphoid organ development, maturation of B cells, and adaptive humoral immunity, but can also have proliferative and antiapoptotic effects. *IKK*, IκB kinase; *IL*, interleukin; *IκB*, inhibitor of NF-κB; *LPS*, lipopolysaccharide; *TNF*, tumor necrosis factor. *Modified from Karin, M., Yamamoto, Y., & Wang, Q. M. (2004). The IKK NF-kappa B system: A treasure trove for drug development.* Nature Reviews Drug Discovery, *3(1), 17–26. http://doi.org/10.1038/nrd1279.*

the active p52 subunit (Xiao et al., 2004). Ultimately, processing of p100 results in the generation of the noncanonical transcription factor, a p52-RelB dimer, which is then able to move onto available κB DNA-binding sites and control expression of its associated genes (Bonizzi et al., 2004).

2. NF-κB TARGET GENE SPECIFICITY

The inducible family of NF-κB transcription factors regulates a broad spectrum of target genes involved in a number of critical cellular processes (Gaur & Aggarwal, 2003; O'Neill & Kaltschmidt, 1997; Pahl, 1999). In order to achieve specificity, cells utilize a number of mechanisms to regulate the transcriptional response downstream of NF-κB pathway activation (Smale, 2011). The NF-κB subunits can form various combinations of distinct homodimers or heterodimers with each of the two subunits contacting one-half of the κB DNA-binding site (Baldwin, 1996; Siebenlist, Franzoso, & Brown, 1994). Slight alterations within a κB consensus site can, therefore, result in different affinities for specific dimer combinations (Kunsch et al., 1992). Moreover, composition of the activated dimeric complex is determined by the particular activating stimulus as observed in noncanonical pathway activation (Sun, 2011). Target gene specificity can also be achieved through regulation of access to κB sites via chromatin modifications (De Santa et al., 2009, 2007; Levy et al., 2011; Saccani & Natoli, 2002; van Essen, Zhu, & Saccani, 2010). Epigenetic changes during development, reprogramming, and differentiation can create cell-type-specific open or closed regions of chromatin, which will determine the κB sites available for transcription factor binding (Ramirez-Carrozzi et al., 2009). Different stimuli can affect the duration of NF-κB activation leading to stimulus-specific activation kinetics (Ashall et al., 2009; Hoffmann & Baltimore, 2006; Hoffmann, Natoli, & Ghosh, 2006; Lee & Covert, 2010; Paszek, Jackson, & White, 2010; Sen & Smale, 2010; Tay et al., 2010; Wang et al., 2011). Target gene expression can depend on stimulus-dependent posttranslational modifications of the NF-κB subunits, which can reduce autoinhibition or support interaction with cofactors required for efficient induction (Oeckinghaus, Hayden, & Ghosh, 2011; Perkins, 2006). These and other mechanisms for selective gene induction confer context-dependent specificity upon NF-κB activation, despite the pathway's broad range of targets.

3. NF-κB IN INFLAMMATION, SURVIVAL, AND PROLIFERATION

The NF-κB pathway has been well established as a critical regulator of immunity. It is one of the few pathways responsible for the inducible expression of genes like ICAM, IL-1, IL-6, IL-8, and TNF-α, which are crucial for the acute-phase response during inflammation (Ghosh et al., 1998). Cytokines like IL-1 and TNF-α can activate the pathway, potentially leading to a deleterious feed-forward loop (Ling et al., 2012; Yang et al., 2009). However, the NF-κB transcription factor upregulates IκBα expression thereby maintaining transient induction via the appropriate inhibitory feedback (Brown, Park, Kanno, Franzoso, & Siebenlist, 1993; Chiao, Miyamoto, & Verma, 1994; Scott, Fujita, Liou, Nolan, & Baltimore, 1993; Sun, Ganchi, Ballard, & Greene, 1993). Long noncoding RNAs (Liu et al., 2015) and ubiquitin-editing enzymes such as A20 (Ruland, 2011) can also act as negative-feedback regulators. The inextricable relationship between NF-κB and inflammation has made the pathway an obvious culprit in chronic inflammatory diseases (Lawrence, 2009). Antiinflammatory drugs such as aspirin, dexamethasone, and prednisone can indeed inhibit pathway activation (Auphan, DiDonato, Rosette, Helmberg, & Karin, 1995; Kopp & Ghosh, 1994; Ray & Prefontaine, 1994; Scheinman, Cogswell, Lofquist, & Baldwin, 1995; Yamamoto & Gaynor, 2001), placing NF-κB as a prime target of therapies against inflammatory disease.

In addition to regulating inflammatory cytokines, NF-κB also regulates important genes associated with cell survival, such as cIAP1/2, Bcl-2, and Bcl-xL (Piva, Belardo, & Santoro, 2006). The role of NF-κB in cell survival was made clear in the November issue of Science in 1996 (Beg & Baltimore, 1996; Van Antwerp, Martin, Kafri, Green, & Verma, 1996; Wang, Mayo, & Baldwin, 1996). A series of studies lead by Baltimore, Verma, and our group showed that the loss, inhibition, or lack of NF-κB activation significantly increased stimulus-induced cell death. Wang et al. found that the pathway is activated by ionizing radiation and some chemotherapeutic compounds and that NF-κB could be reducing the efficacy of cancer therapies by promoting cell survival (Wang, et al., 1996).

NF-κB influences cell proliferation via its regulation of cyclin D1 (Guttridge, Albanese, Reuther, Pestell, & Baldwin, 1999; Joyce et al., 2001) and growth factors (Oeckinghaus & Ghosh, 2009). In B cells, the pathway promotes proliferation and differentiation, indispensible processes in mounting an adaptive immune response (Ghosh et al., 1998). However, because NF-κB and IKK proteins are expressed ubiquitously, aberrant cells of almost any kind can hijack this proliferative signal and, coupled with its prosurvival target genes, drive uncontrolled growth leading to diseases such as cancer (Ben-Neriah & Karin, 2011).

4. IKK AND NF-κB SIGNALING IN DISEASE

The NF-κB pathways have been functionally implicated in a variety of diseases including muscular dystrophy (Acharyya et al., 2007; Li, Malhotra, & Kumar, 2008), asthma (Barnes & Adcock, 1997), bone diseases (Boyce, Yao, & Xing, 2010), age-related diseases (Salminen et al., 2008), and a number of cancers (Ben-Neriah & Karin, 2011; Gilmore & Garbati, 2011). The crucial role of NF-κB in such ailments has led to promising outcomes in studies using specific inhibitors against the pathway. An early example of this was in an animal model of rheumatoid arthritis, where intraarticular administration of NF-κB decoys prevented the recurrence of streptococcal cell wall-induced arthritis in rats (Miagkov et al., 1998). In another model of rheumatoid arthritis, the allosteric IKK inhibitor BMS-345541 was used to treat collagen-induced arthritis in mice. When administered as a prophylactic (before disease), BMS-345541 was effective, in a dose-dependent manner, in reducing the incidence and clinical signs of disease. Histologic evaluation of the joints showed that the IKK inhibitor blocked both inflammation and joint destruction. Dose-dependent reduction in disease severity was also observed with the therapeutic (after disease) dosing regimen, with the 100 mg/kg dose resulting in resolution of disease (McIntyre et al., 2003). Since then, more than a decade of research has explored mechanisms to inhibit the NF-κB pathway in a number of different settings. More recently, efforts have been directed at blocking IKK and NF-κB in cancer.

NF-κB was first implicated in cancer when the c-Rel subunit was identified as the homolog of v-REL (Kieran et al., 1990; Wilhelmsen, Eggleton, & Temin, 1984), the transforming gene of the avian REL retrovirus strain T (REV-T) (Gilmore, 1999a). The c-Rel subunit has since been found to play important roles in several different cancers including lymphoma, breast, and pancreatic cancer (Gilmore & Gerondakis, 2011; Hunter, Leslie, & Perkins, 2016). The RelA/p65 subunit is required for transformation induced by oncogenes such as BCR-ABL (Reuther, Reuther, Cortez, Pendergast, & Baldwin, 1998) and RAS (Finco et al., 1997). Inhibiting the canonical pathway by expressing the dominant-negative form of IκBα (Meylan et al., 2009) or via gene knockouts of RelA/p65 (Bassères, Ebbs, Levantini, & Baldwin, 2010) and IKKβ (Xia et al., 2012) blocks tumor progression in KRAS-driven murine cancer models.

More recent evidence has shown that both canonical and noncanonical NF-κB pathways can promote oncogenesis, potentially through control of tumor-initiating cells (TICs) (Bang, Wilson, Ryan, Yeh, & Baldwin, 2013; Kendellen, Bradford, Lawrence, Clark, & Baldwin, 2014; Rinkenbaugh et al., 2016). In ER-negative basal-like breast cancer cells, EZH2 interacts with RelA/p65 and RelB to confer constitutive activation of NF-κB target gene expression independent of its histone methyltransferase activity. This appears to be context dependent, since EZH2 represses NF-κB target genes in ER-positive luminal-like breast cancer cells (Lee et al., 2011). Recently, our group has shown that EZH2, through a methyltransferase-independent mechanism, promotes the transcriptional activation of RelB to drive self-renewal and the TIC phenotype of triple-negative breast cancer cells (Lawrence & Baldwin, 2016).

RelA/p65 has also been shown to interact with signal transducer and activator 3 (STAT3) in cancer (Fan, Mao, & Yang, 2013). Association with STAT3 can prolong NF-κB nuclear retention by altering p300-mediated RelA/p65 acetylation (Lee et al., 2009). Nuclear RelA/p65-STAT3 interaction is significantly elevated by radiation in human glioma xenograft cell lines. Radiation, part of the standard treatment in glioblastoma multiforme, also increases migration, invasion, and the levels of intercellular adhesion molecule-1 (ICAM-1) (Kesanakurti, Chetty, Rajasekhar Maddirela, Gujrati, & Rao, 2013). Glioma cancer stem cells also show nuclear RelA/p65–STAT3 interaction, upregulation of NF-κB and STAT3 target genes, and sensitivity to inhibitors of STAT3 and NF-κB (Garner et al., 2013; Rinkenbaugh et al., 2016). Therefore, STAT3 appears to cooperate with NF-κB to promote the progression of certain cancers (Grivennikov & Karin, 2010). However, in some breast cancer cells, depending on the methylation at the miR-146b promoter, STAT3 induces the expression of this miRNA, a negative regulator of NF-κB (Xiang et al., 2014). Innate immune cells lacking STAT3 also display elevated NF-κB activity in response to LPS and receptor activator of NF-κB ligand (RANKL) (Zhang et al., 2014). Together these findings suggest that STAT3 is an important and context-dependent modulator of inflammation and NF-κB activity.

Due to its ability to self-regulate, dysregulation of the NF-κB pathway can initiate perpetual activation in cancer cells (Baldwin, 2001). Constitutive activation of KRAS and IKK/NF-κB are signature alterations in pancreatic ductal adenocarcinoma (PDAC). In a mouse model of PDAC, $KRAS^{G12D}$-activated AP-1 induces IL-1α, which activates RelA/p65 and its target genes IL-1α and p62. This initiates IL-1α and p62 feed-forward loops that sustain

NF-κB activity. KRAS mutation correlates with NF-κB activity, IL-1α overexpression, and poor survival in PDAC patients (Ling et al., 2012).

Interest in the noncanonical NF-κB pathway continues to grow as its role in cancer becomes more and more apparent. Polyps isolated from an experimental colitis mouse model deficient for a protein called NLRP12 shows an increased expression of cancer-associated target genes, including *Cxcl13* and *Cxcl12*. *Nlrp12* knockout mice showed elevated noncanonical NF-κB signaling and were highly susceptible to gastrointestinal inflammation and colitis-associated colon cancer (Allen et al., 2012). Inhibition of immunosuppressive myeloid-derived suppressor cells (MDSCs) is important for the efficacy of cancer immunotherapy. A subset of MDSCs in breast cancer can inhibit T-cell function through STAT3-dependent IDO upregulation. Increased NIK levels and RelB-p52 binding to the IDO promoter accompany increased STAT3 activation in these cells, while the STAT3 inhibitor JSI-124 and NIK knockdown are able to inhibit IDO expression (Yu et al., 2014). NIK signaling is also important in pancreatic cancer (Döppler, Liou, & Storz, 2013; Nishina, Yamaguchi, Gohda, Semba, & Inoue, 2009; Prabhu, Mundade, Korc, Loehrer, & Lu, 2014; Storz, 2013).

The efficacy of an inhibitor will rely on the cell-type-dependent and stimulus-specific responses of the NF-κB pathways. IKK inhibition in melanoma cells leads to apoptosis, likely due to the NF-κB-dependent regulation of the programmed cell death machinery. Inhibition of constitutive IKK activity by BMS-345541 reduces NF-κB activity and melanoma cell survival in vitro and in vivo. The effect of BMS-345541 on tumor cell growth was through mitochondria-mediated apoptosis, based on the release of apoptosis-inducing factor, dissipation of mitochondrial membrane potential, and the reduced ratio of B-cell lymphoma gene-2 (Bcl-2)/Bcl-associated X protein (Bax) in mitochondria (Yang, Amiri, Burke, Schmid, & Richmond, 2006).

Both IKKα and IKKβ have been linked to oncogenic pathways that are NF-κB independent. For instance, IKKα was shown by our group to promote both mTORC1 and mTORC2 signaling (Dan, Antonia, & Baldwin, 2016; Dan et al., 2014). IKKβ activates mTORC1 downstream of TNF-α signaling via phosphorylation and suppression of TSC1 (Lee et al., 2007). IKKβ also phosphorylates FOXO3a to promote oncogenesis (Hu et al., 2004). Finally, Karin and colleagues showed that IKKα phosphorylates the CDK inhibitor p27 in HER2+ brain cancer to promote removal from the nucleus and drive TICs (Zhang et al., 2013).

Taken together, an important role for the IKK/NF-κB pathway emerges in many different types of cancer. The field has made great strides in developing inhibitors of IKK/NF-κB activity. However, the quest to reliably and selectively inhibit the NF-κB pathway in all contexts continues.

5. TARGETING IKK AND NF-κB SIGNALING

Numerous inhibitors of IKK and NF-κB activity have been reported, and these have been reviewed extensively (Gilmore & Garbati, 2011; Gilmore & Herscovitch, 2006; Karin, Yamamoto, & Wang, 2004). Here, we will provide an updated perspective of the general strategies for pathway inhibition and discuss therapeutic uses of IKK/NF-κB inhibitors.

Some natural products including certain antioxidants and thiol-reactive compounds can be classified as inhibitors of IKK/NF-κB activity (Folmer, Jaspars, Dicato, & Diederich, 2008; Khanna et al., 2007; Ríos, Recio, Escandell, & Andújar, 2009). The natural products epigallocatechin gallate (found in green tea) and resveratrol (found in grapes, berries, peanuts, etc.) are able to inhibit NF-κB activity, which may partly explain the antiinflammatory and anticancer properties of these compounds (Bhardwaj et al., 2007; Park et al., 2012; Ren, Wang, Cui, Huoc, & Xue, 2013; Syed et al., 2007). Extensive data indicate that the natural product curcumin also has antiinflammatory and anticancer properties, partly through its ability to suppress NF-κB activation (Deguchi, 2015). Data indicate that curcumin induces S-nitrosylation of IKKβ to block its activity (Kao, Hu, Wu, & Kong, 2016). Triptolide, a diterpenoid epoxide endogenously produced by the thunder god vine (*Tripterygium wilfordii*) (Ziaei & Halaby, 2016) and several other natural products with known antiinflammatory properties have been shown to suppress NF-κB activation (Gilmore & Herscovitch, 2006).

Biomolecular and peptide IKK/NF-κB inhibitors include decoy oligonucleotides containing the κB site, ribozymes, siRNAs, the IκB super-repressor, dominant-negative molecules, and interfering peptides like the NEMO-binding domain (Habineza Ndikuyeze et al., 2014). Synthetic, small-molecule inhibitors are usually designed against the IKK complex (Edwards et al., 2009; Karin et al., 2004; Lee & Hung, 2008), because of its upstream regulatory role and the clinical success that other protein kinase inhibitors have shown. These inhibitors (natural, biomolecular, and synthetic drugs) can target the pathway at different stages in the multistep activation of NF-κB. The following sections have been divided based on the steps of activation that are commonly targeted (Fig. 3).

5.1 Receptor Inhibition

Upon ligand binding, the TNFR family members activate the NF-κB pathway; therefore, inhibiting receptor engagement prevents signal transduction. Research into blocking this stage of pathway activation has spawned anti-TNF antibodies. These inhibitors include monoclonal antibodies such as infliximab (Remicade) (Liu et al., 2016; Tsang, Dvorakova, Enright, Murphy, & Gleeson, 2016), adalimumab (Humira) (Asahina et al., 2016; Colombel et al., 2007; Yokoyama, Yamazaki, Katafuchi, & Ferchichi, 2016), golimumab (Simponi) (Sandborn et al., 2014a, 2014b), certolizumab pegol (Cimzia) (Mittal & Raychaudhuri, 2010; Patel & Moreland, 2010; Sandborn et al., 2007), and a circulating receptor fusion protein called etanercept (Enbrel) (Bozkurt et al., 2001; Mease et al., 2000; Taylor, 2010; van Schie et al., 2016). This strategy has been one of the more clinically successful efforts and is used to counteract the effects of TNF in a number of chronic inflammatory diseases (Taylor & Feldmann, 2009). Despite numerous side effects (Scheinfeld, 2004), anti-TNF antibodies provide significant clinical benefit.

5.2 Adaptor Inhibition as a Potential Strategy

Once an NF-κB-activating transmembrane receptor is engaged by its ligand, its associated adaptors and signal transducing complexes are recruited to the plasma membrane. These intracellular complexes include the TRAF proteins, which are required for IKK activation. The TRAF proteins function as E3 ligases and form a complex with the E2 ubiquitin-conjugating UBC13 and UVE1A proteins (Skaug, Jiang, & Chen, 2009; Wang et al., 2012). NEMO interacts with the K63 ubiquitin oligomers deposited by the TRAF-containing complex, which leads to activation of the canonical IKK complex (Clark, Nanda, & Cohen, 2013). Loss-of-function studies have emphasized the crucial role for these ubiquitinating adaptor proteins in NF-κB signal transduction (Chen, 2005). Moreover, mutations in the ubiquitin-editing enzymes A20 and CYLD, which usually function as inhibitors of NF-κB activity, have been found in human cancers (Courtois & Gilmore, 2006). Even though ubiquitinase inhibitors have been discovered (Tsukamoto et al., 2008), the TRAF and UBC proteins are involved in too

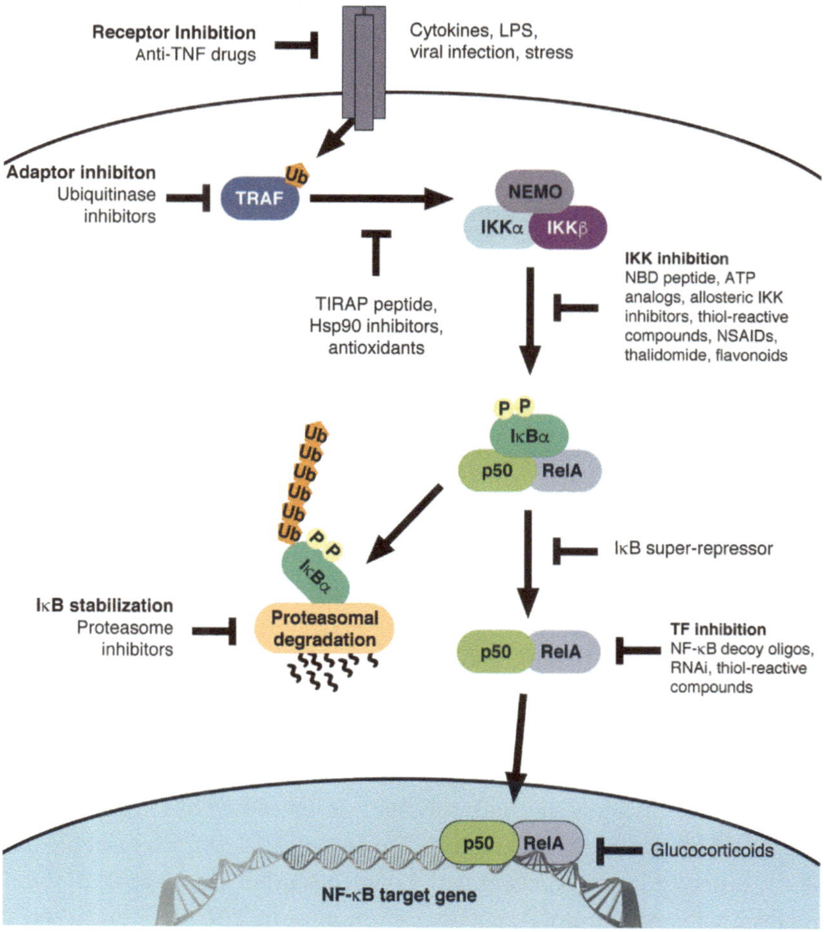

Fig. 3 Inhibition of the NF-κB pathway. The majority of research into developing inhibitors of the canonical pathway has been focused on selectively targeting IKK. However, there are a number of other possible therapeutic approaches. NF-κB decoy oligos and RNAi can target binding to DNA and regulation of NF-κB protein expression; the NEMO-binding domain (NBD) peptide can interfere with the formation of the IKK complex; the TIRAP inhibitor peptide, antioxidants, and Hsp90 inhibitors can block IKK activation (Hsp90 stabilizes RIP protein components of the TNF-α receptor signaling complex); proteasome inhibitors can stabilize IκB; IκBα super-repressor can inhibit NF-κB nuclear translocation; and glucocorticoids can inhibit NF-κB transcriptional activity. Due to the complex and variable mechanisms in regulating NF-κB signaling, more research is needed to develop new, more effective, and clinically useful inhibitors. *TIRAP*, toll-interleukin-1 receptor domain-containing adapter protein. *Modified from Gilmore, T. D., & Gerondakis, S. (2011). The c-Rel transcription factor in development and disease.* Genes & Cancer, *2(7), 695–711. http://doi.org/10.1177/1947601911421925; Karin, M., Yamamoto, Y., & Wang, Q. M. (2004). The IKK NF-kappa B system: A treasure trove for drug development.* Nature Reviews Drug Discovery, *3(1), 17–26. http://doi.org/10.1038/nrd1279.*

many different signaling pathways to be considered as promising targets for specific inhibition of NF-κB activity.

5.3 IKK Inhibition

IKK has been the main focus in the effort to target the NF-κB pathway with numerous small molecules specifically designed to inhibit its activity (Gilmore & Herscovitch, 2006). Pharmaceutical companies have developed a number of synthetic compounds capable of competing for the ATP-binding pocket of IKK. Some of these ATP analogs are highly selective, such as SPC-839, which has 200-fold higher affinity for IKKβ than IKKα (Karin et al., 2004; Lee & Hung, 2008). Bay 65–1942 (Compound A) is another highly selective ATP-competitive IKKβ inhibitor that shows broad antiinflammatory and anticancer activity (Bassères et al., 2010; Ziegelbauer et al., 2005). Some studies indicate that β-carboline alkaloid, a natural product widely expressed in plants, may be an ATP analog inhibitor of IKK activity (Yoon et al., 2005). Allosteric IKK inhibition can also be very effective. BMS-345541 is a selective IKK inhibitor that binds similar allosteric sites on IKKβ and IKKα and serves as an effective tool compound in studies of inflammation and cancer (Burke et al., 2003; McIntyre et al., 2003; Yang et al., 2006).

Thiol-reactive compounds like parthenolide, arsenite, and certain epoxyquinoids can interact with the thiol group of Cys179 in the activation loop of IKKβ and block its activity (Kapahi et al., 2000; Liang et al., 2003, 2006). Epoxyquinone A monomer (EqM), a synthetic derivative of the natural product epoxyquinol A, is a potent inhibitor of TNF-α-induced NF-κB activation. EqM inhibits IκBα phosphorylation and degradation by targeting Cys179 of IKKβ. EqM may even inhibit DNA binding of the canonical transcription factor via interaction with the thiol group of Cys38 on the RelA/p65 subunit. Finally, inhibition of the NF-κB pathway by EqM significantly reduces growth of human leukemia, kidney, and colon cancer cell lines (Liang et al., 2006). Withaferin A, a natural product found in the Ayurvedic plant *Withania somnifera*, has also been shown to inhibit IKKβ and inflammation in vitro and in vivo through its interaction with Cys179 (Heyninck, Lahtela-Kakkonen, Van der Veken, Haegeman, & Vanden Berghe, 2014). The Ser177 and Ser181 residues of IKKβ are part of the kinase activation loop and must be phosphorylated for IKKβ to be activated in response to upstream signals (Perkins, 2006; Scheidereit, 2006). The fact that these important serine residues are adjacent to Cys179 resulted in researchers assuming that these thiol-reactive compounds simply worked through steric hindrance. However, the mechanism proved to be more complicated, when some thiol-reactive compounds were able to inhibit the constitutively active S171/181E IKKβ mutant (Kwok, Koh, Ndubuisi, Elofsson, & Crews, 2001; Liang et al., 2003).

Dominant-negative mutants of the kinases in the NF-κB pathway can limit activation. Dominant-negative kinases can contain a mutated ATP-binding site, a mutated kinase activation loop, or a deleted kinase domain (Scheidereit, 2006). IKKβ, IKKα, and NIK mutants exhibit stimulus-specific effects, due to their distinct roles in the canonical and noncanonical pathways (Fig. 2).

Another drug target within the pathway is the NEMO subunit of the IKK complex. Synthetic inhibitors of NEMO have not yet been developed; however, the cell-permeable NEMO-binding domain (NBD) peptide, derived from the NBD of IKKβ, prevents IKK from binding to NEMO, and inhibits TNF-induced activation of the canonical pathway (Strickland & Ghosh, 2006). In chronic inflammatory diseases like rheumatoid arthritis, NF-κB can lead to increased osteoclast differentiation and inflammatory bone resorption. Treatment with the NBD peptide inhibits RANKL-stimulated NF-κB activation and osteoclastogenesis in vitro and in vivo. The peptide significantly reduces the severity of collagen-induced arthritis in mice by reducing the levels of TNF-α and IL-1β, abrogating joint swelling and reducing destruction of bone and cartilage (Jimi et al., 2004). The efficacy of the NBD peptide was tested in inflammatory bowel disease using a colitis mouse model established by adding dextran sulfate sodium salt to the drinking water and a mouse model established by a trinitrobenzene sulfonic acid (TNBS) enema. The peptide was able to reduce colonic inflammatory injury in both models by downregulating proinflammatory cytokines mediated by NF-κB (Shibata et al., 2007). Activated B-cell (ABC) diffuse large B-cell lymphoma (DLBCL) is characterized by constitutive canonical NF-κB signaling. In a phase I trial, dogs with spontaneous newly diagnosed or relapsed ABC-like DLBCL were treated with NBD peptide and biopsies of malignant lymph nodes were taken before and 24 h after treatment. Intravenous delivery of less than 2 mg/kg of NBD peptide inhibited constitutive canonical NF-κB activity in 6/10 dogs and showed no significant hematological, biochemical, or long-term toxicities. The results suggest that systemic administration of NBD peptide is safe, effective, and reduces malignant B-cell proliferation in some dogs and could be relevant for human ABC-DLBCL (Habineza Ndikuyeze et al., 2014).

Plevy and colleagues showed that NBD peptide reduced colitis in an IL-10-deficient murine model of inflammatory bowel disease (Davé et al., 2007). Guttridge and colleagues used NBD treatment to demonstrate the importance of IKK and NF-κB signaling in muscle degeneration in the murine model of Duchenne muscular dystrophy (Acharyya et al., 2007). Furthermore, NBD peptide improved disease markers in the golden retriever model of Duchenne muscular dystrophy (Kornegay et al., 2014). Peptides corresponding to the region of NEMO responsible for oligomerization can also inhibit NF-κB activity (Agou et al., 2004).

The IKK-related kinases, IKKε and TBK1, can drive context-specific NF-κB signaling and induce NF-κB-independent antiviral inflammatory responses (Hiscott, Nguyen, Arguello, Nakhaei, & Paz, 2006). Amlexanox (Aphthasol), a small-molecule compound used in the clinic to treat aphthous ulcers and asthma, is an inhibitor of these kinases. Remarkably, treatment of obese mice with amlexanox elevates energy expenditure through increased thermogenesis, producing weight loss, improved insulin sensitivity, and decreased hepatic steatosis (Reilly et al., 2013). Experiments performed in double-knockout MEFs supported these findings. Together, these results and its approved status make amlexanox a potential therapeutic for obesity and related disorders.

IKKε and TBK1 may also promote cell proliferation and cell survival in cancer (Barbie et al., 2009; Boehm et al., 2007; Clément, Meloche, & Servant, 2008; Eddy et al., 2005; Li et al., 2014; Muvaffak et al., 2014). Some studies have shown that the IKK-related kinases promote KRAS-driven tumorigenesis. TBK1/IKKε promotes tumor survival by activating CCL5 and IL-6. CYT387, a potent JAK/TBK1/IKKε inhibitor, uniquely disrupts a RAS-associated cytokine circuit involving CCL5, IL-6, and STAT3, and impairs KRAS-driven murine lung cancer growth. Treatment in combination with mitogen-activated protein kinase (MAPK) pathway inhibition led to the regression of aggressive KRAS mutant and p53 null lung adenocarcinomas in mice. The efficacy of CYT387-based treatment in murine KRAS-driven lung cancer suggests that concurrent inhibition of TBK1/IKKε, Janus-activated kinase, and MEK signaling could be an effective treatment in oncogenic KRAS-driven lung adenocarcinoma (Zhu et al., 2014). Although some compounds have displayed selective TBK1 inhibition in vitro and in vivo, the development of more efficient and selective TBK1 inhibitors is still required (Hasan et al., 2015; Yu et al., 2015).

5.4 IκB Stabilization

While there are multiple IκB proteins (Fig. 1), IκBα is the major NF-κB inhibitor protein in most cells. Some studies indicate that the induced upregulation of IκBα expression can be therapeutic in some settings. The natural peptides deposited in brain plaques of Alzheimer's disease patients, called β-amyloid peptides, induce IκBα expression. Treatment of rat cortical cultures with these peptides increases IκBα expression and decreases activation of NF-κB (Bales et al., 1998). Other studies show that antiinflammatory cytokines like IL-10, IL-11, and IL-13 increase IκBα mRNA levels and limit NF-κB nuclear localization (Ehrlich, Hu, Peterson, & Chao, 1998; Lentsch, Shanley, Sarma, & Ward, 1997; Trepicchio & Dorner, 1998).

Activated IKKs can rapidly phosphorylate IκBα, leading to its ubiquitination and proteasomal degradation. The ubiquitin proteasome system plays an important regulatory role in the NF-κB pathway (Chen, 2005). To limit proteasome-mediated degradation of IκBα and prevent NF-κB activation, several strategies may be employed. Microinjection of phosphopeptides corresponding to Ser32/36 of IκBα can inhibit IκBα ubiquitination (Yaron et al., 1997). The SCF-βTrCP E3 ubiquitin ligase complex, which leads to rapid degradation of IκBα, can be inhibited with RNAi against βTrCP or by overexpressing a dominant-negative βTrCP mutant (Tang et al., 2005). The YopJ protein, from the *Yersinia* bacteria, can deubiquitinate IκBα and prevent NF-κB nuclear translocation (Zhou et al., 2005). Although the mechanism is unknown, the small-molecule Ro 106–9920 can also prevent IκBα ubiquitination and inhibit inflammation (Swinney et al., 2002).

Several proteasome inhibitors are able to suppress NF-κB activity. MG101, MG115, and MG132 are peptide aldehydes capable of inhibiting the chymotrypsin-like activity of the proteasome complex (Grisham et al., 1999; Jobin, Hellerbrand, Licato, Brenner, & Sartor, 1998; Palombella, Rando, Goldberg, & Maniatis, 1994). Lactacystin and its precursor β-lactone are microbial products that block proteasome activity by acylating an active site threonine residue in the β subunits of the 20S proteasome (Fenteany & Schreiber, 1998; Hayes, Sherlock, & Selby, 2006). PS-262, PS-273, PS-341, and PS-402 are peptide boronic acids (or dipeptidyl boronates) that also block the chymotrypsin-like activity in the 20S proteasome (Adams, 2004). PS-341, commonly known as bortezomib, is the first drug to be approved for clinical application that targets the ubiquitin proteasome system (Orlowski et al., 2002). Bortezomib has significant efficacy against multiple myeloma, as well as several other hematologic and solid tumors (Jagannath et al., 2010), and works in part by stabilizing IκBα and preventing the activation of NF-κB (Hideshima et al., 2001; Richardson, Hideshima, & Anderson, 2003). However, the therapeutic window of bortezomib is relatively narrow due to its side effects (Nowis et al., 2010) and the high incidence of intrinsic and acquired resistance (Lü & Wang, 2013). Therefore, inhibitors of the ubiquitin proteasome system acting through mechanisms distinct from bortezomib could serve as effective therapeutic alternatives (Eldridge & O'Brien, 2010; Ruschak, Slassi, Kay, & Schimmer, 2011).

E1 inhibitors, such as PYR-41, target the ubiquitin-activating enzyme (UBE1) at the first step in the ubiquitination cascade (da Silva, Paiva, Lukkarila, & Gunning, 2013). PYR-41 has been reported to kill tumor cells by inhibiting NF-κB activation, and promoting p53 accumulation (Yang et al., 2007). NSC697923, an E2 inhibitor of the UBC13-UEV1A E2 enzyme, blocks the formation of the E2-ubiquitin thioester conjugate (Landré, Rotblat, Melino, Bernassola, & Melino, 2014). Since UBC13-UEV1A catalyzes the formation of K63-linked polyubiquitin chains, this compound inhibits NF-κB activation, and leads to reduced proliferation and viability of some cancer cells (Pulvino et al., 2012). E3 ubiquitin ligases represent promising targets for therapeutic intervention as key regulators of protein degradation, activation, and localization (Berndsen & Wolberger, 2014; Landré et al., 2014). One example is the family of E3 ligases composed of inhibitor of apoptosis (IAP) proteins such as cIAP1/2. These IAPs are interesting targets as they promote antiapoptotic NF-κB activity while inhibiting proapoptotic caspases and SMAC proteins (Fulda & Vucic, 2012). However, it will be hard to predict the specificity of drugs targeting E3 ligases, since they will affect the ubiquitination of several different proteins.

5.5 Blocking Nuclear Translocation

IκB degradation frees the NF-κB transcription factor to move into the nucleus via a subset of importin-α subunits (Fagerlund, Melén, Cao, & Julkunen, 2008). Peptides corresponding to the nuclear localization sequence of the p50 subunit have been used as a decoy to saturate the nuclear import machinery required by NF-κB dimers containing the p50 subunit (Letoha et al.,

2005; Lin, Yao, Veach, Torgerson, & Hawiger, 1995). However, some of these peptides, like SN50, can be nonspecific and can also block the nuclear localization of other factors including AP-1, STAT, and NFAT (Boothby, 2001; Torgerson, Colosia, Donahue, Lin, & Hawiger, 1998).

A small molecule, called PBS-1086, has been reported as a dual inhibitor of both canonical and noncanonical NF-κB pathways that inhibits NF-κB translocation into the nucleus (Fabre et al., 2012; Oh et al., 2011). Dehydroxymethylepoxyquinomicin (DHMEQ) is a fungal compound with antiinflammatory and anticancer effects that inhibits nuclear translocation of RelA/p65 and c-Rel, but not p50 (Umezawa & Chaicharoenpong, 2002). DHMEQ, a thiol-reactive compound, targets the conserved Cys38 residue of RelA/p65, conferring its specificity (Watanabe et al., 2008).

5.6 Transcription Factor Inhibition

In the final phase of NF-κB activation, the transcription factor binds to a stimulus-dependent suite of target genes. Transcription factors have not traditionally been considered optimal drug targets, but their ability to bind DNA and activate transcription can be inhibited. Estrogen and glucocorticoid receptors inhibit NF-κB-dependent transcription through several mechanisms, including cofactor competition and repressor protein recruitment (De Bosscher, Vanden Berghe, & Haegeman, 2006; Garside et al., 2004; Kalaitzidis & Gilmore, 2005; Ling & Kumar, 2012). Several of the thiol-reactive compounds used to target Cys179 of IKKβ can also prevent DNA binding by interacting with a conserved cysteine residue in the DNA-binding loop of the NF-κB/Rel subunits (Pande, Sousa, & Ramos, 2009). Using this mechanism, some natural products and their derivatives like EqM and parthenolide can target two stages of the pathway and inhibit both IKKβ activation and transcription factor binding (García-Piñeres, Lindenmeyer, & Merfort, 2004; Kwok et al., 2001; Liang et al., 2006; Straus et al., 2000). However, these compounds may lack specificity and are likely to interact with the reactive cysteine residues in many different proteins.

Dimethyl fumarate (DMF) is an antiinflammatory drug in clinical use for multiple sclerosis that also inhibits the NF-κB pathway (Cross et al., 2011; Gillard et al., 2015; Nicolay et al., 2016; Seidel et al., 2009). In breast cancer cells, Kastrati and colleagues showed that DMF effectively abrogates NF-κB-dependent mammosphere formation, inhibits cell proliferation, and significantly impairs xenograft tumor growth. In addition, their data suggest that DMF blocks NF-κB activity by covalently modifying RelA/p65 at Cys38, which prevents RelA/p65 nuclear translocation and attenuates its DNA-binding activity (Kastrati et al., 2016).

Decoy oligonucleotides corresponding to the NF-κB DNA-binding site and oligopeptide analogs can compete for binding of activated NF-κB dimers (Edwards et al., 2009). Some of these decoy oligonucleotides have had positive results in animal models (Desmet et al., 2004; Matsuda et al., 2004) and have gone on to clinical trials.

6. IKK/NF-κB AND ANTICANCER DRUG RESISTANCE

Radiation and conventional chemotherapies activate NF-κB and lead to increased expression of target genes that promote proliferation and survival (e.g., cIAP1/2, Bcl-2, and Bcl-xL). In this setting, the cancer cells will acquire NF-κB-dependent resistance to anticancer treatment (Chuang et al., 2002; Kesanakurti et al., 2013; Wang et al., 1996; Wang, Cusack, Liu, & Baldwin, 1999). The NF-κB inhibitor DHMEQ is able to resensitize prostate cancer cells to chemotherapy by decreasing the expression of Snail and Bcl-xL and increasing the levels of RKIP, a negative regulator of NF-κB (Baritaki, Yeung, Palladino, Berenson, & Bonavida, 2009). In BRCA mutant cancers cells, PARP inhibitors have shown anticancer activity. However, these tumors eventually develop drug resistance with a preferential upregulation of the NF-κB pathway, suggesting an NF-κB-dependent mechanism for acquired resistance. The PARP inhibitor-resistant cells are also sensitive to the inhibitors of NF-κB, Bay 11–7082, or bortezomib (Nakagawa et al., 2015).

In mouse models of KRAS-driven lung adenocarcinoma, studies have identified NF-κB as a promising therapeutic target important for tumor initiation and maintenance (Bassères et al., 2010; Meylan et al., 2009). Synthetic small-molecule NF-κB inhibitors bortezomib (Velcade/PS-341) and Bay 11–7082 induce tumor regression and prolong survival in tumor-bearing $KRAS^{LSL\text{-}G12D/wt}$; $p53^{flox/flox}$ mice. However, sustained bortezomib or Bay 11–7082 treatment inevitably confers drug resistance to the tumors (Xue et al., 2011). Since some cancer cells can achieve resistance to chemotherapy and radiation through activation of NF-κB, inhibiting the pathway in the right context and in combination with conventional therapies could significantly improve the efficacy of cancer treatment (Nakanishi & Toi, 2005).

7. CLINICAL TRIALS

While drugs such as aspirin exhibit clinical benefits related to blocking NF-κB activity (Yin, Yamamoto, & Gaynor, 1998), trials testing the specific inhibition of NF-κB remain scarce. A comprehensive database of clinical studies curated by the US National Institutes of Health can be found at the ClinicalTrials.gov website. One clinical trial testing the antiinflammatory effects of indirect IKK/NF-κB inhibition uses Denosumab, a monoclonal antibody against RANKL in Crohn's disease. This study's sponsor (Identifier: NCT02321280), University of Manitoba, is conducting phase I and phase II trials. In another phase II trial (Identifier: NCT02280785), sponsored by the Samsung Medical Center, researchers are testing the anticancer effects of indirect IKK/NF-κB inhibition using brentuximab vedotin (Adcetris), an antibody–drug conjugate targeting CD30, one of the surface antigens expressed in lymphoma cells. CD30 (TNFRSF8) is a transmembrane glycoprotein of the TNFR superfamily involved in the activation of the NF-κB pathway and the MAPKs, ultimately modulating cell growth, proliferation, and apoptosis. In January 2016, AnGes MG, a Japanese biotechnology company, announced the completion of patient enrollment for the phase III clinical trial of its NF-κB decoy oligonucleotide drug for atopic dermatitis.

8. CONCLUDING REMARKS

As a crucial regulator of inflammation and cell survival, the importance of the NF-κB pathway in disease is clear. The challenge lies in how to optimize therapeutic strategies for target specificity, disease context, dosage, toxicity, schedules, adjuvants, drug combinations, side effects, and resistance. Some strategies have been more effective than others. Anti-TNF drugs continue to provide significant clinical benefit, but these are accompanied by side effects including lymphomas, infection, autoimmunity, and heart failure. The NBD peptide has shown promise in animal trials, but whether it can be used safely and effectively in humans remains to be seen. Then, there are the IKK inhibitors, which have served researchers for years as reliable tool compounds, but have not been able to move beyond animal studies and into the clinic. Even with numerous IKK inhibitor compounds, very few can reliably distinguish between IKKα and IKKβ, and none are able to selectively target IKKα (Tian et al., 2015).

Ideally an IKK/NF-κB inhibitor should be highly specific and prevent activation without affecting other signaling pathways, at least not directly.

This ideal inhibitor should be more active in diseased cells than in normal cells or it should be amenable to targeted delivery (drug-loaded nanoparticles) (Zhang et al., 2016). It should be efficient, but transient and reversible since long-term immunosuppression would be unfavorable, considering the important role of NF-κB in the innate immune system. Finally, the efficacy of this ideal inhibitor as a single agent and in combination with other therapeutic agents would have to be thoroughly understood.

Elucidating the specific sequence of events that shapes the context-dependent NF-κB response is an important first step in designing a clinically relevant and truly specific strategy for NF-κB inhibition in disease. In addition to immunosuppression, a major caveat associated with long-term IKK/NF-κB inhibition would be increased susceptibility to endotoxin-induced shock. Studies have shown that sustained inhibition of IKKβ in mice results in enhanced production of IL-1β and related cytokines and increases susceptibility to endotoxin-induced shock due to upregulated inflammasome activation during bacterial infections (Greten et al., 2007). If this scenario were unavoidable, inhibitors would have to be limited to short durations and paired with antibiotics (Baud & Karin, 2009).

In September 2016, the Cancer Moonshot Blue Ribbon Panel led by the United States presented a report describing plans for making a decade's worth of progress in cancer research by the year 2020. Called the Cancer Moonshot 2020, it is one of the most comprehensive collaborative initiatives ever assembled for cancer research. The initiative hopes to accelerate the implementation of immunotherapy as the new standard of care for cancer patients—transitioning from therapies that decimate the immune system (radiation and chemotherapy) to an approach that would empower the immune system. This would potentially allow patients to manage cancer in a manner similar to diseases such as diabetes, asthma, or other chronic diseases. The IKK/NF-κB pathway is essential to immune response, but the pathway is also a major culprit in cancer. It will be important to determine if progress can be made toward successfully blocking IKK or NF-κB in cancer without impairing the immune system, and whether there is still a place for these strategies in a future where immunotherapy is the standard of care.

ACKNOWLEDGMENTS

The authors thank the Baldwin lab members for discussion and feedback.

Conflicts of interest: The authors declare no conflicts of interest.

REFERENCES

Acharyya, S., Villalta, S. A., Bakkar, N., Bupha-Intr, T., Janssen, P. M. L., Carathers, M., et al. (2007). Interplay of IKK/NF-kappaB signaling in macrophages and myofibers promotes muscle degeneration in Duchenne muscular dystrophy. *Journal of Clinical Investigation, 117*(4), 889–901. http://doi.org/10.1172/JCI30556.

Adams, J. (2004). The development of proteasome inhibitors as anticancer drugs. *Cancer Cell, 5*(5), 417–421.

Agou, F., Courtois, G., Chiaravalli, J., Baleux, F., Coïc, Y.-M., Traincard, F., et al. (2004). Inhibition of NF-kappa B activation by peptides targeting NF-kappa B essential modulator (nemo) oligomerization. *The Journal of Biological Chemistry, 279*(52), 54248–54257. http://doi.org/10.1074/jbc.M406423200.

Allen, I. C., Wilson, J. E., Schneider, M., Lich, J. D., Roberts, R. A., Arthur, J. C., et al. (2012). NLRP12 suppresses colon inflammation and tumorigenesis through the negative regulation of noncanonical NF-κB signaling. *Immunity, 36*(5), 742–754. http://doi.org/10.1016/j.immuni.2012.03.012.

Antonaki, A., Demetriades, C., Polyzos, A., Banos, A., Vatsellas, G., Lavigne, M. D., et al. (2011). Genomic analysis reveals a novel nuclear factor-κB (NF-κB)-binding site in Alu-repetitive elements. *Journal of Biological Chemistry, 286*(44), 38768–38782. http://doi.org/10.1074/jbc.M111.234161.

Asahina, A., Torii, H., Ohtsuki, M., Tokimoto, T., Hase, H., Tsuchiya, T., et al. (2016). Safety and efficacy of adalimumab treatment in Japanese patients with psoriasis: Results of SALSA study. *The Journal of Dermatology, 43*(11), 1257–1266. http://doi.org/10.1111/1346-8138.13409.

Ashall, L., Horton, C. A., Nelson, D. E., Paszek, P., Harper, C. V., Sillitoe, K., et al. (2009). Pulsatile stimulation determines timing and specificity of NF-kappaB-dependent transcription. *Science, 324*(5924), 242–246. http://doi.org/10.1126/science.1164860.

Auphan, N., DiDonato, J. A., Rosette, C., Helmberg, A., & Karin, M. (1995). Immunosuppression by glucocorticoids: Inhibition of NF-kappa B activity through induction of I kappa B synthesis. *Science, 270*(5234), 286–290.

Baeuerle, P. A., & Baltimore, D. (1988a). Activation of DNA-binding activity in an apparently cytoplasmic precursor of the NF-kappa B transcription factor. *Cell, 53*(2), 211–217.

Baeuerle, P. A., & Baltimore, D. (1988b). I kappa B: A specific inhibitor of the NF-kappa B transcription factor. *Science, 242*(4878), 540–546.

Baldwin, A. S. (1996). The NF-kappa B and I kappa B proteins: New discoveries and insights. *Annual Review of Immunology, 14*(1), 649–683. http://doi.org/10.1146/annurev.immunol.14.1.649.

Baldwin, A. S. (2001). Control of oncogenesis and cancer therapy resistance by the transcription factor NF-kappaB. *Journal of Clinical Investigation, 107*(3), 241–246. http://doi.org/10.1172/JCI11991.

Bales, K. R., Du, Y., Dodel, R. C., Yan, G. M., Hamilton-Byrd, E., & Paul, S. M. (1998). The NF-kappaB/Rel family of proteins mediates Abeta-induced neurotoxicity and glial activation. *Brain Research. Molecular Brain Research, 57*(1), 63–72.

Bang, D., Wilson, W., Ryan, M., Yeh, J. J., & Baldwin, A. S. (2013). GSK-3α promotes oncogenic KRAS function in pancreatic cancer via TAK1-TAB stabilization and regulation of noncanonical NF-κB. *Cancer Discovery, 3*(6), 690–703. http://doi.org/10.1158/2159-8290.CD-12-0541.

Barbie, D. A., Tamayo, P., Boehm, J. S., Kim, S. Y., Moody, S. E., Dunn, I. F., et al. (2009). Systematic RNA interference reveals that oncogenic KRAS-driven cancers require TBK1. *Nature, 462*(7269), 108–112. http://doi.org/10.1038/nature08460.

Baritaki, S., Yeung, K., Palladino, M., Berenson, J., & Bonavida, B. (2009). Pivotal roles of snail inhibition and RKIP induction by the proteasome inhibitor NPI-0052 in tumor cell

chemoimmunosensitization. *Cancer Research, 69*(21), 8376–8385. http://doi.org/10.1158/0008-5472.CAN-09-1069.
Barnes, P. J., & Adcock, I. M. (1997). NF-kappa B: A pivotal role in asthma and a new target for therapy. *Trends in Pharmacological Sciences, 18*(2), 46–50.
Bassères, D. S., Ebbs, A., Levantini, E., & Baldwin, A. S. (2010). Requirement of the NF-kappaB subunit p65/RelA for K-Ras-induced lung tumorigenesis. *Cancer Research, 70*(9), 3537–3546. http://doi.org/10.1158/0008-5472.CAN-09-4290.
Baud, V., & Karin, M. (2009). Is NF-kappaB a good target for cancer therapy? Hopes and pitfalls. *Nature Reviews. Drug Discovery, 8*(1), 33–40. http://doi.org/10.1038/nrd2781.
Beg, A. A., & Baltimore, D. (1996). An essential role for NF-kappaB in preventing TNF-alpha-induced cell death. *Science, 274*(5288), 782–784.
Beinke, S., Belich, M. P., & Ley, S. C. (2002). The death domain of NF-kappa B1 p105 is essential for signal-induced p105 proteolysis. *The Journal of Biological Chemistry, 277*(27), 24162–24168. http://doi.org/10.1074/jbc.M201576200.
Ben-Neriah, Y., & Karin, M. (2011). Inflammation meets cancer, with NF-κB as the matchmaker. *Nature Immunology, 12*(8), 715–723. http://doi.org/10.1038/ni.2060.
Berndsen, C. E., & Wolberger, C. (2014). New insights into ubiquitin E3 ligase mechanism. *Nature Structural & Molecular Biology, 21*(4), 301–307. http://doi.org/10.1038/nsmb.2780.
Betts, J. C., & Nabel, G. J. (1996). Differential regulation of NF-kappaB2(p100) processing and control by amino-terminal sequences. *Molecular and Cellular Biology, 16*(11), 6363–6371.
Bhardwaj, A., Sethi, G., Vadhan-Raj, S., Bueso-Ramos, C., Takada, Y., Gaur, U., et al. (2007). Resveratrol inhibits proliferation, induces apoptosis, and overcomes chemoresistance through down-regulation of STAT3 and nuclear factor-kappaB-regulated antiapoptotic and cell survival gene products in human multiple myeloma cells. *Blood, 109*(6), 2293–2302. http://doi.org/10.1182/blood-2006-02-003988.
Boehm, J. S., Zhao, J. J., Yao, J., Kim, S. Y., Firestein, R., Dunn, I. F., et al. (2007). Integrative genomic approaches identify IKBKE as a breast cancer oncogene. *Cell, 129*(6), 1065–1079. http://doi.org/10.1016/j.cell.2007.03.052.
Bonizzi, G., Bebien, M., Otero, D. C., Johnson-Vroom, K. E., Cao, Y., Vu, D., et al. (2004). Activation of IKKalpha target genes depends on recognition of specific kappaB binding sites by RelB:p52 dimers. *The EMBO Journal, 23*(21), 4202–4210. http://doi.org/10.1038/sj.emboj.7600391.
Boothby, M. (2001). Specificity of sn50 for NF-kappa B? *Nature Immunology, 2*(6), 471–472. http://doi.org/10.1038/88652.
Boyce, B. F., Yao, Z., & Xing, L. (2010). Functions of nuclear factor kappaB in bone. *Annals of the New York Academy of Sciences, 1192*, 367–375. http://doi.org/10.1111/j.1749-6632.2009.05315.x.
Bozkurt, B., Torre-Amione, G., Warren, M. S., Whitmore, J., Soran, O. Z., Feldman, A. M., et al. (2001). Results of targeted anti-tumor necrosis factor therapy with etanercept (ENBREL) in patients with advanced heart failure. *Circulation, 103*(8), 1044–1047.
Brach, M. A., Hass, R., Sherman, M. L., Gunji, H., Weichselbaum, R., & Kufe, D. (1991). Ionizing radiation induces expression and binding activity of the nuclear factor kappa B. *Journal of Clinical Investigation, 88*(2), 691–695. http://doi.org/10.1172/JCI115354.
Brown, K., Gerstberger, S., Carlson, L., Franzoso, G., & Siebenlist, U. (1995). Control of I kappa B-alpha proteolysis by site-specific, signal-induced phosphorylation. *Science, 267*(5203), 1485–1488.
Brown, K., Park, S., Kanno, T., Franzoso, G., & Siebenlist, U. (1993). Mutual regulation of the transcriptional activator NF-kappa B and its inhibitor, I kappa B-alpha. *Proceedings of the National Academy of Sciences of the United States of America, 90*(6), 2532–2536.

Burke, J. R., Pattoli, M. A., Gregor, K. R., Brassil, P. J., MacMaster, J. F., McIntyre, K. W., et al. (2003). BMS-345541 is a highly selective inhibitor of I kappa B kinase that binds at an allosteric site of the enzyme and blocks NF-kappa B-dependent transcription in mice. *The Journal of Biological Chemistry*, *278*(3), 1450–1456. http://doi.org/10.1074/jbc.M209677200.

Chen, Z. J. (2005). Ubiquitin signalling in the NF-kappaB pathway. *Nature Cell Biology*, 7(8), 758–765. http://doi.org/10.1038/ncb0805-758.

Chen, F. E., & Ghosh, G. (1999). Regulation of DNA binding by Rel/NF-kappaB transcription factors: Structural views. *Oncogene*, *18*(49), 6845–6852. http://doi.org/10.1038/sj.onc.1203224.

Chen, Z., Hagler, J., Palombella, V. J., Melandri, F., Scherer, D., Ballard, D., et al. (1995). Signal-induced site-specific phosphorylation targets I kappa B alpha to the ubiquitin-proteasome pathway. *Genes & Development*, *9*(13), 1586–1597.

Chen, Z. J., Parent, L., & Maniatis, T. (1996). Site-specific phosphorylation of IkappaBalpha by a novel ubiquitination-dependent protein kinase activity. *Cell*, *84*(6), 853–862.

Chiao, P. J., Miyamoto, S., & Verma, I. M. (1994). Autoregulation of I kappa B alpha activity. *Proceedings of the National Academy of Sciences of the United States of America*, *91*(1), 28–32.

Chow, J. C., Young, D. W., Golenbock, D. T., Christ, W. J., & Gusovsky, F. (1999). Toll-like receptor-4 mediates lipopolysaccharide-induced signal transduction. *The Journal of Biological Chemistry*, *274*(16), 10689–10692.

Chuang, S. E., Yeh, P. Y., Lu, Y. S., Lai, G. M., Liao, C. M., Gao, M., et al. (2002). Basal levels and patterns of anticancer drug-induced activation of nuclear factor-kappaB (NF-kappaB), and its attenuation by tamoxifen, dexamethasone, and curcumin in carcinoma cells. *Biochemical Pharmacology*, *63*(9), 1709–1716.

Clark, K., Nanda, S., & Cohen, P. (2013). Molecular control of the NEMO family of ubiquitin-binding proteins. *Nature Publishing Group*, *14*(10), 673–685. http://doi.org/10.1038/nrm3644.

Claudio, E., Brown, K., Park, S., Wang, H., & Siebenlist, U. (2002). BAFF-induced NEMO-independent processing of NF-kappa B2 in maturing B cells. *Nature Immunology*, *3*(10), 958–965. http://doi.org/10.1038/ni842.

Clément, J.-F., Meloche, S., & Servant, M. J. (2008). The IKK-related kinases: From innate immunity to oncogenesis. *Cell Research*, *18*(9), 889–899. http://doi.org/10.1038/cr.2008.273.

Cohen, S., Ciechanover, A., Kravtsova-Ivantsiv, Y., Lapid, D., & Lahav-Baratz, S. (2009). ABIN-1 negatively regulates NF-kappaB by inhibiting processing of the p105 precursor. *Biochemical and Biophysical Research Communications*, *389*(2), 205–210. http://doi.org/10.1016/j.bbrc.2009.08.074.

Colombel, J. F., Sandborn, W. J., Rutgeerts, P., Enns, R., Hanauer, S. B., Panaccione, R., et al. (2007). Adalimumab for maintenance of clinical response and remission in patients with Crohn's disease: The CHARM trial. *Gastroenterology*, *132*(1), 52–65. http://doi.org/10.1053/j.gastro.2006.11.041.

Coope, H. J., Atkinson, P. G. P., Huhse, B., Belich, M., Janzen, J., Holman, M. J., et al. (2002). CD40 regulates the processing of NF-kappaB2 p100 to p52. *The EMBO Journal*, *21*(20), 5375–5385.

Courtois, G., & Gilmore, T. D. (2006). Mutations in the NF-kappaB signaling pathway: Implications for human disease. *Oncogene*, *25*(51), 6831–6843. http://doi.org/10.1038/sj.onc.1209939.

Cross, S. A., Cook, D. R., Chi, A. W. S., Vance, P. J., Kolson, L. L., Wong, B. J., et al. (2011). Dimethyl fumarate, an immune modulator and inducer of the antioxidant response, suppresses HIV replication and macrophage-mediated neurotoxicity:

A novel candidate for HIV neuroprotection. *The Journal of Immunology, 187*(10), 5015–5025. http://doi.org/10.4049/jimmunol.1101868.
da Silva, S. R., Paiva, S.-L., Lukkarila, J. L., & Gunning, P. T. (2013). Exploring a new frontier in cancer treatment: Targeting the ubiquitin and ubiquitin-like activating enzymes. *Journal of Medicinal Chemistry, 56*(6), 2165–2177. http://doi.org/10.1021/jm301420b.
Dan, H. C., Antonia, R. J., & Baldwin, A. S. (2016). PI3K/Akt promotes feedforward mTORC2 activation through IKKα. *Oncotarget*, 7(16), 21064–21075. http://doi.org/10.18632/oncotarget.8383.
Dan, H. C., Ebbs, A., Pasparakis, M., Van Dyke, T., Basseres, D. S., & Baldwin, A. S. (2014). Akt-dependent activation of mTORC1 complex involves phosphorylation of mTOR (mammalian target of rapamycin) by IκB kinase α (IKKα). *Journal of Biological Chemistry, 289*(36), 25227–25240. http://doi.org/10.1074/jbc.M114.554881.
Davé, S. H., Tilstra, J. S., Matsuoka, K., Li, F., Karrasch, T., Uno, J. K., et al. (2007). Amelioration of chronic murine colitis by peptide-mediated transduction of the IkappaB kinase inhibitor NEMO binding domain peptide. *Journal of Immunology (Baltimore, Md: 1950), 179*(11), 7852–7859. http://doi.org/10.4049/jimmunol.179.11.7852.
De Bosscher, K., Vanden Berghe, W., & Haegeman, G. (2006). Cross-talk between nuclear receptors and nuclear factor kappaB. *Oncogene, 25*(51), 6868–6886. http://doi.org/10.1038/sj.onc.1209935.
De Santa, F., Narang, V., Yap, Z. H., Tusi, B. K., Burgold, T., Austenaa, L., et al. (2009). Jmjd3 contributes to the control of gene expression in LPS-activated macrophages. *The EMBO Journal, 28*(21), 3341–3352. http://doi.org/10.1038/emboj.2009.271.
De Santa, F., Totaro, M. G., Prosperini, E., Notarbartolo, S., Testa, G., & Natoli, G. (2007). The histone H3 lysine-27 demethylase Jmjd3 links inflammation to inhibition of polycomb-mediated gene silencing. *Cell, 130*(6), 1083–1094. http://doi.org/10.1016/j.cell.2007.08.019.
Deguchi, A. (2015). Curcumin targets in inflammation and cancer. *Endocrine, Metabolic & Immune Disorders Drug Targets, 15*(2), 88–96.
Dejardin, E. (2006). The alternative NF-kappaB pathway from biochemistry to biology: Pitfalls and promises for future drug development. *Biochemical Pharmacology, 72*(9), 1161–1179. http://doi.org/10.1016/j.bcp.2006.08.007.
Dejardin, E., Droin, N. M., Delhase, M., Haas, E., Cao, Y., Makris, C., et al. (2002). The lymphotoxin-beta receptor induces different patterns of gene expression via two NF-kappaB pathways. *Immunity, 17*(4), 525–535.
Desmet, C., Gosset, P., Pajak, B., Cataldo, D., Bentires-Alj, M., Lekeux, P., et al. (2004). Selective blockade of NF-kappa B activity in airway immune cells inhibits the effector phase of experimental asthma. *Journal of Immunology (Baltimore, Md: 1950), 173*(9), 5766–5775. http://doi.org/10.4049/jimmunol.173.9.5766.
DiDonato, J. A., Hayakawa, M., Rothwarf, D. M., Zandi, E., & Karin, M. (1997). A cytokine-responsive IkappaB kinase that activates the transcription factor NF-kappaB. *Nature, 388*(6642), 548–554. http://doi.org/10.1038/41493.
DiDonato, J., Mercurio, F., Rosette, C., Wu-Li, J., Suyang, H., Ghosh, S., et al. (1996). Mapping of the inducible IkappaB phosphorylation sites that signal its ubiquitination and degradation. *Molecular and Cellular Biology, 16*(4), 1295–1304.
Döppler, H., Liou, G.-Y., & Storz, P. (2013). Downregulation of TRAF2 mediates NIK-induced pancreatic cancer cell proliferation and tumorigenicity. *PloS One, 8*(1), e53676. http://doi.org/10.1371/journal.pone.0053676.
Eddy, S. F., Guo, S., Demicco, E. G., Romieu-Mourez, R., Landesman-Bollag, E., Seldin, D. C., et al. (2005). Inducible IkappaB kinase/IkappaB kinase epsilon expression is induced by CK2 and promotes aberrant nuclear factor-kappaB activation in breast cancer cells. *Cancer Research, 65*(24), 11375–11383. http://doi.org/10.1158/0008-5472.CAN-05-1602.

Edwards, M. R., Bartlett, N. W., Clarke, D., Birrell, M., Belvisi, M., & Johnston, S. L. (2009). Targeting the NF-kappaB pathway in asthma and chronic obstructive pulmonary disease. *Pharmacology & Therapeutics*, *121*(1), 1–13. http://doi.org/10.1016/j.pharmthera.2008.09.003.

Ehrlich, L. C., Hu, S., Peterson, P. K., & Chao, C. C. (1998). IL-10 down-regulates human microglial IL-8 by inhibition of NF-kappaB activation. *Neuroreport*, *9*(8), 1723–1726.

Eldridge, A. G., & O'Brien, T. (2010). Therapeutic strategies within the ubiquitin proteasome system. *Cell Death & Differentiation*, *17*(1), 4–13. http://doi.org/10.1038/cdd.2009.82.

Fabre, C., Mimura, N., Bobb, K., Kong, S.-Y., Gorgun, G., Cirstea, D., et al. (2012). Dual inhibition of canonical and noncanonical NF-κB pathways demonstrates significant antitumor activities in multiple myeloma. *Clinical Cancer Research*, *18*(17), 4669–4681. http://doi.org/10.1158/1078-0432.CCR-12-0779.

Fagerlund, R., Melén, K., Cao, X., & Julkunen, I. (2008). NF-kappaB p52, RelB and c-Rel are transported into the nucleus via a subset of importin alpha molecules. *Cellular Signalling*, *20*(8), 1442–1451. http://doi.org/10.1016/j.cellsig.2008.03.012.

Fan, C. M., & Maniatis, T. (1991). Generation of p50 subunit of NF-kappa B by processing of p105 through an ATP-dependent pathway. *Nature*, *354*(6352), 395–398. http://doi.org/10.1038/354395a0.

Fan, Y., Mao, R., & Yang, J. (2013). NF-κB and STAT3 signaling pathways collaboratively link inflammation to cancer. *Protein & Cell*, *4*(3), 176–185. http://doi.org/10.1007/s13238-013-2084-3.

Fenteany, G., & Schreiber, S. L. (1998). Lactacystin, proteasome function, and cell fate. *The Journal of Biological Chemistry*, *273*(15), 8545–8548.

Finco, T. S., Westwick, J. K., Norris, J. L., Beg, A. A., Der, C. J., & Baldwin, A. S. (1997). Oncogenic Ha-Ras-induced signaling activates NF-kappaB transcriptional activity, which is required for cellular transformation. *The Journal of Biological Chemistry*, *272*(39), 24113–24116.

Folmer, F., Jaspars, M., Dicato, M., & Diederich, M. (2008). Marine natural products as targeted modulators of the transcription factor NF-kappaB. *Biochemical Pharmacology*, *75*(3), 603–617. http://doi.org/10.1016/j.bcp.2007.07.044.

Fulda, S., & Vucic, D. (2012). Targeting IAP proteins for therapeutic intervention in cancer. *Nature Reviews. Drug Discovery*, *11*(2), 109–124. http://doi.org/10.1038/nrd3627.

García-Piñeres, A. J., Lindenmeyer, M. T., & Merfort, I. (2004). Role of cysteine residues of p65/NF-kappaB on the inhibition by the sesquiterpene lactone parthenolide and N-ethyl maleimide, and on its transactivating potential. *Life Sciences*, *75*(7), 841–856. http://doi.org/10.1016/j.lfs.2004.01.024.

Garner, J. M., Fan, M., Yang, C. H., Du, Z., Sims, M., Davidoff, A. M., et al. (2013). Constitutive activation of signal transducer and activator of transcription 3 (STAT3) and nuclear factor κB signaling in glioblastoma cancer stem cells regulates the Notch pathway. *Journal of Biological Chemistry*, *288*(36), 26167–26176. http://doi.org/10.1074/jbc.M113.477950.

Garside, H., Stevens, A., Farrow, S., Normand, C., Houle, B., Berry, A., et al. (2004). Glucocorticoid ligands specify different interactions with NF-kappaB by allosteric effects on the glucocorticoid receptor DNA binding domain. *The Journal of Biological Chemistry*, *279*(48), 50050–50059. http://doi.org/10.1074/jbc.M407309200.

Gaur, U., & Aggarwal, B. B. (2003). Regulation of proliferation, survival and apoptosis by members of the TNF superfamily. *Biochemical Pharmacology*, *66*(8), 1403–1408. http://doi.org/10.1016/S0006-2952(03)00490-8.

Ghosh, S., May, M. J., & Kopp, E. B. (1998). NF-kappa B and Rel proteins: Evolutionarily conserved mediators of immune responses. *Annual Review of Immunology*, *16*(1), 225–260. http://doi.org/10.1146/annurev.immunol.16.1.225.

Gil, J., Alcamí, J., & Esteban, M. (2000). Activation of NF-kappa B by the dsRNA-dependent protein kinase, PKR involves the I kappa B kinase complex. *Oncogene*, *19*(11), 1369–1378. http://doi.org/10.1038/sj.onc.1203448.

Gillard, G. O., Collette, B., Anderson, J., Chao, J., Scannevin, R. H., Huss, D. J., et al. (2015). DMF, but not other fumarates, inhibits NF-κB activity in vitro in an Nrf2-independent manner. *Journal of Neuroimmunology*, *283*, 74–85. http://doi.org/10.1016/j.jneuroim.2015.04.006.

Gilmore, T. D. (1999a). Multiple mutations contribute to the oncogenicity of the retroviral oncoprotein v-Rel. *Oncogene*, *18*(49), 6925–6937. http://doi.org/10.1038/sj.onc.1203222.

Gilmore, T. D. (1999b). The Rel/NF-kappaB signal transduction pathway: Introduction. *Oncogene*, *18*(49), 6842–6844. http://doi.org/10.1038/sj.onc.1203237.

Gilmore, T. D., & Garbati, M. R. (2011). Inhibition of NF-κB signaling as a strategy in disease therapy. *Current Topics in Microbiology and Immunology*, *349*, 245–263. Chapter 105. http://doi.org/10.1007/82_2010_105.

Gilmore, T. D., & Gerondakis, S. (2011). The c-Rel transcription factor in development and disease. *Genes & Cancer*, *2*(7), 695–711. http://doi.org/10.1177/1947601911421925.

Gilmore, T. D., & Herscovitch, M. (2006). Inhibitors of NF-kappaB signaling: 785 and counting. *Oncogene*, *25*(51), 6887–6899. http://doi.org/10.1038/sj.onc.1209982.

Greten, F. R., Arkan, M. C., Bollrath, J., Hsu, L.-C., Goode, J., Miething, C., et al. (2007). NF-kappaB is a negative regulator of IL-1beta secretion as revealed by genetic and pharmacological inhibition of IKKbeta. *Cell*, *130*(5), 918–931. http://doi.org/10.1016/j.cell.2007.07.009.

Grisham, M. B., Palombella, V. J., Elliott, P. J., Conner, E. M., Brand, S., Wong, H. L., et al. (1999). Inhibition of NF-kappa B activation in vitro and in vivo: Role of 26S proteasome. *Methods in Enzymology*, *300*, 345–363.

Grivennikov, S. I., & Karin, M. (2010). Dangerous liaisons: STAT3 and NF-kappaB collaboration and crosstalk in cancer. *Cytokine & Growth Factor Reviews*, *21*(1), 11–19. http://doi.org/10.1016/j.cytogfr.2009.11.005.

Guttridge, D. C., Albanese, C., Reuther, J. Y., Pestell, R. G., & Baldwin, A. S. (1999). NF-kappaB controls cell growth and differentiation through transcriptional regulation of cyclin D1. *Molecular and Cellular Biology*, *19*(8), 5785–5799.

Habineza Ndikuyeze, G., Gaurnier-Hausser, A., Patel, R., Baldwin, A. S., May, M. J., Flood, P., et al. (2014). A phase I clinical trial of systemically delivered NEMO binding domain peptide in dogs with spontaneous activated B-cell like diffuse large B-cell lymphoma. *PloS One*, *9*(5), e95404. http://doi.org/10.1371/journal.pone.0095404.

Hasan, M., Dobbs, N., Khan, S., White, M. A., Wakeland, E. K., Li, Q.-Z., et al. (2015). Cutting edge: Inhibiting TBK1 by compound II ameliorates autoimmune disease in mice. *The Journal of Immunology*, *195*(10), 4573–4577. http://doi.org/10.4049/jimmunol.1500162.

Hayden, M. S., & Ghosh, S. (2004). Signaling to NF-kappaB. *Genes & Development*, *18*(18), 2195–2224. http://doi.org/10.1101/gad.1228704.

Hayes, C. J., Sherlock, A. E., & Selby, M. D. (2006). Enantioselective total syntheses of (−)-clasto-lactacystin beta-lactone and 7-epi-(−)-clasto-lactacystin beta-lactone. *Organic & Biomolecular Chemistry*, *4*(2), 193–195. http://doi.org/10.1039/b516311k.

Heyninck, K., Lahtela-Kakkonen, M., Van der Veken, P., Haegeman, G., & Vanden Berghe, W. (2014). Withaferin A inhibits NF-kappaB activation by targeting cysteine 179 in IKKβ. *Biochemical Pharmacology*, *91*(4), 501–509. http://doi.org/10.1016/j.bcp.2014.08.004.

Hideshima, T., Richardson, P., Chauhan, D., Palombella, V. J., Elliott, P. J., Adams, J., et al. (2001). The proteasome inhibitor PS-341 inhibits growth, induces apoptosis, and overcomes drug resistance in human multiple myeloma cells. *Cancer Research*, *61*(7), 3071–3076.

Hiscott, J., Nguyen, T.-L. A., Arguello, M., Nakhaei, P., & Paz, S. (2006). Manipulation of the nuclear factor-kappaB pathway and the innate immune response by viruses. *Oncogene*, *25*(51), 6844–6867. http://doi.org/10.1038/sj.onc.1209941.

Hoffmann, A., & Baltimore, D. (2006). Circuitry of nuclear factor kappaB signaling. *Immunological Reviews*, *210*(1), 171–186. http://doi.org/10.1111/j.0105-2896.2006.00375.x.

Hoffmann, A., Natoli, G., & Ghosh, G. (2006). Transcriptional regulation via the NF-kappaB signaling module. *Oncogene*, *25*(51), 6706–6716. http://doi.org/10.1038/sj.onc.1209933.

Hoshino, K., Takeuchi, O., Kawai, T., Sanjo, H., Ogawa, T., Takeda, Y., et al. (1999). Cutting edge: Toll-like receptor 4 (TLR4)-deficient mice are hyporesponsive to lipopolysaccharide: Evidence for TLR4 as the Lps gene product. *Journal of Immunology (Baltimore, Md: 1950)*, *162*(7), 3749–3752.

Hu, M. C.-T., Lee, D.-F., Xia, W., Golfman, L. S., Ou-Yang, F., Yang, J.-Y., et al. (2004). IkappaB kinase promotes tumorigenesis through inhibition of forkhead FOXO3a. *Cell*, *117*(2), 225–237.

Hunter, J. E., Leslie, J., & Perkins, N. D. (2016). c-Rel and its many roles in cancer: An old story with new twists. *British Journal of Cancer*, *114*(1), 1–6. http://doi.org/10.1038/bjc.2015.410.

Jagannath, S., Kyle, R. A., Palumbo, A., Siegel, D. S., Cunningham, S., & Berenson, J. (2010). The current status and future of multiple myeloma in the clinic. *Clinical Lymphoma, Myeloma & Leukemia*, *10*(1), 28–43. http://doi.org/10.3816/CLML.2010.n.003.

Jimi, E., Aoki, K., Saito, H., D'Acquisto, F., May, M. J., Nakamura, I., et al. (2004). Selective inhibition of NF-κB blocks osteoclastogenesis and prevents inflammatory bone destruction in vivo. *Nature Medicine*, *10*(6), 617–624. http://doi.org/10.1038/nm1054.

Jobin, C., Hellerbrand, C., Licato, L. L., Brenner, D. A., & Sartor, R. B. (1998). Mediation by NF-kappa B of cytokine induced expression of intercellular adhesion molecule 1 (ICAM-1) in an intestinal epithelial cell line, a process blocked by proteasome inhibitors. *Gut*, *42*(6), 779–787.

Joyce, D., Albanese, C., Steer, J., Fu, M., Bouzahzah, B., & Pestell, R. G. (2001). NF-kappaB and cell-cycle regulation: The cyclin connection. *Cytokine & Growth Factor Reviews*, *12*(1), 73–90.

Kalaitzidis, D., & Gilmore, T. D. (2005). Transcription factor cross-talk: The estrogen receptor and NF-kappaB. *Trends in Endocrinology and Metabolism: TEM*, *16*(2), 46–52. http://doi.org/10.1016/j.tem.2005.01.004.

Kao, N.-J., Hu, J.-Y., Wu, C.-S., & Kong, Z.-L. (2016). Curcumin represses the activity of inhibitor-κB kinase in dextran sulfate sodium-induced colitis by S-nitrosylation. *International Immunopharmacology*, *38*, 1–7. http://doi.org/10.1016/j.intimp.2016.05.015.

Kapahi, P., Takahashi, T., Natoli, G., Adams, S. R., Chen, Y., Tsien, R. Y., et al. (2000). Inhibition of NF-kappa B activation by arsenite through reaction with a critical cysteine in the activation loop of Ikappa B kinase. *The Journal of Biological Chemistry*, *275*(46), 36062–36066. http://doi.org/10.1074/jbc.M007204200.

Karin, M., Yamamoto, Y., & Wang, Q. M. (2004). The IKK NF-kappa B system: A treasure trove for drug development. *Nature Reviews. Drug Discovery*, *3*(1), 17–26. http://doi.org/10.1038/nrd1279.

Kastrati, I., Siklos, M. I., Calderon-Gierszal, E. L., El-Shennawy, L., Georgieva, G., Thayer, E. N., et al. (2016). Dimethyl fumarate inhibits the nuclear factor κB pathway in breast cancer cells by covalent modification of p65 protein. *Journal of Biological Chemistry*, *291*(7), 3639–3647. http://doi.org/10.1074/jbc.M115.679704.

Kayagaki, N., Yan, M., Seshasayee, D., Wang, H., Lee, W., French, D. M., et al. (2002). BAFF/BLyS receptor 3 binds the B cell survival factor BAFF ligand through a discrete surface loop and promotes processing of NF-kappaB2. *Immunity*, *17*(4), 515–524.

Kendellen, M. F., Bradford, J. W., Lawrence, C. L., Clark, K. S., & Baldwin, A. S. (2014). Canonical and non-canonical NF-κB signaling promotes breast cancer tumor-initiating cells. *Oncogene*, *33*(10), 1297–1305. http://doi.org/10.1038/onc.2013.64.

Kesanakurti, D., Chetty, C., Rajasekhar Maddirela, D., Gujrati, M., & Rao, J. S. (2013). Essential role of cooperative NF-κB and Stat3 recruitment to ICAM-1 intronic consensus elements in the regulation of radiation-induced invasion and migration in glioma. *Oncogene*, *32*(43), 5144–5155. http://doi.org/10.1038/onc.2012.546.

Khanna, D., Sethi, G., Ahn, K. S., Pandey, M. K., Kunnumakkara, A. B., Sung, B., et al. (2007). Natural products as a gold mine for arthritis treatment. *Current Opinion in Pharmacology*, 7(3), 344–351. http://doi.org/10.1016/j.coph.2007.03.002.

Kieran, M., Blank, V., Logeat, F., Vandekerckhove, J., Lottspeich, F., Le Bail, O., et al. (1990). The DNA binding subunit of NF-kappa B is identical to factor KBF1 and homologous to the rel oncogene product. *Cell*, *62*(5), 1007–1018.

Kopp, E., & Ghosh, S. (1994). Inhibition of NF-kappa B by sodium salicylate and aspirin. *Science*, *265*(5174), 956–959.

Kornegay, J. N., Peterson, J. M., Bogan, D. J., Kline, W., Bogan, J. R., Dow, J. L., et al. (2014). NBD delivery improves the disease phenotype of the golden retriever model of Duchenne muscular dystrophy. *Skeletal Muscle*, *4*, 18. http://doi.org/10.1186/2044-5040-4-18.

Kunsch, C., Ruben, S. M., & Rosen, C. A. (1992). Selection of optimal kappa B/Rel DNA-binding motifs: Interaction of both subunits of NF-kappa B with DNA is required for transcriptional activation. *Molecular and Cellular Biology*, *12*(10), 4412–4421.

Kwok, B. H., Koh, B., Ndubuisi, M. I., Elofsson, M., & Crews, C. M. (2001). The anti-inflammatory natural product parthenolide from the medicinal herb Feverfew directly binds to and inhibits IkappaB kinase. *Chemistry & Biology*, *8*(8), 759–766. http://doi.org/10.1016/S1074-5521(01)00049-7.

Landré, V., Rotblat, B., Melino, S., Bernassola, F., & Melino, G. (2014). Screening for E3-ubiquitin ligase inhibitors: Challenges and opportunities. *Oncotarget*, *5*(18), 7988–8013. http://doi.org/10.18632/oncotarget.2431.

Lawrence, T. (2009). The nuclear factor NF-kappaB pathway in inflammation. *Cold Spring Harbor Perspectives in Biology*, *1*(6), a001651. http://doi.org/10.1101/cshperspect.a001651.

Lawrence, C. L., & Baldwin, A. S. (2016). Non-canonical EZH2 transcriptionally activates RelB in triple negative breast cancer. *PloS One*, *11*(10), e0165005. http://doi.org/10.1371/journal.pone.0165005.

Lee, T. K., & Covert, M. W. (2010). High-throughput, single-cell NF-κB dynamics. *Current Opinion in Genetics & Development*, *20*(6), 677–683. http://doi.org/10.1016/j.gde.2010.08.005.

Lee, H., Herrmann, A., Deng, J.-H., Kujawski, M., Niu, G., Li, Z., et al. (2009). Persistently activated Stat3 maintains constitutive NF-kappaB activity in tumors. *Cancer Cell*, *15*(4), 283–293. http://doi.org/10.1016/j.ccr.2009.02.015.

Lee, D.-F., & Hung, M.-C. (2008). Advances in targeting IKK and IKK-related kinases for cancer therapy. *Clinical Cancer Research*, *14*(18), 5656–5662. http://doi.org/10.1158/1078-0432.CCR-08-0123.

Lee, D.-F., Kuo, H.-P., Chen, C.-T., Hsu, J.-M., Chou, C.-K., Wei, Y., et al. (2007). IKK beta suppression of TSC1 links inflammation and tumor angiogenesis via the mTOR pathway. *Cell*, *130*(3), 440–455. http://doi.org/10.1016/j.cell.2007.05.058.

Lee, S. T., Li, Z., Wu, Z., Aau, M., Guan, P., Karuturi, R. K. M., et al. (2011). Context-specific regulation of NF-κB target gene expression by EZH2 in breast cancers. *Molecular Cell*, *43*(5), 798–810. http://doi.org/10.1016/j.molcel.2011.08.011.

Lentsch, A. B., Shanley, T. P., Sarma, V., & Ward, P. A. (1997). In vivo suppression of NF-kappa B and preservation of I kappa B alpha by interleukin-10 and interleukin-13. *Journal of Clinical Investigation*, *100*(10), 2443–2448. http://doi.org/10.1172/JCI119786.

Letoha, T., Somlai, C., Takacs, T., Szabolcs, A., Jarmay, K., Rakonczay, Z., et al. (2005). A nuclear import inhibitory peptide ameliorates the severity of cholecystokinin-induced acute pancreatitis. *World Journal of Gastroenterology: WJG, 11*(7), 990–999.

Levy, D., Kuo, A. J., Chang, Y., Schaefer, U., Kitson, C., Cheung, P., et al. (2011). Lysine methylation of the NF-κB subunit RelA by SETD6 couples activity of the histone methyltransferase GLP at chromatin to tonic repression of NF-κB signaling. *Nature Publishing Group, 12*(1), 29–36. http://doi.org/10.1038/ni.1968.

Li, J., Huang, J., Jeong, J.-H., Park, S.-J., Wei, R., Peng, J., et al. (2014). Selective TBK1/IKKi dual inhibitors with anticancer potency. *International Journal of Cancer, 134*(8), 1972–1980. http://doi.org/10.1002/ijc.28507.

Li, H., Malhotra, S., & Kumar, A. (2008). Nuclear factor-kappa B signaling in skeletal muscle atrophy. *Journal of Molecular Medicine (Berlin, Germany), 86*(10), 1113–1126. http://doi.org/10.1007/s00109-008-0373-8.

Liang, M.-C., Bardhan, S., Li, C., Pace, E. A., Porco, J. A., & Gilmore, T. D. (2003). Jesterone dimer, a synthetic derivative of the fungal metabolite jesterone, blocks activation of transcription factor nuclear factor kappaB by inhibiting the inhibitor of kappaB kinase. *Molecular Pharmacology, 64*(1), 123–131. http://doi.org/10.1124/mol.64.1.123.

Liang, M.-C., Bardhan, S., Pace, E. A., Rosman, D., Beutler, J. A., Porco, J. A., et al. (2006). Inhibition of transcription factor NF-kappaB signaling proteins IKKbeta and p65 through specific cysteine residues by epoxyquinone A monomer: Correlation with its anti-cancer cell growth activity. *Biochemical Pharmacology, 71*(5), 634–645. http://doi.org/10.1016/j.bcp.2005.11.013.

Liao, G., Zhang, M., Harhaj, E. W., & Sun, S.-C. (2004). Regulation of the NF-kappaB-inducing kinase by tumor necrosis factor receptor-associated factor 3-induced degradation. *The Journal of Biological Chemistry, 279*(25), 26243–26250. http://doi.org/10.1074/jbc.M403286200.

Lin, L., & Kobayashi, M. (2003). Stability of the Rel homology domain is critical for generation of NF-kappa B p50 subunit. *The Journal of Biological Chemistry, 278*(34), 31479–31485. http://doi.org/10.1074/jbc.M304140200.

Lin, Y. Z., Yao, S. Y., Veach, R. A., Torgerson, T. R., & Hawiger, J. (1995). Inhibition of nuclear translocation of transcription factor NF-kappa B by a synthetic peptide containing a cell membrane-permeable motif and nuclear localization sequence. *The Journal of Biological Chemistry, 270*(24), 14255–14258.

Ling, L., Cao, Z., & Goeddel, D. V. (1998). NF-kappaB-inducing kinase activates IKK-alpha by phosphorylation of Ser-176. *Proceedings of the National Academy of Sciences of the United States of America, 95*(7), 3792–3797.

Ling, J., Kang, Y., Zhao, R., Xia, Q., Lee, D.-F., Chang, Z., et al. (2012). KrasG12D-induced IKK2/β/NF-κB activation by IL-1α and p62 feedforward loops is required for development of pancreatic ductal adenocarcinoma. *Cancer Cell, 21*(1), 105–120. http://doi.org/10.1016/j.ccr.2011.12.006.

Ling, J., & Kumar, R. (2012). Crosstalk between NFkB and glucocorticoid signaling: A potential target of breast cancer therapy. *Cancer Letters, 322*(2), 119–126. http://doi.org/10.1016/j.canlet.2012.02.033.

Liu, F., Ai, F., Tian, L., Liu, S., Zhao, L., & Wang, X. (2016). Infliximab enhances the therapeutic effects of 5-fluorouracil resulting in tumor regression in colon cancer. *OncoTargets and Therapy, 9*, 5999–6008. http://doi.org/10.2147/OTT.S109342.

Liu, B., Sun, L., Liu, Q., Gong, C., Yao, Y., Lv, X., et al. (2015). A cytoplasmic NF-κB interacting long noncoding RNA blocks IκB phosphorylation and suppresses breast cancer metastasis. *Cancer Cell, 27*(3), 370–381. http://doi.org/10.1016/j.ccell.2015.02.004.

Lü, S., & Wang, J. (2013). The resistance mechanisms of proteasome inhibitor bortezomib. *Biomarker Research, 1*(1), 13. http://doi.org/10.1186/2050-7771-1-13.

Marienfeld, R. B., Palkowitsch, L., & Ghosh, S. (2006). Dimerization of the I kappa B kinase-binding domain of NEMO is required for tumor necrosis factor alpha-induced NF-kappa B activity. *Molecular and Cellular Biology*, *26*(24), 9209–9219. http://doi.org/10.1128/MCB.00478-06.

Matsuda, N., Hattori, Y., Takahashi, Y., Nishihira, J., Jesmin, S., Kobayashi, M., et al. (2004). Therapeutic effect of in vivo transfection of transcription factor decoy to NF-kappaB on septic lung in mice. *AJP: Lung Cellular and Molecular Physiology*, *287*(6), L1248–L1255. http://doi.org/10.1152/ajplung.00164.2004.

McIntyre, K. W., Shuster, D. J., Gillooly, K. M., Dambach, D. M., Pattoli, M. A., Lu, P., et al. (2003). A highly selective inhibitor of I kappa B kinase, BMS-345541, blocks both joint inflammation and destruction in collagen-induced arthritis in mice. *Arthritis and Rheumatism*, *48*(9), 2652–2659. http://doi.org/10.1002/art.11131.

Mease, P. J., Goffe, B. S., Metz, J., VanderStoep, A., Finck, B., & Burge, D. J. (2000). Etanercept in the treatment of psoriatic arthritis and psoriasis: A randomised trial. *Lancet (London, England)*, *356*(9227), 385–390. http://doi.org/10.1016/S0140-6736(00)02530-7.

Mercurio, F., Zhu, H., Murray, B. W., Shevchenko, A., Bennett, B. L., Li, J., et al. (1997). IKK-1 and IKK-2: Cytokine-activated IkappaB kinases essential for NF-kappaB activation. *Science*, *278*(5339), 860–866.

Meylan, E., Dooley, A. L., Feldser, D. M., Shen, L., Turk, E., Ouyang, C., et al. (2009). Requirement for NF-kappaB signalling in a mouse model of lung adenocarcinoma. *Nature*, *462*(7269), 104–107. http://doi.org/10.1038/nature08462.

Miagkov, A. V., Kovalenko, D. V., Brown, C. E., Didsbury, J. R., Cogswell, J. P., Stimpson, S. A., et al. (1998). NF-kappaB activation provides the potential link between inflammation and hyperplasia in the arthritic joint. *Proceedings of the National Academy of Sciences of the United States of America*, *95*(23), 13859–13864.

Mittal, M., & Raychaudhuri, S. P. (2010). Golimumab and certolizumab: The two new anti-tumor necrosis factor kids on the block. *Indian Journal of Dermatology, Venereology and Leprology*, *76*(6), 602–608. quiz 609. http://doi.org/10.4103/0378-6323.72445.

Miyamoto, S., Chiao, P. J., & Verma, I. M. (1994). Enhanced I kappa B alpha degradation is responsible for constitutive NF-kappa B activity in mature murine B-cell lines. *Molecular and Cellular Biology*, *14*(5), 3276–3282.

Moorthy, A. K., Savinova, O. V., Ho, J. Q., Wang, V. Y.-F., Vu, D., & Ghosh, G. (2006). The 20S proteasome processes NF-kappaB1 p105 into p50 in a translation-independent manner. *The EMBO Journal*, *25*(9), 1945–1956. http://doi.org/10.1038/sj.emboj.7601081.

Muvaffak, A., Pan, Q., Yan, H., Fernandez, R., Lim, J., Dolinski, B., et al. (2014). Evaluating TBK1 as a therapeutic target in cancers with activated IRF3. *Molecular Cancer Research*, *12*(7), 1055–1066. http://doi.org/10.1158/1541-7786.MCR-13-0642.

Nakagawa, Y., Sedukhina, A. S., Okamoto, N., Nagasawa, S., Suzuki, N., Ohta, T., et al. (2015). NF-κB signaling mediates acquired resistance after PARP inhibition. *Oncotarget*, *6*(6), 3825–3839. http://doi.org/10.18632/oncotarget.2868.

Nakanishi, C., & Toi, M. (2005). Nuclear factor-kappaB inhibitors as sensitizers to anticancer drugs. *Nature Reviews. Cancer*, *5*(4), 297–309. http://doi.org/10.1038/nrc1588.

Nicolay, J. P., Müller-Decker, K., Schroeder, A., Brechmann, M., Möbs, M., Géraud, C., et al. (2016). Dimethyl fumarate restores apoptosis sensitivity and inhibits tumor growth and metastasis in CTCL by targeting NF-κB. *Blood*, *128*(6), 805–815. http://doi.org/10.1182/blood-2016-01-694117.

Nishina, T., Yamaguchi, N., Gohda, J., Semba, K., & Inoue, J.-I. (2009). NIK is involved in constitutive activation of the alternative NF-kappaB pathway and proliferation of pancreatic cancer cells. *Biochemical and Biophysical Research Communications*, *388*(1), 96–101. http://doi.org/10.1016/j.bbrc.2009.07.125.

Novack, D. V., Yin, L., Hagen-Stapleton, A., Schreiber, R. D., Goeddel, D. V., Ross, F. P., et al. (2003). The IkappaB function of NF-kappaB2 p100 controls stimulated osteoclastogenesis. *The Journal of Experimental Medicine, 198*(5), 771–781. http://doi.org/10.1084/jem.20030116.

Nowis, D., Mączewski, M., Mackiewicz, U., Kujawa, M., Ratajska, A., Wieckowski, M. R., et al. (2010). Cardiotoxicity of the anticancer therapeutic agent bortezomib. *The American Journal of Pathology, 176*(6), 2658–2668. http://doi.org/10.2353/ajpath.2010.090690.

Oeckinghaus, A., & Ghosh, S. (2009). The NF-kappaB family of transcription factors and its regulation. *Cold Spring Harbor Perspectives in Biology, 1*(4), a000034. http://doi.org/10.1101/cshperspect.a000034.

Oeckinghaus, A., Hayden, M. S., & Ghosh, S. (2011). Crosstalk in NF-κB signaling pathways. *Nature Immunology, 12*(8), 695–708. http://doi.org/10.1038/ni.2065.

Oh, U., McCormick, M. J., Datta, D., Turner, R. V., Bobb, K., Monie, D. D., et al. (2011). Inhibition of immune activation by a novel nuclear factor-kappa B inhibitor in HTLV-I-associated neurologic disease. *Blood, 117*(12), 3363–3369. http://doi.org/10.1182/blood-2010-10-306571.

O'Neill, L. A., & Kaltschmidt, C. (1997). NF-kappa B: A crucial transcription factor for glial and neuronal cell function. *Trends in Neurosciences, 20*(6), 252–258.

Orlowski, R. Z., Stinchcombe, T. E., Mitchell, B. S., Shea, T. C., Baldwin, A. S., Stahl, S., et al. (2002). Phase I trial of the proteasome inhibitor PS-341 in patients with refractory hematologic malignancies. *Journal of Clinical Oncology: Official Journal of the American Society of Clinical Oncology, 20*(22), 4420–4427. http://doi.org/10.1200/JCO.2002.01.133.

Osborn, L., Kunkel, S., & Nabel, G. J. (1989). Tumor necrosis factor alpha and interleukin 1 stimulate the human immunodeficiency virus enhancer by activation of the nuclear factor kappa B. *Proceedings of the National Academy of Sciences of the United States of America, 86*(7), 2336–2340.

Pahl, H. L. (1999). Activators and target genes of Rel/NF-kappaB transcription factors. *Oncogene, 18*(49), 6853–6866. http://doi.org/10.1038/sj.onc.1203239.

Palombella, V. J., Rando, O. J., Goldberg, A. L., & Maniatis, T. (1994). The ubiquitin-proteasome pathway is required for processing the NF-kappa B1 precursor protein and the activation of NF-kappa B. *Cell, 78*(5), 773–785.

Pande, V., Sousa, S. F., & Ramos, M. J. (2009). Direct covalent modification as a strategy to inhibit nuclear factor-kappa B. *Current Medicinal Chemistry, 16*(32), 4261–4273.

Park, H. J., Lee, J.-Y., Chung, M.-Y., Park, Y.-K., Bower, A. M., Koo, S. I., et al. (2012). Green tea extract suppresses NFκB activation and inflammatory responses in diet-induced obese rats with nonalcoholic steatohepatitis. *The Journal of Nutrition, 142*(1), 57–63. http://doi.org/10.3945/jn.111.148544.

Paszek, P., Jackson, D. A., & White, M. R. (2010). Oscillatory control of signalling molecules. *Current Opinion in Genetics & Development, 20*(6), 670–676. http://doi.org/10.1016/j.gde.2010.08.004.

Patel, A. M., & Moreland, L. W. (2010). Certolizumab pegol: A new biologic targeting rheumatoid arthritis. *Expert Review of Clinical Immunology, 6*(6), 855–866. http://doi.org/10.1586/eci.10.69.

Perkins, N. D. (2006). Post-translational modifications regulating the activity and function of the nuclear factor kappa B pathway. *Oncogene, 25*(51), 6717–6730. http://doi.org/10.1038/sj.onc.1209937.

Perkins, N. D. (2007). Integrating cell-signalling pathways with NF-kappaB and IKK function. *Nature Reviews. Molecular Cell Biology, 8*(1), 49–62. http://doi.org/10.1038/nrm2083.

Pierce, J. W., Lenardo, M., & Baltimore, D. (1988). Oligonucleotide that binds nuclear factor NF-kappa B acts as a lymphoid-specific and inducible enhancer element. *Proceedings of the National Academy of Sciences of the United States of America, 85*(5), 1482–1486.

Piva, R., Belardo, G., & Santoro, M. G. (2006). NF-kappaB: A stress-regulated switch for cell survival. *Antioxidants & Redox Signaling, 8*(3–4), 478–486. http://doi.org/10.1089/ars.2006.8.478.

Prabhu, L., Mundade, R., Korc, M., Loehrer, P. J., & Lu, T. (2014). Critical role of NF-κB in pancreatic cancer. *Oncotarget, 5*(22), 10969–10975. http://doi.org/10.18632/oncotarget.2624.

Pulvino, M., Liang, Y., Oleksyn, D., DeRan, M., Van Pelt, E., Shapiro, J., et al. (2012). Inhibition of proliferation and survival of diffuse large B-cell lymphoma cells by a small-molecule inhibitor of the ubiquitin-conjugating enzyme Ubc13-Uev1A. *Blood, 120*(8), 1668–1677. http://doi.org/10.1182/blood-2012-02-406074.

Ramirez-Carrozzi, V. R., Braas, D., Bhatt, D. M., Cheng, C. S., Hong, C., Doty, K. R., et al. (2009). A unifying model for the selective regulation of inducible transcription by CpG islands and nucleosome remodeling. *Cell, 138*(1), 114–128. http://doi.org/10.1016/j.cell.2009.04.020.

Ray, A., & Prefontaine, K. E. (1994). Physical association and functional antagonism between the p65 subunit of transcription factor NF-kappa B and the glucocorticoid receptor. *Proceedings of the National Academy of Sciences of the United States of America, 91*(2), 752–756.

Reilly, S. M., Chiang, S.-H., Decker, S. J., Chang, L., Uhm, M., Larsen, M. J., et al. (2013). An inhibitor of the protein kinases TBK1 and IKK-ε improves obesity-related metabolic dysfunctions in mice. *Nature Medicine, 19*(3), 313–321. http://doi.org/10.1038/nm.3082.

Ren, Z., Wang, L., Cui, J., Huoc, Z., & Xue, J. (2013). Resveratrol inhibits NF-κB signaling through suppression of p65 and IκB kinase activities. *Die Pharmazie, 68*(8), 689–694. http://doi.org/10.1691/ph.2013.2916.

Reuther, J. Y., Reuther, G. W., Cortez, D., Pendergast, A. M., & Baldwin, A. S. (1998). A requirement for NF-kappaB activation in Bcr-Abl-mediated transformation. *Genes & Development, 12*(7), 968–981.

Richardson, P. G., Hideshima, T., & Anderson, K. C. (2003). Bortezomib (PS-341): A novel, first-in-class proteasome inhibitor for the treatment of multiple myeloma and other cancers. *Cancer Control: Journal of the Moffitt Cancer Center, 10*(5), 361–369.

Rinkenbaugh, A. L., Cogswell, P. C., Calamini, B., Dunn, D. E., Persson, A. I., Weiss, W. A., et al. (2016). IKK/NF-κB signaling contributes to glioblastoma stem cell maintenance. *Oncotarget*, 7(43), 69173–69187. http://doi.org/10.18632/oncotarget.12507.

Ríos, J. L., Recio, M. C., Escandell, J. M., & Andújar, I. (2009). Inhibition of transcription factors by plant-derived compounds and their implications in inflammation and cancer. *Current Pharmaceutical Design, 15*(11), 1212–1237.

Rothwarf, D. M., Zandi, E., Natoli, G., & Karin, M. (1998). IKK-gamma is an essential regulatory subunit of the IkappaB kinase complex. *Nature, 395*(6699), 297–300. http://doi.org/10.1038/26261.

Ruland, J. (2011). Return to homeostasis: Downregulation of NF-κB responses. *Nature Immunology, 12*(8), 709–714. http://doi.org/10.1038/ni.2055.

Ruschak, A. M., Slassi, M., Kay, L. E., & Schimmer, A. D. (2011). Novel proteasome inhibitors to overcome bortezomib resistance. *JNCI Journal of the National Cancer Institute, 103*(13), 1007–1017. http://doi.org/10.1093/jnci/djr160.

Saccani, S., & Natoli, G. (2002). Dynamic changes in histone H3 Lys 9 methylation occurring at tightly regulated inducible inflammatory genes. *Genes & Development, 16*(17), 2219–2224. http://doi.org/10.1101/gad.232502.

Salminen, A., Huuskonen, J., Ojala, J., Kauppinen, A., Kaarniranta, K., & Suuronen, T. (2008). Activation of innate immunity system during aging: NF-kB signaling is the molecular culprit of inflamm-aging. *Ageing Research Reviews*, 7(2), 83–105. http://doi.org/10.1016/j.arr.2007.09.002.

Sandborn, W. J., Feagan, B. G., Marano, C., Zhang, H., Strauss, R., Johanns, J., et al. (2014a). Subcutaneous golimumab induces clinical response and remission in patients with moderate-to-severe ulcerative colitis. *Gastroenterology*, *146*(1), 85–95. quiz e14–5. http://doi.org/10.1053/j.gastro.2013.05.048.

Sandborn, W. J., Feagan, B. G., Marano, C., Zhang, H., Strauss, R., Johanns, J., et al. (2014b). Subcutaneous golimumab maintains clinical response in patients with moderate-to-severe ulcerative colitis. *Gastroenterology*, *146*(1), 96–109.e1. http://doi.org/10.1053/j.gastro.2013.06.010.

Sandborn, W. J., Feagan, B. G., Stoinov, S., Honiball, P. J., Rutgeerts, P., Mason, D., et al. (2007). Certolizumab pegol for the treatment of Crohn's disease. *The New England Journal of Medicine*, *357*(3), 228–238. http://doi.org/10.1056/NEJMoa067594.

Scheidereit, C. (2006). IkappaB kinase complexes: Gateways to NF-kappaB activation and transcription. *Oncogene*, *25*(51), 6685–6705. http://doi.org/10.1038/sj.onc.1209934.

Scheinfeld, N. (2004). A comprehensive review and evaluation of the side effects of the tumor necrosis factor alpha blockers etanercept, infliximab and adalimumab. *The Journal of Dermatological Treatment*, *15*(5), 280–294. http://doi.org/10.1080/09546630410017275.

Scheinman, R. I., Cogswell, P. C., Lofquist, A. K., & Baldwin, A. S. (1995). Role of transcriptional activation of I kappa B alpha in mediation of immunosuppression by glucocorticoids. *Science*, *270*(5234), 283–286.

Scherer, D. C., Brockman, J. A., Chen, Z., Maniatis, T., & Ballard, D. W. (1995). Signal-induced degradation of I kappa B alpha requires site-specific ubiquitination. *Proceedings of the National Academy of Sciences of the United States of America*, *92*(24), 11259–11263.

Schmitz, M. L., & Baeuerle, P. A. (1991). The p65 subunit is responsible for the strong transcription activating potential of NF-kappa B. *The EMBO Journal*, *10*(12), 3805–3817.

Schröfelbauer, B., Polley, S., Behar, M., Ghosh, G., & Hoffmann, A. (2012). NEMO ensures signaling specificity of the pleiotropic IKKβ by directing its kinase activity toward IκBα. *Molecular Cell*, *47*(1), 111–121. http://doi.org/10.1016/j.molcel.2012.04.020.

Scott, M. L., Fujita, T., Liou, H. C., Nolan, G. P., & Baltimore, D. (1993). The p65 subunit of NF-kappa B regulates I kappa B by two distinct mechanisms. *Genes & Development*, 7(7A), 1266–1276.

Seidel, P., Merfort, I., Hughes, J. M., Oliver, B. G. G., Tamm, M., & Roth, M. (2009). Dimethylfumarate inhibits NF-{kappa}B function at multiple levels to limit airway smooth muscle cell cytokine secretion. *American Journal of Physiology. Lung Cellular and Molecular Physiology*, *297*(2), L326–L339. http://doi.org/10.1152/ajplung.90624.2008.

Sen, R., & Baltimore, D. (1986). Multiple nuclear factors interact with the immunoglobulin enhancer sequences. *Cell*, *46*(5), 705–716.

Sen, R., & Smale, S. T. (2010). Selectivity of the NF-{kappa}B response. *Cold Spring Harbor Perspectives in Biology*, *2*(4), a000257. http://doi.org/10.1101/cshperspect.a000257.

Senftleben, U., Cao, Y., Xiao, G., Greten, F. R., Krähn, G., Bonizzi, G., et al. (2001). Activation by IKKalpha of a second, evolutionary conserved, NF-kappa B signaling pathway. *Science*, *293*(5534), 1495–1499. http://doi.org/10.1126/science.1062677.

Shibata, W., Maeda, S., Hikiba, Y., Yanai, A., Ohmae, T., Sakamoto, K., et al. (2007). Cutting edge: The IkappaB kinase (IKK) inhibitor, NEMO-binding domain peptide, blocks inflammatory injury in murine colitis. *Journal of Immunology (Baltimore, Md : 1950)*, *179*(5), 2681–2685. http://doi.org/10.4049/jimmunol.179.5.2681.

Siebenlist, U., Brown, K., & Claudio, E. (2005). Control of lymphocyte development by nuclear factor-kappaB. *Nature Reviews. Immunology*, *5*(6), 435–445. http://doi.org/10.1038/nri1629.

Siebenlist, U., Franzoso, G., & Brown, K. (1994). Structure, regulation and function of NF-kappa B. *Annual Review of Cell Biology*, *10*(1), 405–455. http://doi.org/10.1146/annurev.cb.10.110194.002201.

Skaug, B., Jiang, X., & Chen, Z. J. (2009). The role of ubiquitin in NF-kappaB regulatory pathways. *Annual Review of Biochemistry*, *78*, 769–796. http://doi.org/10.1146/annurev.biochem.78.070907.102750.

Smale, S. T. (2011). Hierarchies of NF-κB target-gene regulation. *Nature Immunology*, *12*(8), 689–694. http://doi.org/10.1038/ni.2070.

Storz, P. (2013). Targeting the alternative NF-κB pathway in pancreatic cancer: a new direction for therapy? *Expert Review of Anticancer Therapy*, *13*(5), 501–504. http://doi.org/10.1586/era.13.28.

Straus, D. S., Pascual, G., Li, M., Welch, J. S., Ricote, M., Hsiang, C. H., et al. (2000). 15-deoxy-delta 12,14-prostaglandin J2 inhibits multiple steps in the NF-kappa B signaling pathway. *Proceedings of the National Academy of Sciences of the United States of America*, *97*(9), 4844–4849.

Strickland, I., & Ghosh, S. (2006). Use of cell permeable NBD peptides for suppression of inflammation. *Annals of the Rheumatic Diseases*, *65*(Suppl. 3), iii75–iii82. http://doi.org/10.1136/ard.2006.058438.

Sun, S.-C. (2011). Non-canonical NF-κB signaling pathway. *Cell Research*, *21*(1), 71–85. http://doi.org/10.1038/cr.2010.177.

Sun, S.-C., Ganchi, P. A., Ballard, D. W., & Greene, W. C. (1993). NF-kappa B controls expression of inhibitor I kappa B alpha: Evidence for an inducible autoregulatory pathway. *Science*, *259*(5103), 1912–1915.

Swinney, D. C., Xu, Y.-Z., Scarafia, L. E., Lee, I., Mak, A. Y., Gan, Q.-F., et al. (2002). A small molecule ubiquitination inhibitor blocks NF-kappa B-dependent cytokine expression in cells and rats. *The Journal of Biological Chemistry*, *277*(26), 23573–23581. http://doi.org/10.1074/jbc.M200842200.

Syed, D. N., Afaq, F., Kweon, M.-H., Hadi, N., Bhatia, N., Spiegelman, V. S., et al. (2007). Green tea polyphenol EGCG suppresses cigarette smoke condensate-induced NF-kappaB activation in normal human bronchial epithelial cells. *Oncogene*, *26*(5), 673–682. http://doi.org/10.1038/sj.onc.1209829.

Tang, W., Li, Y., Yu, D., Thomas-Tikhonenko, A., Spiegelman, V. S., & Fuchs, S. Y. (2005). Targeting beta-transducin repeat-containing protein E3 ubiquitin ligase augments the effects of antitumor drugs on breast cancer cells. *Cancer Research*, *65*(5), 1904–1908. http://doi.org/10.1158/0008-5472.CAN-04-2597.

Tay, S., Hughey, J. J., Lee, T. K., Lipniacki, T., Quake, S. R., & Covert, M. W. (2010). Single-cell NF-kappaB dynamics reveal digital activation and analogue information processing. *Nature*, *466*(7303), 267–271. http://doi.org/10.1038/nature09145.

Taylor, P. C. (2010). Pharmacology of TNF blockade in rheumatoid arthritis and other chronic inflammatory diseases. *Current Opinion in Pharmacology*, *10*(3), 308–315. http://doi.org/10.1016/j.coph.2010.01.005.

Taylor, P. C., & Feldmann, M. (2009). Anti-TNF biologic agents: Still the therapy of choice for rheumatoid arthritis. *Nature Reviews. Rheumatology*, *5*(10), 578–582. http://doi.org/10.1038/nrrheum.2009.181.

Tian, F., Zhou, P., Kang, W., Luo, L., Fan, X., Yan, J., et al. (2015). The small-molecule inhibitor selectivity between IKKα and IKKβ kinases in NF-κB signaling pathway. *Journal of Receptor and Signal Transduction Research*, *35*(4), 307–318. http://doi.org/10.3109/10799893.2014.980950.

Torgerson, T. R., Colosia, A. D., Donahue, J. P., Lin, Y. Z., & Hawiger, J. (1998). Regulation of NF-kappa B, AP-1, NFAT, and STAT1 nuclear import in T lymphocytes by noninvasive delivery of peptide carrying the nuclear localization sequence of NF-kappa B p50. *Journal of Immunology (Baltimore, Md: 1950)*, *161*(11), 6084–6092.

Traenckner, E. B., Pahl, H. L., Henkel, T., Schmidt, K. N., Wilk, S., & Baeuerle, P. A. (1995). Phosphorylation of human I kappa B-alpha on serines 32 and 36 controls I kappa B-alpha proteolysis and NF-kappa B activation in response to diverse stimuli. *The EMBO Journal*, *14*(12), 2876–2883.
Trepicchio, W. L., & Dorner, A. J. (1998). The therapeutic utility of Interleukin-11 in the treatment of inflammatory disease. *Expert Opinion on Investigational Drugs*, 7(9), 1501–1504. http://doi.org/10.1517/13543784.7.9.1501.
Tsang, V., Dvorakova, V., Enright, F., Murphy, M., & Gleeson, C. (2016). Successful use of infliximab as first line treatment for severe childhood generalized pustular psoriasis. *Journal of the European Academy of Dermatology and Venereology: JEADV*, *30*(11), e117–e119. http://doi.org/10.1111/jdv.13388.
Tsukamoto, S., Takeuchi, T., Rotinsulu, H., Mangindaan, R. E. P., van Soest, R. W. M., Ukai, K., et al. (2008). Leucettamol A: A new inhibitor of Ubc13-Uev1A interaction isolated from a marine sponge, Leucetta aff. microrhaphis. *Bioorganic & Medicinal Chemistry Letters*, *18*(24), 6319–6320. http://doi.org/10.1016/j.bmcl.2008.10.110.
Umezawa, K., & Chaicharoenpong, C. (2002). Molecular design and biological activities of NF-kappaB inhibitors. *Molecules and Cells*, *14*(2), 163–167.
Urban, M. B., & Baeuerle, P. A. (1990). The 65-kD subunit of NF-kappa B is a receptor for I kappa B and a modulator of DNA-binding specificity. *Genes & Development*, *4*(11), 1975–1984.
Van Antwerp, D. J., Martin, S. J., Kafri, T., Green, D. R., & Verma, I. M. (1996). Suppression of TNF-alpha-induced apoptosis by NF-kappaB. *Science*, *274*(5288), 787–789.
van Essen, D., Zhu, Y., & Saccani, S. (2010). A feed-forward circuit controlling inducible NF-κB target gene activation by promoter histone demethylation. *Molecular Cell*, *39*(5), 750–760. http://doi.org/10.1016/j.molcel.2010.08.010.
van Schie, K. A., Ooijevaar-de Heer, P., Dijk, L., Kruithof, S., Wolbink, G., & Rispens, T. (2016). Therapeutic TNF inhibitors can differentially stabilize trimeric TNF by inhibiting monomer exchange. *Scientific Reports*, *6*, 32747. http://doi.org/10.1038/srep32747.
Wang, C. Y., Cusack, J. C., Liu, R., & Baldwin, A. S. (1999). Control of inducible chemoresistance: Enhanced anti-tumor therapy through increased apoptosis by inhibition of NF-kappaB. *Nature Medicine*, *5*(4), 412–417. http://doi.org/10.1038/7410.
Wang, G., Gao, Y., Li, L., Jin, G., Cai, Z., Chao, J.-I., et al. (2012). K63-linked ubiquitination in kinase activation and cancer. *Frontiers in Oncology*, *2*, 5. http://doi.org/10.3389/fonc.2012.00005.
Wang, C. Y., Mayo, M. W., & Baldwin, A. S. (1996). TNF- and cancer therapy-induced apoptosis: Potentiation by inhibition of NF-kappaB. *Science*, *274*(5288), 784–787.
Wang, Y., Paszek, P., Horton, C. A., Kell, D. B., White, M. R., Broomhead, D. S., et al. (2011). Interactions among oscillatory pathways in NF-kappa B signaling. *BMC Systems Biology*, *5*(1), 23. http://doi.org/10.1186/1752-0509-5-23.
Watanabe, M., Nakashima, M., Togano, T., Higashihara, M., Watanabe, T., Umezawa, K., et al. (2008). Identification of the RelA domain responsible for action of a new NF-kappaB inhibitor DHMEQ. *Biochemical and Biophysical Research Communications*, *376*(2), 310–314. http://doi.org/10.1016/j.bbrc.2008.08.148.
Whiteside, S. T., Ernst, M. K., LeBail, O., Laurent-Winter, C., Rice, N., & Israel, A. (1995). N- and C-terminal sequences control degradation of MAD3/I kappa B alpha in response to inducers of NF-kappa B activity. *Molecular and Cellular Biology*, *15*(10), 5339–5345.
Wilhelmsen, K. C., Eggleton, K., & Temin, H. M. (1984). Nucleic acid sequences of the oncogene v-rel in reticuloendotheliosis virus strain T and its cellular homolog, the proto-oncogene c-rel. *Journal of Virology*, *52*(1), 172–182.
Xia, Y., Yeddula, N., Leblanc, M., Ke, E., Zhang, Y., Oldfield, E., et al. (2012). Reduced cell proliferation by IKK2 depletion in a mouse lung-cancer model. *Nature Publishing Group*, *14*(3), 257–265. http://doi.org/10.1038/ncb2428.

Xiang, M., Birkbak, N. J., Vafaizadeh, V., Walker, S. R., Yeh, J. E., Liu, S., et al. (2014). STAT3 induction of miR-146b forms a feedback loop to inhibit the NF-κB to IL-6 signaling axis and STAT3-driven cancer phenotypes. *Science Signaling*, 7(310), ra11. http://doi.org/10.1126/scisignal.2004497.

Xiao, G., Fong, A., & Sun, S.-C. (2004). Induction of p100 processing by NF-kappaB-inducing kinase involves docking IkappaB kinase alpha (IKKalpha) to p100 and IKKalpha-mediated phosphorylation. *The Journal of Biological Chemistry*, *279*(29), 30099–30105. http://doi.org/10.1074/jbc.M401428200.

Xiao, G., Harhaj, E. W., & Sun, S.-C. (2001). NF-kappaB-inducing kinase regulates the processing of NF-kappaB2 p100. *Molecular Cell*, 7(2), 401–409.

Xue, W., Meylan, E., Oliver, T. G., Feldser, D. M., Winslow, M. M., Bronson, R., et al. (2011). Response and resistance to NF-κB inhibitors in mouse models of lung adenocarcinoma. *Cancer Discovery*, *1*(3), 236–247. http://doi.org/10.1158/2159-8290.CD-11-0073.

Yamamoto, Y., & Gaynor, R. B. (2001). Therapeutic potential of inhibition of the NF-kappaB pathway in the treatment of inflammation and cancer. *Journal of Clinical Investigation*, *107*(2), 135–142. http://doi.org/10.1172/JCI11914.

Yamaoka, S., Courtois, G., Bessia, C., Whiteside, S. T., Weil, R., Agou, F., et al. (1998). Complementation cloning of NEMO, a component of the IkappaB kinase complex essential for NF-kappaB activation. *Cell*, *93*(7), 1231–1240.

Yang, J., Amiri, K. I., Burke, J. R., Schmid, J. A., & Richmond, A. (2006). BMS-345541 targets inhibitor of kappaB kinase and induces apoptosis in melanoma: Involvement of nuclear factor kappaB and mitochondria pathways. *Clinical Cancer Research*, *12*(3 Pt. 1), 950–960. http://doi.org/10.1158/1078-0432.CCR-05-1220.

Yang, Y., Kitagaki, J., Dai, R.-M., Tsai, Y. C., Lorick, K. L., Ludwig, R. L., et al. (2007). Inhibitors of ubiquitin-activating enzyme (E1), a new class of potential cancer therapeutics. *Cancer Research*, *67*(19), 9472–9481. http://doi.org/10.1158/0008-5472.CAN-07-0568.

Yang, J., Park, Y., Zhang, H., Xu, X., Laine, G. A., Dellsperger, K. C., et al. (2009). Feed-forward signaling of TNF-alpha and NF-kappaB via IKK-beta pathway contributes to insulin resistance and coronary arteriolar dysfunction in type 2 diabetic mice. *AJP: Heart and Circulatory Physiology*, *296*(6), H1850–H1858. http://doi.org/10.1152/ajpheart.01199.2008.

Yaron, A., Gonen, H., Alkalay, I., Hatzubai, A., Jung, S., Beyth, S., et al. (1997). Inhibition of NF-kappa-B cellular function via specific targeting of the I-kappa-B-ubiquitin ligase. *The EMBO Journal*, *16*(21), 6486–6494. http://doi.org/10.1093/emboj/16.21.6486.

Yin, M. J., Yamamoto, Y., & Gaynor, R. B. (1998). The anti-inflammatory agents aspirin and salicylate inhibit the activity of I(kappa)B kinase-beta. *Nature*, *396*(6706), 77–80. http://doi.org/10.1038/23948.

Yokoyama, K., Yamazaki, K., Katafuchi, M., & Ferchichi, S. (2016). A retrospective claims database study on drug utilization in Japanese patients with Crohn's disease treated with adalimumab or infliximab. *Advances in Therapy*, *33*(11), 1947–1963. http://doi.org/10.1007/s12325-016-0406-6.

Yoon, J. W., Kang, J. K., Lee, K. R., Lee, H. W., Han, J. W., Seo, D. W., et al. (2005). beta-Carboline alkaloid suppresses NF-kappaB transcriptional activity through inhibition of IKK signaling pathway. *Journal of Toxicology and Environmental Health. Part A*, *68*(23–24), 2005–2017. http://doi.org/10.1080/15287390491009110.

Yu, J., Wang, Y., Yan, F., Zhang, P., Li, H., Zhao, H., et al. (2014). Noncanonical NF-κB activation mediates STAT3-stimulated IDO upregulation in myeloid-derived suppressor cells in breast cancer. *The Journal of Immunology*, *193*(5), 2574–2586. http://doi.org/10.4049/jimmunol.1400833.

Yu, T., Yang, Y., Yin, D. Q., Hong, S., Son, Y.-J., Kim, J.-H., et al. (2015). TBK1 inhibitors: A review of patent literature (2011–2014). *Expert Opinion on Therapeutic Patents*, *25*(12), 1385–1396. http://doi.org/10.1517/13543776.2015.1081168.

Zhang, H., Hu, H., Greeley, N., Jin, J., Matthews, A. J., Ohashi, E., et al. (2014). STAT3 restrains RANK- and TLR4-mediated signalling by suppressing expression of the E2 ubiquitin-conjugating enzyme Ubc13. *Nature Communications*, *5*, 5798. http://doi.org/10.1038/ncomms6798.

Zhang, W., Tan, W., Wu, X., Poustovoitov, M., Strasner, A., Li, W., et al. (2013). A NIK-IKKα module expands ErbB2-induced tumor-initiating cells by stimulating nuclear export of p27/Kip1. *Cancer Cell*, *23*(5), 647–659. http://doi.org/10.1016/j.ccr.2013.03.012.

Zhang, H., Tian, Y., Zhu, Z., Xu, H., Li, X., Zheng, D., et al. (2016). Efficient antitumor effect of co-drug-loaded nanoparticles with gelatin hydrogel by local implantation. *Scientific Reports*, *6*, 26546. http://doi.org/10.1038/srep26546.

Zhou, H., Monack, D. M., Kayagaki, N., Wertz, I., Yin, J., Wolf, B., et al. (2005). Yersinia virulence factor YopJ acts as a deubiquitinase to inhibit NF-kappa B activation. *The Journal of Experimental Medicine*, *202*(10), 1327–1332. http://doi.org/10.1084/jem.20051194.

Zhu, Z., Aref, A. R., Cohoon, T. J., Barbie, T. U., Imamura, Y., Yang, S., et al. (2014). Inhibition of KRAS-driven tumorigenicity by interruption of an autocrine cytokine circuit. *Cancer Discovery*, *4*(4), 452–465. http://doi.org/10.1158/2159-8290.CD-13-0646.

Ziaei, S., & Halaby, R. (2016). Immunosuppressive, anti-inflammatory and anti-cancer properties of triptolide: A mini review. *Avicenna Journal of Phytomedicine*, *6*(2), 149–164.

Ziegelbauer, K., Gantner, F., Lukacs, N. W., Berlin, A., Fuchikami, K., Niki, T., et al. (2005). A selective novel low-molecular-weight inhibitor of IkappaB kinase-beta (IKK-beta) prevents pulmonary inflammation and shows broad anti-inflammatory activity. *British Journal of Pharmacology*, *145*(2), 178–192. http://doi.org/10.1038/sj.bjp.0706176.

CHAPTER FOUR

NF-κB as a Therapeutic Target in Inflammatory-Associated Bone Diseases

T.-h. Lin*, J. Pajarinen*, L. Lu*, A. Nabeshima*, L.A. Cordova*,†, Z. Yao*, S.B. Goodman*,[1]

*Stanford University, Stanford, CA, United States
†Faculty of Dentistry, University of Chile, Santiago, Chile
[1]Corresponding author: e-mail address: goodbone@stanford.edu

Contents

Abstract

Inflammation is a defensive mechanism for pathogen clearance and maintaining tissue homeostasis. In the skeletal system, inflammation is closely associated with many bone disorders including fractures, nonunions, periprosthetic osteolysis (bone loss around orthopedic implants), and osteoporosis. Acute inflammation is a critical step for proper bone-healing and bone-remodeling processes. On the other hand, chronic inflammation with excessive proinflammatory cytokines disrupts the balance of skeletal

Advances in Protein Chemistry and Structural Biology, Volume 107
ISSN 1876-1623
http://dx.doi.org/10.1016/bs.apcsb.2016.11.002

homeostasis involving osteoblastic (bone formation) and osteoclastic (bone resorption) activities. NF-κB is a transcriptional factor that regulates the inflammatory response and bone-remodeling processes in both bone-forming and bone-resorption cells. In vitro and in vivo evidences suggest that NF-κB is an important potential therapeutic target for inflammation-associated bone disorders by modulating inflammation and bone-remodeling process simultaneously. The challenges of NF-κB-targeting therapy in bone disorders include: (1) the complexity of canonical and noncanonical NF-κB pathways; (2) the fundamental roles of NF-κB-mediated signaling for bone regeneration at earlier phases of tissue damage and acute inflammation; and (3) the potential toxic effects on nontargeted cells such as lymphocytes. Recent developments of novel inhibitors with differential approaches to modulate NF-κB activity, and the controlled release (local) or bone-targeting drug delivery (systemic) strategies, have largely increased the translational application of NF-κB therapy in bone disorders. Taken together, temporal modulation of NF-κB pathways with the combination of recent advanced bone-targeting drug delivery techniques is a highly translational strategy to reestablish homeostasis in the skeletal system.

1. INTRODUCTION

Bone is the major component of the skeletal system and provides physical support and protection of the body, calcium metabolism, and endocrine regulation, and it facilitates the hematopoietic system in bone marrow. Bone remodeling is a dynamic process that continues throughout life and involves bone formation and bone resorption activities. The common pathophysiological event in bone disorders is the disruption of bone homeostasis (Theoleyre et al., 2004). Bone homeostasis depends on the functional balance between bone-forming cells (osteoblasts, OBs) and bone-resorptive cells (osteoclasts, OCs). A functional imbalance between these two arms determines either osteosclerotic bone-forming diseases (i.e., osteopetrosis) or osteolytic bone-resorptive diseases (Theoleyre et al., 2004).

Inflammation is a protective mechanism involving the activation of innate and adaptive immune systems in response to exogenous (bacteria, virus, etc.) or endogenous (necrotic cells) stimuli. Immune cells recognize the inflammatory stimuli to activate several cellular signaling including nuclear factor-κB (NF-κB) (Cordova et al., 2014). NF-κB is a master transcriptional factor in regulation of the inflammatory response and bone-remodeling process (Lin, Tamaki, et al., 2014; Novack, 2011). The proinflammatory cytokines driven by NF-κB are powerful signals to modulate OB and OC activities (Purdue, Koulouvaris, Potter, Nestor, & Sculco, 2007). Activation of NF-κB signaling in OCs is crucial for their differentiation and activation (Boyle, Simonet, & Lacey, 2003), whereas the activation in OBs inhibits bone formation (Chang et al., 2009). These unique characteristics imply the great potential of NF-κB as a therapeutic target for the treatment of inflammatory-associated bone disorders.

Acute inflammation is an essential step to initiate tissue repair processes including bone healing (Alexander et al., 2011; Raggatt et al., 2014). Unresolved inflammation progresses into chronic inflammation and leads to pathological conditions in affected organs. This review will focus on the biological significance and therapeutic potential of NF-κB in bone disorders with acute (fracture healing) or chronic (fracture nonunion (FNU), periprosthetic osteolysis (see Section 2.3.2), and senile osteoporosis) inflammation. Tumor, osteoarthritis, rheumatoid arthritis, bone infection, and metabolic bone disorders are excluded because of their complicated pathogenesis involving (in some instances) systemic factors, the adaptive immune system, and factors beyond innate immunity and NF-κB signaling.

2. INFLAMMATION AND BONE DISORDERS

2.1 Inflammation

The major functions of inflammation are clearance of pathogens and reestablishment of tissue homeostasis. In addition to pathogen infection, sterile inflammation is defined as inflammatory responses induced by trauma, ischemia-reperfusion injury, or chemical-induced injury (Chen & Nunez, 2010). The acute inflammatory response in damaged tissue initiates the release of chemical mediators that increase vascular permeability and leukocyte infiltration via activation of the local endothelium. The infiltrated leukocytes, including neutrophils and macrophages, can recognize necrotic cell debris and secrete proinflammatory cytokines and chemokines to further enhance immune cell infiltration. The infiltrated cells engulf the damaged tissue and cell debris, and secrete proteinases and growth factors to facilitate tissue remodeling and reconstruction. Successful clearance of inflammatory stimuli is accompanied by increased antiinflammatory and reparative cytokines to resolve the inflammatory response and reestablish tissue homeostasis (Serhan & Savill, 2005). However, if unresolved, these events may progress to chronic inflammation when inflammatory stimuli persist in damaged tissue. This results in continuous secretion of cytokines that enhance tissue destruction and impair the homeostasis.

2.1.1 Acute vs Chronic Inflammation

Acute inflammation is initiated by recognition of inflammatory stimuli including microorganisms or damaged cell debris via the pattern-recognition receptors (PRRs). There are several classes of PRRs that recognize a variety of stimuli and trigger downstream inflammatory responses, including toll-like receptors (TLRs), NOD-like receptors (NLRs), RIG-I-like receptors (RLRs), C-type lectin receptors (CLRs), and absence in melanoma 2 (AIM2)-like receptors. PRRs can recognize pathogen-associated molecular patterns (PAMPs) from microorganisms. In sterile inflammation, PRRs can recognize damage-associated molecular patterns (DAMPs) from damaged tissue. The recognition of PAMPs and DAMPs and the activation of proinflammatory responses have been well documented by Takeuchi and Akira (2010) and Chen and Nunez (2010). DAMPs are intracellular proteins such as heat-shock proteins (Quintana & Cohen, 2005), nucleic acid (Cavassani et al., 2008; Imaeda et al., 2009), and intracellular cytokines (Eigenbrod, Park, Harder, Iwakura, & Nunez, 2008). Release of DAMPs from necrotic cells with impaired membrane integrity is an indicator of tissue damage. Recognition of DAMPs via PRRs in tissue macrophages leads to the secretion of chemoattractants and local recruitment of neutrophils and circulating monocytes/macrophages.

The inflammatory response may switch from acute to chronic inflammation when the stimuli persist (tissue damage, infection, etc.). Chronic inflammation is marked by infiltration of macrophages and lymphocytes, as well as ongoing attempts at repair. Excessive production of inflammatory cytokines from macrophages during chronic inflammation may cause tissue damage and fibrosis. For example, in patients with joint replacement surgery, continued release of wear particles from implanted biomaterials induce chronic inflammation and periprosthetic osteolysis in confined regions (Lin, Tamaki, et al., 2014).

2.1.2 Proinflammatory and Antiinflammatory Functions of Macrophages

Macrophages are crucial regulators of initiation, progression, and resolution of inflammation (Martinez, Sica, Mantovani, & Locati, 2008). Polarized macrophages may acquire distinct phenotypes with proinflammatory (M1) or antiinflammatory (M2) behaviors (Mantovani, Biswas, Galdiero, Sica, & Locati, 2013; Martinez et al., 2008). Classical activation of macrophages with interferon-γ and/or lipopolysaccharide leads to M1 macrophage polarization. M1 macrophages secrete proinflammatory cytokines (tumor necrosis factor-α (TNF-α), interleukin-1β (IL-1β), etc.) and chemokines (monocyte chemotactic protein-1 (MCP-1), macrophage inflammatory protein-1α (MIP-1α), etc.) in a NF-κB-dependent pathway, which can result in tissue damage with additional leukocytes infiltration. Alternatively, macrophages exposed to IL-4 or IL-13 are polarized into M2 macrophages, marked by increased arginase-1 and antiinflammatory cytokines such as IL-10 and IL-1 receptor antagonist (IL-1Ra). The interaction between the NF-κB pathway and M2 macrophage polarization remains unclear. M2 or M2-like macrophages are capable of modulating and terminating the inflammatory response and are crucial for tissue remodeling and repair. Crosstalk between polarized macrophages and bone remodeling has been reviewed comprehensively elsewhere (Loi et al., 2016).

Thus, the biological roles of inflammation are to eliminate pathogens and foreign bodies, as well as initiate tissue repair and remodeling. The crosstalk of macrophages and others cells in the tissue microenvironment via modulation of inflammatory status and tissue remodeling may determine the status of inflammatory-associated conditions such as bone remodeling and repair.

2.2 Acute Inflammation-Associated Bone Disorders

Traumatic fracture and fragility fracture secondary to osteoporosis are closely associated with acute inflammatory responses. Infiltration of immune cells, especially macrophages, is critical for bone-healing processes (Loi et al., 2016). Fracture healing involves four consecutive phases: (a) acute inflammation, (b) cell proliferation and progenitor recruitment, (c) a stabilization step characterized by the consecutive formation of a fibrous, cartilaginous, and immature bony callus, and then (d) remodeling of the immature callus (Loi et al., 2016). The acute inflammation phase is driven by phagocytic cells including macrophages and polymorphonuclear neutrophils (PMNs), which are recruited from the hematopoietic niche and then attracted to the fractured site (Purdue et al., 2007). Macrophages and PMNs recognize DAMPs and PAMPs through TLRs and other receptors (see Section 2.1.1) and mediate NF-κB-dependent induction of proinflammatory and pro-osteoclastogenic cytokine secretion including TNF-α and IL-1β (Lin et al., 2015; Lin, Tamaki, et al., 2014). Transient elevated TNF-α levels during the acute inflammatory stage (d1–d3) are critical to the mediation of mesenchymal stem cell (MSC) migration into the fracture site and the differentiation into osteoblastic lineage cells (Karnes, Daffner, & Watkins, 2015). Supplementation of TNF-α (1 ng/ml) accelerates fracture healing rate and mineralization of callus (Glass et al., 2011). Nevertheless, persisted proinflammatory responses at the callus formation stage (days 3–7) could impair fracture healing. Martensson et al. reported that TNF-α and IL-1β synergistically decrease chondrocyte proliferation and survival at fracture callus, leading to reduction of bone formation in a rat model (Martensson, Chrysis, & Savendahl, 2004).

2.3 Chronic Inflammation-Associated Bone Disorders

2.3.1 Fracture Nonunion

FNU consists in incomplete consolidation of the fracture, with an absence of progressive radiographic signs of healing over three consecutive months (Loi et al., 2016; Panteli, Pountos, Jones, & Giannoudis, 2015). Based on the radiographic findings, FNU can be classified into either hypertrophic, exhibiting an oversized soft callus around the fracture site; or atrophic, featured by the absence of visible soft callus (Marsell & Einhorn, 2011). While hypertrophic FNU is associated with mechanical instability, atrophic FNU involves an intrinsic deficit of host immune and/or bone-healing responses (Karnes et al., 2015). Histologically, FNU demonstrates persistence of disorganized fibrous tissue, woven bone, and cartilage at the nonunion site (Karnes et al., 2015).

2.3.2 Periprosthetic Osteolysis

Total joint replacement (TJR) is an effective surgical procedure to treat patients with end-stage arthritis. In 2011, approximately 1 million of TJRs were performed in the United States (AAOS, 2013). Wear particles and other orthopedic byproducts generated from the bearing surfaces of TJRs induce chronic inflammation and bone loss around the implant, leading to aseptic loosening, and revision surgery in nearly 10% of TJR patients (Purdue et al., 2007). Wear particles less than 10 μm are either phagocytized by macrophages, or enclosed by foreign body giant cells (FBGCs) (if >10 μm) (Cobelli, Scharf, Crisi, Hardin, & Santambrogio, 2011). Recognition of wear particles or DAMPs adhering to the particles via TLR2 and TLR4 induces the secretion of proinflammatory cytokines in the NF-κB-dependent pathway (Pearl et al., 2011). The proinflammatory cytokines including TNF-α and IL-1β further activate NF-κB in macrophages as a positive regulatory loop during the chronic inflammation, leading to OC activation, and osteolytic processes (Lin, Pajarinen, et al., 2016; Lin, Tamaki, et al., 2014). Inhibition of NF-κB activity suppresses PMMA and UHMWPE wear particle induced OC activation in vitro and in vivo, suggesting the therapeutic potential of targeting NF-κB pathway in the particle disease (Clohisy, Hirayama, Frazier, Han, & Abu-Amer, 2004; Lin, Pajarinen, et al., 2016).

Exposure of wear particles also impairs osteoblastic phenotypes and paracrine regulation functions in MSCs and OB-lineage cells. Reduced expression of collagen type 1, bone sialoprotein, and decreased bone mineralization was found in human and murine MSCs exposed to titanium or polyethylene wear particles (Chiu, Ma, Smith, & Goodman, 2009; Lin et al., 2015; Wang et al., 2002). Inhibition of NF-κB activity in the MSCs exposed to wear particles mitigated the reduced bone formation (Lin et al., 2015). The paracrine regulators including IL-8, GM-CSF (Haleem-Smith et al., 2012), M-CSF, and RANKL (Pioletti & Kottelat, 2004) are also upregulated in OB or MSC exposed to wear particles, which could further enhance inflammation and the osteolytic process.

2.3.3 Senile Osteoporosis

Aging is associated with chronic inflammation and increased reactive oxygen species. The concept of "inflamm-aging" has been suggested to accelerate the aging process (Franceschi et al., 2000). Aging-associated bone loss is marked by reduced bone formation ability (compared to increased OC activity in postmenopause osteoporosis) and is often referred to as "senile osteoporosis" (Dobbs, Buckwalter, & Saltzman, 1999). Senile osteoporosis is associated with increased risk of fracture in 44–65% of women and 25–42% of men during the lifetime (Nguyen, Ahlborg, Center, Eisman, & Nguyen, 2007).

Though NF-κB activation could be associated with inflamm-aging process (Salminen et al., 2008), direct evidence of a correlation with senile osteoporosis in humans remain unclear. In one study using a murine model, NF-κB activation (increased phosphorylation of RelA) has been reported in the trabecular bone in natural aging mice (Yu et al., 2014). Furthermore, increased NF-κB activity was found in the MSCs isolated from aged mice (15 months) compared with younger mice (Lin, Gibon, et al., 2016). Inhibition of the NF-κB pathway partially rescued the reduction of osteogenesis in aged MSCs. Increased RANKL and decreased OPG expression (thus leading to increased RANKL/OPG ratio and OC activation) was observed in aged MSCs. Further investigation is essential to clarify the correlation of NF-κB pathway in the process of senile osteoporosis.

3. INFLAMMATION AND NF-κB SIGNALING

NF-κB is one of the best characterized transcription factors that regulate inflammation and innate and adaptive immune reactions (Hayden & Ghosh, 2011; Lawrence, 2009). NF-κB signaling is activated by a variety of proinflammatory and danger signals as well as cell stress sensed by multiple receptors expressed on the cell membrane, endosomal compartment, and cytoplasm. Activation of NF-κB signaling leads to the production of various inflammatory cytokines, chemokines, adhesion molecules, transcription factors, and antimicrobial effector molecules that initiate and module the inflammatory reaction, and orchestrates the immediate host response against pathogens and tissue damage. NF-κB signaling is also involved in lymphoid organogenesis; activation and proliferation of $CD4^+$ T cells; T cell polarization into various different effector T cell populations; and to the maturation of B cells. The role of NF-κB signaling in these adaptive immune functions has comprehensively been reviewed elsewhere (Hayden & Ghosh, 2011). In the context of bone, NF-κB signaling is directly involved in the differentiation and activation of bone resorbing OCs (Novack, 2011; Soysa & Alles, 2009). Chronic NF-κB activation has also been shown to impair both the differentiation of MSCs along the osteogenic pathway and OB-mediated bone formation (Lin, Tamaki, et al., 2014). Thus NF-κB signaling plays a role both in physiological bone remodeling as well as pathological bone loss occurring in such inflammatory conditions as rheumatoid arthritis and periimplant osteolysis of TJRs. Indeed, blocking the chronic NF-κB signaling in these and other inflammatory conditions limits inflammation and prevents bone loss (Lin, Tamaki, et al., 2014; Xu et al., 2009).

3.1 The NF-κB Protein Family

The NF-κB transcription factor family consists of five members p50 (NF-κB1), p52 (NF-κB2), RelA (p65), RelB, and c-Rel (Huxford, Hoffmann, & Ghosh, 2011; Lawrence, 2009). All of the NF-κB subunits share a structurally conserved N-terminal sequence spanning 300 amino acid residues called the Rel homology domain (RHD) (Ghosh, May, & Kopp, 1998; Huxford et al., 2011). The RHD is responsible for DNA binding, dimerization, and nuclear translocation of the NF-κB subunits (Courtois & Gilmore, 2006; Huxford, Huang, Malek, & Ghosh, 1998). As described by Huxford, Hoffman, and Ghosh, the RHD can be divided into three structural components—the N-terminal domain (NTD), dimerization domain

(DD), and the nuclear localization sequence (NLS) polypeptide—all of which mediate the various activities of the RHD and subsequent NF-κB signaling (Huxford et al., 2011). At the C-terminal end of the RHD, the DD folds so that two antiparallel β-sheets form an immunoglobulin-like (Ig-like) structure; one of the sheets forms an interface for subunit dimer formation while the other mediates nonspecific DNA contacts (Chen & Ghosh, 1999). Similarly, the NTD contains an Ig-like fold that binds DNA both base-specifically and backbone nonspecifically. Subunits RelA, RelB, and c-Rel are produced as a mature protein while the p50 subunit is produced from an inactive precursor p105 and the p52 from precursor p100 via posttranslational processing in proteasome (Courtois & Gilmore, 2006; Gilmore, 2006; Lawrence, 2009). The production of p50 is constitutive while the production of p52 is regulated and induced upon the activation of the noncanonical NF-κB pathway (Fig. 1). In addition, the NF-κB p105 and p100 proteins are distinguished from Rel proteins by IκB-like inhibitory domains that play a role in the binding of inhibitory IκB proteins (Karin, Yamamoto, & Wang, 2004).

In resting state, NF-κB subunits reside in the cell cytoplasm noncovalently bound by a group of IκB proteins that maintain the NF-κB subunits in inactive form (Huxford et al., 2011; Whiteside & Israel, 1997). The mechanism underlying the inhibitory effects of IκB binding has been elucidated by determining the X-ray crystal structure of IκBα bound to p50:RelA dimers (Jacobs & Harrison, 1998). It is thought that the long-range electrostatic interactions between the C-terminal PEST region of IκBα and the NTD of RelA causes a conformational change that does not allow for RelA to bind DNA (Huxford et al., 1998; Jacobs & Harrison, 1998). Moreover, IκBα conceals the NLS peptide of p65, preventing its nuclear localization.

The degradation of these inhibiting proteins is regulated by a group of IκB kinase complexes IKK1 (IKKα), IKK2 (IKKβ), and IKKγ (NEMO) (Chen, 2005; Karin & Ben-Neriah, 2000). Once activated by upstream signaling cascades, IKKs phosphorylate the inhibitory IκB proteins leading to their ubiquitination by E3 ubiquitin-protein ligase followed by degradation in 26S proteasome. Once released the NF-κB subunits home to the nucleus as hetero- or homodimers and regulate the transcription of multitude of genes by binding to the gene promoter regions know as κB sites; currently more than 500 NF-κB target genes have been recognized (Boston_University). The dimers bind to DNA via 10 flexible loops that extend from the Ig-like fold; this mechanism is unlike most transcription factors, which utilize alpha helices (Chen & Ghosh, 1999). Beyond this unique

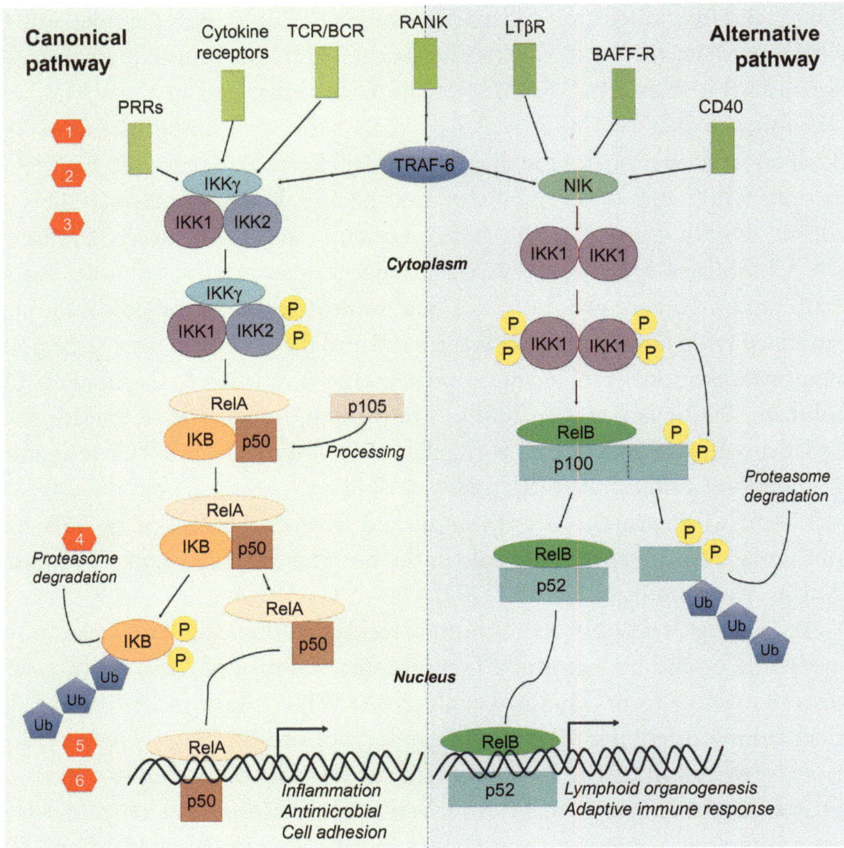

Fig. 1 Canonical (*left*) and alternative (*right*) NF-κB pathway and the target strategies: (1) inhibition of cell-surface receptor binding, (2) inhibition of receptor adaptor proteins, (3) IKK inhibition, (4) blocking IκB degradation/upregulating IκB/inhibiting proteasome, (5) inhibit nuclear translocation, and (6) block DNA binding/transcriptional activation.

DNA-binding scheme, the NF-κB dimer arranges along the major groove for a full turn to create a "butterfly" structure.

Depending on the composition of the NF-κB dimer, the NF-κB can either induce or inhibit gene transcription; complexes containing RelA, RelB, and c-Rel function as transcription promoters while p50 and p52 lack the C-terminal transcription activation domains necessary to induce gene reading and can activate transcription only when paired with other NF-κB family members (Hayden & Ghosh, 2011); correspondingly p50 homodimers lacking activation domains function as transcription suppressors (Bohuslav et al., 1998). The signaling pathway leading to the degradation of IκB and the subsequent release of NF-κB subunits is known as the canonical (classical) NF-κB pathway, while the pathway culminating to the cleavage of p100 to active p52 is known as noncanonical (alternative) NF-κB pathway (Fig. 1).

3.2 Activators and Targets of Canonical Pathway

The canonical NF-κB signaling pathway is primarily involved in the sensing of danger due to tissue damage or infection, and is followed by rapid initiation and progression of an inflammatory reaction and antimicrobial functions (Huxford et al., 2011; Lawrence, 2009; Lin, Tamaki, et al., 2014). In the context of innate immunity, the pathway is activated by signals originating from two broad groups of receptors; the receptors for proinflammatory cytokines and the PRRs for various danger signal molecules (Hayden & Ghosh, 2014; Kawai & Akira, 2010). Downstream signals from these receptors convene to active a kinase complex formed by IKK1, IKK2, and IKKγ, with IKK2 playing the key role, leading to the release and nuclear translocation of mainly p50-RelA and p50-cRel dimers (Hayden & Ghosh, 2011). These factors promote the transcription of multiple proinflammatory cytokines, prostaglandins, chemokines, endothelial and leucocyte adhesion molecules as well as proteinases leading to recruitment and activation of further inflammatory cells, mainly neutrophils and macrophages. Antimicrobial effector molecules, such as defensing and reactive oxygen and nitrogen species, are also produced and the antigen presenting machinery induced for the subsequent activation of the adaptive immune system. In addition to transcription of proinflammatory signals, antiinflammatory cytokines such as IL-10 and IL-1rA as well as multiple inhibitors of NF-κB pathway, e.g., IκB proteins, are produced thus limiting the inflammatory reaction in a manner of an autocrine feedback loop (Lawrence, 2009).

Proinflammatory cytokines including TNF-α and IL-1 are among the best-known inducers and also target of the canonical NF-κB pathway (Hayden & Ghosh, 2014). These archetypal proinflammatory cytokines are abundantly produced during the inflammatory reaction and play a key role in multiple chronic inflammatory conditions. Binding of these cytokines to corresponding receptors TNF receptor 1 (TNFR1, widely expressed) and TNF receptor 2 (TNFR2, expressed on immune cells), and type I IL-1 receptor (IL-1R1) activates canonical NF-κB pathway signaling but also the MAP kinase/AP-1 pathway amplifying the inflammatory reaction.

The second set of receptors that activate the canonical NF-κB pathway are PRRs that sense danger signals originating from tissue damage and invading pathogens (Matzinger, 2002). The first recognized and the best-known PRR family is the TLRs (Kawai & Akira, 2010; Kumar, Kawai, & Akira, 2009). This set of receptors recognizes various evolutionally well-conserved molecular repeat structures expressed on various pathogens; the best-known examples include TLR2 ligand lipoteichoic acid (LTA) and TLR4 ligand lipopolysaccharide (LPS) both of which are fundamental structural components of Gram-positive or Gram-negative bacteria cell walls, respectively. TLRs also recognize conserved viral (e.g., single- and double-stranded RNA, recognized by TLR3, TLR7, and TLR8) and fungal (e.g., zymosan, recognized by TLR2) structures. In addition to these pathogen-derived molecules, or PAMPs, it has been suggested that several TLRs recognize endogenous ligands collectively known as alarmins (Bianchi, 2007; Kono & Rock, 2008). Other families of PPRs, including NLRs and RLRs, activate canonical NF-κB pathway and IRF3-mediated type 1 interferon production (Kawai & Akira, 2010, 2011). Unlike TLRs that are restricted to cell and endosomal membranes, NLRs and RLRs are located to the cell cytoplasm thus being optimally located to recognize viral structures and complementing the cells danger signal sensing machinery.

3.3 Activators and Targets of the Alternative Pathway

While the canonical NF-κB pathway is related to the rapid initiation and amplification of an inflammatory reaction, the best-known functions of the noncanonical pathway are related to lymphoid organogenesis and the activation and progression of adaptive immune response (Hayden & Ghosh, 2011; Lawrence, 2009). The activation of the pathway culminates in the induction of NF-κB-inducing kinase (NIK) and formation of IKK1 dimers. IKK1 phosphorylates p100 leading to its proteosomal processing into p52 and nuclear translocation of mainly p52/RelB dimers. The ligands that activate the noncanonical pathway include several TNF family members related to adaptive immune functions including lymphotoxin b, CD40 ligand, and B-cell activating factor (BAFF) but not TNF-α itself.

In the context of bone, the best-known function for the noncanonical NF-κB pathway is related to the bone resorption (Abu-Amer, 2013; Boyce, Yao, & Xing, 2010; Novack, 2011). The formation and function of bone resorbing OC are dependent on the receptor RANK expressed on circulating OC precursors (derived from monocytes) and its ligand, RANKL, produced by OBs, other MSCs, and activated T cells (Lacey et al., 1998; Suda et al., 1999). Both RANK and RANKL are necessary for OC formation. RANK signaling activates both the canonical and non-canonical NF-κB pathway via TRAF6 (Novack, 2011; Soysa & Alles, 2009). RANK signaling also activates the MAP kinase pathway with induction of the transcription factor c-Fos that is also necessary for OC formation (Grigoriadis et al., 1994). All of these pathways synergize to induce the expression of transcription factor NFATc1 that directly activates the transcription of OC-related genes and is considered as the master regulator of OC formation and sustained function (Kim & Kim, 2014; Takayanagi et al., 2002). The activity of RANKL is regulated on the one hand by a secreted decoy receptor, OPG that inhibits RANKL activity, and on the other hand by various inflammatory cytokines such as TNF and IL-1 and TLR ligands that enhance OC formation by synergizing with RANKL in activating the canonical NF-κB pathway (Osta, Benedetti, & Miossec, 2014; Simonet et al., 1997). Furthermore, the balance of RANKL and OPG production is regulated by these inflammatory signals typically increasing the RANKL/OPG ratio (Abu-Amer, 2013). Thus multiple mechanisms drive the increased OC formation and bone loss seen in the context of chronic inflammatory conditions.

4. NF-κB AND BONE REMODELING

4.1 Bone Remodeling

Bone remodeling is a necessary process to repair damaged bone and involves the resorption and formation of hard tissue. OCs, OBs, and osteocytes are the cells maintaining bone matrix homeostasis in bone. Immune cells, especially macrophages, are critical in the bone-remodeling process in response to damage and other inflammatory stimuli (Alexander et al., 2011; Cho et al., 2014; Guihard et al., 2012; Lin, Tamaki, et al., 2014). Under normal physiological conditions, osteocytes secrete sclerostin (van Bezooijen et al., 2004) and transforming growth factor β1 (TGF-β1) (Heino, Hentunen, & Vaananen, 2002) to suppress the activities of OB and OC, respectively. Osteocyte apoptosis caused by damage or inflammation is associated with a reduction of the suppressive paracrine regulators and thus, initiates bone remodeling at the damaged site. The basic multicellular unit (BMU) is a specialized structure formed in the bone-remodeling process. In an active

BMU, OBs and OCs line the bone surface, covered by a canopy-like structure consisting of bone resident macrophages (known as osteomacs) (Chang et al., 2008), and acting in bone resorption and bone formation by a coupling reaction. The process is then terminated when the bone homeostasis is reestablished.

The crosstalk between immune cells and bone cells is tightly associated with mineral homeostasis and tissue repair in bone. However, inflammation-associated NF-κB activation is not limited to immune cells. The signals can be activated in OBs and OCs via direct exposure to the stimuli like PAMPs or DAMPs, or indirect regulation by the immune cells (Xu et al., 2009; Chang et al., 2013; Lin, Tamaki, et al., 2014; Lin et al., 2015). The fundamental roles of NF-κB activation in OC differentiation and activation are well defined (Boyle et al., 2003). The biological functions of NF-κB activation in OBs and MSCs are recently being clarified (Chang et al., 2013, 2009; Cho et al., 2010; Lin et al., 2015), although the detailed regulation of their bone-forming ability remains unclear. In this section, we summarize the current findings of NF-κB activation effects on bone-remodeling process (Fig. 2), including the studies using transgenic animal models or the stimulation by proinflammatory cytokines such as TNF-α and IL-1β. In addition, several reports have indicated that the TLR agonists can modulate the biological activities in OC (Itoh et al., 2003; Takami, Kim, Rho, & Choi, 2002) and OB (Hwa Cho, Bae, & Jung, 2006; Lombardo et al., 2009; Mo et al., 2008).

4.2 NF-κB in Osteoclasts

OCs are terminal differentiated myeloid lineage cells, which can be characterized by the unique multinuclear morphology and the expression of OCs-specific markers including tartrate-resistant acid phosphatase (TRAP), cathepsin K, calcitonin receptor, and β3 integrin (Boyle et al., 2003). Myeloid precursor cells differentiate into macrophages when exposed to macrophage colony-stimulating factor (M-CSF), or differentiate into OCs in the presence of M-CSF and RANKL.

RANKL can induce both classical and alternative NF-κB activation in OCs and their precursors (Novack, 2011). Classical activation of NF-κB requires IKK2 to phosphorylate and degrade IκB, and releases RelA/p50 or cRel/p50 heterodimer to translocate into the nucleus. Deletion of IKK2 ($IKK2^{-/-}$) in transgenic mice caused defective osteoclastogenesis in OC precursors in response to RANKL, TNF-α, or IL-1β, which results

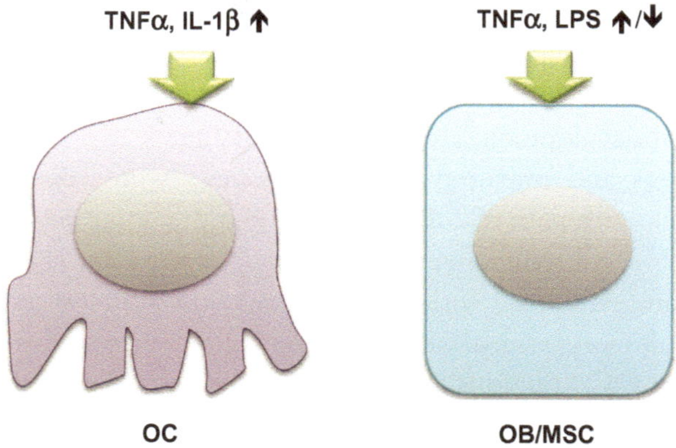

	Phenotypes with target gene silencing	
IKK1	*Normal OC*	–
IKK2	*OC defect*	*Increased osteogenesis*
***IKKγ**	*Embryonic lethal*	*Reduced trabecular bone*
IkB	*No phenotypes in bone*	–
p50/p52	*Osteopetrosis*	–
***RelA**	*Embryonic lethal*	–
RelB	*Defective response to RANKL*	*Increase osteogenesis*
c-Rel	*No phenotypes in bone*	–
NIK	*Defective response to RANKL*	–

Fig. 2 The roles of NF-κB pathway in OC and OB/MSC biological activity. The upstream NF-κB stimulators (TNFα, IL-1β) can enhance OC activity. The effect of NF-κB activation to OB/MSC is associated with timing, dose, duration, and the composition of the stimulators (TNFα, LPS, etc.). The table at the bottom summarized the specific role of NF-κB protein family reported in the studies using transgenic mice or pharmacological inhibitors. *IKKγ- and RelB-deficient mice are embryonic lethal, and their roles in OC have not been reported. The specific role of IKKγ in OB was identified using OB-specific transgenic mice model.

in osteopetrosis (excessive bone formation) and resistance to inflammatory-associated bone loss in vivo (Ruocco et al., 2005). In contrast, mutation of IKK1 ($IKK1^{AA}$) shows a defect of osteoclastogenesis in OC precursors in response to RANKL but not TNF-α or IL-1β in vitro. Notably, the defect of osteoclastogenesis was not observed in $IKK1^{AA}$ mice in vivo, which could be explained by compensatory effects from the paracrine regulation of OB. The role of IKKγ in skeletal development remains unclear due to the lethal phenotypes of severe liver degeneration in IKKγ-deficient mice (Rudolph et al., 2000). No phenotypes in bone have been reported in the transgenic mice lacking IκB or cRel expression (Gerondakis et al., 1996; Klement et al., 1996). Further studies are required to clarify their roles in skeletal development and bone remodeling.

Osteopetrosis and impaired osteoclastogenesis were reported in the transgenic mice with p50/p52 deletion, and the phenotypes were rescued by bone marrow cell transplantation (Franzoso et al., 1997; Iotsova et al., 1997). The transgenic mice with the single deletion in p50 or p52 showed no significant phenotypes in bone, suggesting the redundant roles of p50 and p52 to regulate OC functions and bone homeostasis. The RelA-deficient mice are embryonically lethal (Beg, Sha, Bronson, Ghosh, & Baltimore, 1995), while the mice with deficiency in RelA and TNFR1 live for 2–3 weeks (Rosenfeld, Prichard, Shiojiri, & Fausto, 2000). In the radiated wild-type mice transplanted with bone marrow cells from RelA/TNFR1-deficient mice, the OC numbers at the basal level or upon RANKL induction are significantly decreased (Vaira, Alhawagri, et al., 2008). RANKL-induced significant cellular apoptosis in RelA/TNFR1-deficient precursor cells in vitro through JNK/Bid/caspase 3 pathway, and blocking of proapoptotic Bid signaling protect the cells from RANKL-induced cell deaths and rescued the defect of osteoclastogenesis in p65-deficient cells. Taken together, the results suggested that RelA is essential for the anti-apoptotic signaling in OC in response to RANKL stimulation, but is not required for OC activation.

The alternative NF-κB pathway can be activated by RANKL but not the proinflammatory cytokines. In this process, NF-κB-inducing kinase (NIK) is stabilized in response to the stimulation and activates IKKα, which then process p100 into p52 to form RelB/p52 heterodimer (Sun, 2011). The nuclear translocation of RelB is specifically regulated by p100, and thus unique downstream signaling of NIK. Mild increased trabecular bone volume and normal OC numbers at the basal line are observed in both NIK- and RelB-deficient mice (Novack et al., 2003; Vaira, Johnson, et al., 2008). Nevertheless, administration of RANKL failed to induce osteoclastogenesis in the OC precursors with the defect in NIK or RelB in vitro (Novack, 2011). Overexpression of RelA cannot rescue defective osteoclastogenesis in RelB-deficient precursor cells induced by RANKL, suggested that RelB may be the key regulator of OC differentiation (Vaira, Johnson, et al., 2008).

The proinflammatory cytokines including TNF-α and IL-1β induce osteoclastogenesis via direct activation of OC precursor cells, or indirect induction of RANKL secretion in bone marrow stromal cells. Nevertheless, the dependence of RANKL/RANK signaling in the OC activation process is still in debate. Kobayashi et al. (2000) first reported that murine bone marrow myeloid cells differentiated into TRAP+ OCs in the presence of M-CSF and TNF-α. The bone-resorption ability in OCs and RANKL secretion in OBs stimulated by TNF-α was dependent on IL-1β (Kobayashi et al., 2000; Wei, Kitaura, Zhou, Ross, & Teitelbaum, 2005; Zwerina et al., 2007). Antibodies against TNF-α receptors but not RANKL inhibited this process, suggesting that the TNF-α-mediated OC activation is independent of RANKL/RANK signaling (Kobayashi et al., 2000). Lam et al. showed that TNF-α and M-CSF-mediated OC activation and required a permissive level of RANKL (Lam et al., 2000). Inhibition of RANKL by OPG at an earlier stage of OC differentiation (exposed to M-CSF alone) augmented TNF-α-mediated OC activation. Interestingly, bone marrow myeloid cells deficient in RANK can differentiate into OCs by TNF-α stimulation; this demonstrated the existence of RANKL/RANK-independent pathway of OC activation (Kim et al., 2005).

The classical NF-κB activation can be stimulated by TNF-α, whereas the alternative NF-κB activation could be inhibited in OC precursors. TNF-α induced the accumulation of p100 in OC precursors, which inhibited the alternative NF-κB activation (Yao, Xing, & Boyce, 2009). A recent study further demonstrated that the limitation of TNF-α-mediated OC activation was only observed in macrophage/OC precursors stimulated by M-CSF but not in combination with TNF-α (Zhao et al., 2015). Mechanistic studies showed that TNF-α-induced RelB expression and enhanced OC activation, but this process was self-limited by suppression of NFATc1.

4.3 NF-κB in MSCs/Osteoblasts

OBs are specialized bone-forming cells differentiated from MSCs, which have multilineage differentiation abilities including bone, cartilage, adipose tissue, etc. Runx2 is a master regulator of osteoblastogenesis which initiates the commitment of osteogenic differentiation into osteoprogenitors. The expression of Runx2, followed by osterix, induces alkaline phosphatase (ALP) and type I collagen secretion, and turn the cells into mature OBs located on the bone surface (Wu, Scadden, & Kronenberg, 2009). Mature OBs deposit the organic matrix for bone mineralization and, once surrounded by bone matrix, become osteocytes, which account for almost 95% of all bone cells. Osteoblastogenesis is guided by paracrine or endocrine factors including parathyroid hormone, bone morphogenic proteins

(BMPs), Wnt signaling, and growth factors such as TGF-β1 (Novack, 2011). In addition, differentiated OBs secrete M-CSF, RANKL, and OPG to regulate OC activity in a coupling reaction during the bone-remodeling process.

The role of NF-κB in OC activation and differentiation has been well defined. However, the biological effects of NF-κB activation in OB differentiation and bone formation are still in debate. The first direct evidence for the suppressive role of canonical NF-κB signaling in osteogenesis was demonstrated by using transgenic mice expressing a dominant negative form of IKKγ in differentiated OBs as controlled by osteocalcin promoter. Inhibition of NF-κB activity in OBs increased trabecular bone mass and bone mineral density without affecting OC function in young (2–4 weeks old) mice. In addition, inhibition of NF-κB activity prevented ovariectomy-induced bone loss in adult mice by maintaining bone-forming ability in OBs. In a study using human, murine, and rat MSCs, inhibition of IKK2 by a small molecular inhibitor or gene deletion increased osteogenic ability at the basal level or in the presence of TNF-α or IL-17. Local administration of IKK2 inhibitor enhanced MSC-mediated bone repair in a murine calvarial bone defect model. This mechanistic study demonstrated that NF-κB activation induced the expression of Smurf1/2, the ubiquitin ligase controlling the degradation of β-catenin, and inhibits osteogenic differentiation (Chang et al., 2013). Expression of constitutive active IKK2 in OBs and chondrocytes in transgenic mice controlled by Col2α1 promoter exhibits abnormal skeletal development with impeded bone formation and reduced bone mineral density. A heterozygous missense mutation on RelA in an osteopetrosis patient was revealed by trio-based whole exome sequencing, demonstrating the critical role of NF-κB signaling in human skeletal homeostasis.

In studies of noncanonical NF-κB signaling on bone formation, increased OB numbers and increased bone formation rate were found in the transgenic mice with NIK mutation. However, the effects of mutant NIK on osteogenesis had not been examined in MSCs and/or OBs. RelB-deficient mice developed age-related increased trabecular bone mass associated with increased bone formation. RelB-deficient MSCs had increased bone-forming ability in both in vitro studies and in the murine tibia defect model. These studies suggest that the noncanonical NF-κB signaling also plays a significant role in osteogenic differentiation.

Compared to the evidence in transgenic animal studies, the effects of NF-κB activation induced by extracellular stimulation in OB differentiation remain paradoxical (Osta et al., 2014). Bone mineral density and bone formation were elevated in the TNF-α or TNF-α receptor 1-deficient mice, whereas bone-resorption activity was not changed (Li et al., 2007). Direct injection of TNF-α into wild-type mice reduced osteogenic differentiation in MSCs, which was blocked in the ubiquitin ligase WWP1-deficient mice (Zhao et al., 2011). In vitro studies with primary bone marrow stromal cells or MC3T3 E1 (clone 14) cells demonstrated that 10 ng/ml TNF-α potently suppressed osteogenesis via suppressing Runx2 expression (Abbas, Zhang, Clohisy, & Abu-Amer, 2003; Gilbert, Rubin, & Nanes, 2005; Li et al., 2007). Interestingly, TNF-α-mediated inhibition of Runx2 is NF-κB dependent, whereas the suppression of osterix is depended on MEK1/ERK1 signaling but independent of NF-κB (Lu, Gilbert, He, Rubin, & Nanes, 2006).

There is increasing evidence indicating an important induction role of TNF-α in osteogenic differentiation. Hess et al. demonstrated that treatment of 20 ng/ml TNF-α during osteogenic differentiation enhanced BMP2 expression, followed by Runx2 and osterix expression, and increased mineralization in human MSCs (Hess, Ushmorov, Fiedler, Brenner, & Wirth, 2009). Induction of constitutive active IKK2 by using retroviral vectors enhanced, whereas IκB impaired TNF-α-mediated osteogenesis induction. In studies of rat MSCs cultured in poly(ε-carprolactone) scaffold, the lower dose (0.1–5.0 ng/ml) inhibited, whereas the higher dose (50 ng/ml) of TNF-α enhanced mineralization in dexamethasone-pretreated cells (Mountziaris et al., 2013; Mountziaris, Tzouanas, & Mikos, 2010). The high dose of TNF-α mediating osteogenic induction was only observed with continuous (days 1–16) or early (days 1–4) treatments, but not with intermediate or later treated cells (Mountziaris et al., 2013). Early treatments of extracellular NF-κB stimulators including TNF-α, LPS, or peptidoglycan (the agonist for TLR2) enhanced osteogenesis in human adipose tissue-derived (Cho et al., 2010) or bone marrow-derived MSCs (Croes et al., 2015). Inhibition of NF-κB activity or silence of its downstream target TAZ impaired the induction of osteogenic differentiation, suggesting the direct regulation of NF-κB on TNF-α-mediated osteogenesis (Cho et al., 2010). In addition, Lu et al. demonstrated that human OBs and MSCs preconditioned with 1 ng/ml TNF-α for 1–3 days enhanced osteogenic differentiation via induction of BMP2 expression (Lu et al., 2013; Lu, Wang, Dunstan, & Zreiqat, 2012). Mechanistic studies have shown that inhibition of ERK1/2 or Wnt signaling impeded TNF-α-induced osteogenesis (Briolay et al., 2013; Lu et al., 2013); however, these studies did not address the role of NF-κB induced by TNF-α. In vivo studies have demonstrated that local injection of TNF-α at the early stage (24 h) augmented fracture healing by recruitment of muscle-derived stromal cells and infiltrated macrophages. Low and continuous levels of TNF-α were associated with proosteogenic effects, contributing to the formation of bone spurs (enthesis) by upregulating the activity of ALP (Ding et al., 2009; Lencel et al., 2011). However, the involvement of NF-κB signaling during the inflammation and healing processes remains unclear (Chan et al., 2015; Glass et al., 2011).

Taken together, although studies using transgenic animals demonstrated that NF-κB activation impairs normal skeletal development and bone-forming ability in OB-lineage cells, their effects on the osteogenic differentiation stimulated by extracellular signals are still in debate. The controversial finding in these reports could be due to the dose and the exposure times of NF-κB inducer, cell type used, and the conditions of osteogenic induction (DelaRosa & Lombardo, 2010; Osta et al., 2014).

5. PHARMACEUTICAL APPROACHES TO MODULATE NF-κB ACTIVITY

In this section, we will broadly discuss the structural aspects of NF-κB transcription factors, their interactions with select regulatory proteins, and how this knowledge can help in targeting NF-κB signaling for therapeutic purposes.

As aberrant NF-κB signaling is implicated in many disease processes, the development of NF-κB inhibitors has been widespread. To date, there are over 800 inhibitors that have been reported with, undoubtedly, many more to come (Gilmore & Garbati, 2011). There are step-wise approaches to inhibiting the NF-κB transduction pathway by: receptor inhibition, adaptor inhibition, IKK inhibition, IκB stabilization, cytoplasmic retention, and transcription factor inhibition (Gilmore & Garbati, 2011).

Since ligand binding to cell-surface receptors activates NF-κB signaling, inhibition of receptor binding can be used to block NF-κB activation. Most notably, anti-TNF-α antibodies such as etanercept and infliximab have been used to block activation of the canonical NF-κB pathway and have been used in several chronic inflammatory diseases (Gilmore & Garbati, 2011; Taylor & Feldmann, 2009). Similarly, denosumab, an anti-RANKL antibody, has been used to treat osteoporosis, and its role in preventing osteolytic lesions following total hip arthroplasty is in Phase 2 clinical trial (Clinicaltrials.gov, 2012). Once receptors are engaged, adaptor proteins such as TRAF are recruited to the cell membrane. Though these receptor or adaptor inhibitors are effective, their off-target effects on multiple signaling pathways and the compensatory effects from other NF-κB upstream regulators have made them less desirable as specific NF-κB inhibitors.

Most of the developments in NF-κB inhibitors have focused on targeting the IKK protein family since they are the central integrator of the NF-κB pathway without involvement of other cellular signaling. There are a variety of inhibitors that exert their effects via different mechanisms (Gilmore & Herscovitch, 2006): (1) The ATP analog that specifically interacts with IKK. For example, β-carboline, a natural ATP analog, specifically binds IKK2 to inhibit NF-κB signaling (Karin et al., 2004). Similarly, NSAIDs provide COX-independent antiinflammatory effects by competitively inhibiting the ATP-binding site of IKK2 (Karin et al., 2004). (2) Compounds bind to IKK proteins and mediate conformational changes. The synthetic BMS-345541 has been shown to inhibit IKK2 activity through allosteric effects (Karin et al., 2004). (3) Compounds bind to the Cys-179 residue on the activation loop on IKK protein and blocking the kinase activity. Several thiol-reactive compounds were shown to interact with IKK2 and inhibit the kinase activity, although the detailed mechanism remains unclear (Kwok, Koh, Ndubuisi, Elofsson, & Crews, 2001). In the canonical NF-κB pathway, NEMO and IKK1 are also comprised in the IKK complex. Therefore, dominant negative mutants or specific inhibitors to these proteins could also result in NF-κB signaling inhibition (Gilmore & Garbati, 2011; Scheidereit, 2006). For bone-related disease, the IKK inhibitor SAR113945 recently completed Phase 1 of clinical trials as a treatment for knee osteoarthritis (Clinicaltrials.gov, 2011). Notably, most of the current developed IKK inhibitors have targeted IKKβ and thus only inhibit the canonical NF-κB pathway. Inhibition of the noncanonical pathway via targeting IKK1 activity could be particularly important in osteolytic bone diseases, regarding the fact that IKK1 mutant in OC precursors showed defective osteoclastogenesis but normal TNF-α and IL1β signaling (Ruocco et al., 2005). As the IKK complex is not completely understood in itself, it is almost certain that more IKK inhibitors will be developed and/or discovered to match future understanding.

The next important step in the NF-κB signaling is the degradation of IκB. There are three main strategies for blocking IκB degradation: (1) promoting IκB synthesis, (2) blocking IκB ubiquitination, and (3) inhibition of the proteasome (Gilmore & Garbati, 2011). The most prominent proteasome inhibitor is bortezomib, which has shown efficacy in treating multiple myeloma as well as other hematologic and solid tumors (Gilmore & Garbati, 2011; Jagannath et al., 2010). Similarly, sulfasalazine, a medication used to treat inflammatory bowel disease, prevents NF-κB activation by blocking IκB degradation in response to TNFα and LPS (Karin et al., 2004; Wahl, Liptay, Adler, & Schmid, 1998).

Once IκB is degraded, the NF-κB dimer must translocate into the nucleus, so inhibitors of nuclear localization can block NF-κB signaling. As nuclear entry appears to be mediated by importin α, cell-permeable peptides with the NLS of p50 have been used to saturate importin α, and thus, block NF-κB dimer nuclear entry (Gilmore & Garbati, 2011; Letoha et al., 2005; Lin, Yao, Veach, Torgerson, & Hawiger, 1995; Torgerson, Colosia, Donahue, Lin, & Hawiger, 1998). However, these peptides were not specific to NF-κB and have not been used applied beyond the laboratory.

Finally, NF-κB signaling can be blocked at the level of DNA by directly blocking DNA binding or competitively inhibiting binding through the introduction of NF-κB decoy oligodeoxynucleotides (ODNs). For example, there are several compounds that target both IKK2 and NF-κB DNA binding such as parthenolide (Garcia-Pineres, Lindenmeyer, & Merfort, 2004; Gilmore & Garbati, 2011). However, these compounds likely affect other protein targets. Alternatively, NF-κB decoy oligonucleotides (ODN) specifically compete for binding of NF-κB dimers to their DNA targets, and could simultaneously inhibit both canonical and noncanonical NF-κB pathway. It has been shown that NF-κB decoy ODN significantly suppress cytokine and chemokine expression in macrophages, especially in the setting of periprosthetic osteolysis (Lin, Pajarinen, et al., 2016; Lin et al., 2015; Lin, Yao, et al., 2014). Moreover, there are several NF-κB decoy ODN-based therapies that have entered clinical trials for dermatitis and psoriasis (Gilmore & Garbati, 2011).

6. DRUG DELIVERY STRATEGIES IN BONE DISORDERS

6.1 Classification

Over 800 NF-κB inhibitors have been reported, and the number continues to increase (Gilmore & Garbati, 2011; Gilmore & Herscovitch, 2006). These inhibitors can be broadly divided into three categories including (1) proteins and peptides, (2) small molecules, and (3) nucleic acids. Proteins and peptides inhibitors include antibodies and growth factors and are highly specific with less off-target side effects (Craik, Fairlie, Liras, & Price, 2013). The common limitations in protein-based treatment are instability when applied in vivo, high dose requirement, high cost of manufacturing, undesirable immunogenic effects, contamination, and the limitations for oral administration. Small molecules are natural or synthetic nonpeptide molecules with small molecular size (<1000 Da) and have several advantages in clinical application. These molecules are characterized by greater tissue penetration and are less immunogenic due to their smaller size compared to the protein-based compounds (Lo, Ashe, Kan, & Laurencin, 2012). Other advantages include lower manufacturing costs and application for oral administration. However, these compounds could be less selective compared to the protein or nucleic acid-based drugs, raising the concern of nonspecific toxicity, limiting their clinical application (Balmayor, 2015; Torchilin, 2000). Nucleic acid-based molecules include such as oligonucleotides, small interfering RNAs (siRNAs), and microRNAs (miRNAs). These approaches are highly specific and directly silence protein translation via siRNA or miRNA, or inhibit gene transactivation via decoy ODN. In particular, modulating the transcriptional process such as NF-κB by using the decoy ODN is an emerging approach due to the complexity of the signal transduction pathway (Rad et al., 2015). Synthetic decoy ODN is a short fragment of DNA which has the consensus sequence of the binding site of the target transcription factor, thus decoy ODN has the ability to bind to free target transcription factor subsequently preventing these factors from binding to the specific promoter regions (Bielinska, Shivdasani, Zhang, & Nabel, 1990). Like protein-based therapy, nucleic acids are also limited by low cellular uptake and rapid degradation in biological fluids (De Stefano, 2011).

6.2 Drug Delivery to the Skeletal System

Local drug delivery is applicable to many of the bone disorders in a confined region including fractures and periprosthetic osteolysis. Though local delivery strategy has the advantage of reduced systemic adverse effects, systemic drug delivery is still desired for the long-term management during the disease process, and treatment for systemic bone diseases such as osteoporosis. Recent developments in bone cell-specific targeting vehicles allow gene modulation and transcriptional regulation in OC, OB, or MSC in skeletal system. Thus, the adverse effects of NF-κB therapy to nontargeted cell populations (i.e., lymphocytes) in the bone marrow can be reduced and advance its translational application. Here we summarize the current advanced techniques for local and systemic drug delivery in the skeletal system (Fig. 3).

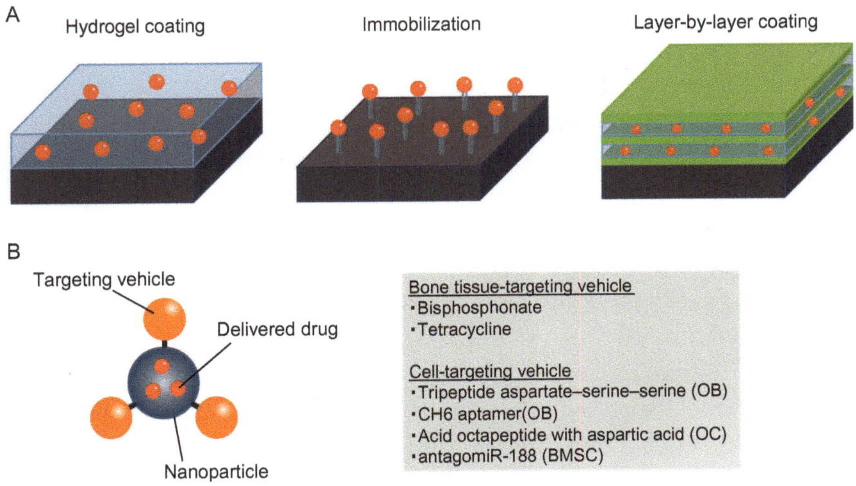

Fig. 3 The strategies of drug delivery to the skeletal system. (A) Three surface coating procedures of the implants. (B) Systemic drug delivery conjugated with bone-targeting vehicle. OB, osteoblast; OC, osteoclast; BMSC, bone marrow mesenchymal stem cell.

6.2.1 Local Drug Delivery

Local drug delivery systems are designed to keep the localized and controlled release of biomolecules in a confined region. The advantages of local drug delivery include reduced systemic toxicity; avoidance of the risk of overdose; and upholding effective concentrations at target regions. The aims of drug delivery in orthopedics are to improve bone-healing processes, increase osseointegration, mitigate inflammation, and prevent infection. Implantable devices with controlled drug release are some of the most commonly used strategies for local drug delivery in the orthopedic field, and the techniques have also been applied to other diseases including cardiovascular diseases, cancer, periodontal diseases, and ophthalmologic diseases (Arruebo, Vilaboa, & Santamaria, 2010). Orthopedic implants are widely used for fracture fixation, TJR, spinal reconstruction, and other orthopedic applications. Several different strategies have been developed for surface coating of local drug release of orthopedic implants (Goodman, Yao, Keeney, & Yang, 2013; Raphel, Holodniy, Goodman, & Heilshorn, 2016), including the use of hydrogel, immobilization (Maia, Bidarra, Granja, & Barrias, 2013), and layer-by-layer coating (de Villiers, Otto, Strydom, & Lvov, 2011).

The layer-by-layer coating method is based on the alternating deposition of positively and negatively charged polyelectrolytes onto the implant surface. Charged biomolecules can be loaded onto the surface, and the amount of encapsulated biomolecules and release rate can be tuned by changing the concentration of the biomolecules, the number of layers, and the type of polyelectrolytes. Coating can be performed onto any shape and dimension of different material surfaces. Furthermore, due to simple process of layer-by-layer coating, various biomolecules have been applied for drug release from implants including proteins (Keeney et al., 2013; Min, Braatz, & Hammond, 2014), nucleic acids (Miyake et al., 2014; Tahara et al., 2011), and small molecules (Min et al., 2014). The limitations of layer-by-layer coating techniques include the fact that many layers are required to prevent rapid diffusion and burst release of biomolecules. Thus, long production times are needed, which may lead high batch-to-batch variability. Also, the influence of the layers on the mechanical stability after a press fit implantation into bone must be considered. Further studies are required to address such limitations.

Combination of drug delivery systems have been used to overcome the instability of target drugs such as ODN. De Rosa and La Rotonda reviewed the details of delivery systems including cationic liposomes, polyethylenimine (PEI), chitosan, and poly(lactide-*co*-glycolide) (PLGA) nanospheres which have the ability to increase the cellular uptake of ODN, prolong their stability, and localize the effect of ODN to target cells and tissues (De Rosa & La Rotonda, 2009). Furthermore, coating with chitosan showed a positive charge, which can penetrate easily through the negative charged cell membrane and enhance cellular uptake of ODN (Mao et al., 2001). Additionally, PLGA is thought to have the proton sponge mechanism to escape from the endosomal pathway (Panyam, Zhou, Prabha, Sahoo, & Labhasetwar, 2002). Taken together, combinations of multiple drug delivery systems could improve the drug release profile and keep the biological stability of target drugs in a confined region.

6.2.2 Systemic Delivery (Bone-Targeting Vehicle)

Early development of drug delivery to the skeletal system was focused on targeting the bone environment. The synthetic compounds including bisphosphonates and tetracycline have high affinity for calcium ions and hydroxyapatite (Pazianas & Abrahamsen, 2011; Russell, 2011). Therefore, the concept of bone-targeted delivery system is the use of biomolecules conjugated with bisphosphonates as ligands targeting bone mineral (Cole, Vargo-Gogola, & Roeder, 2016). Clinical applications investigating bisphosphonates as a targeting ligand include metabolic bone diseases such as osteoporosis (Bhandari, Newa, Chapman, & Doschak, 2012; Fujisaki et al., 1997;

Morioka et al., 2010), bone infection (Houghton et al., 2008), rheumatoid arthritis (Hirabayashi et al., 2001), osteosarcoma, and cancer metastases to bone (Klenner et al., 1990; Reinholz et al., 2010; Torres Martin de Rosales, Finucane, Mather, & Blower, 2009). Intravenous injection or oral administration of bisphosphonate-conjugated vehicle delivered the desired compounds to bone due to their high affinity for calcium crystals (Widler, Jahnke, & Green, 2012). Also, bisphosphonates can regulate calcium homeostasis and inhibit OC activity, further enhancing the clinical application for several bone disorders. Conjugation of bisphosphonates with PLGA nanoparticles (Thamake, Raut, Gryczynski, Ranjan, & Vishwanatha, 2012) or N-(2-hydroxypropyl)methacrylamide (HPMA) (Miller et al., 2011) successfully delivered the small molecule compounds and siRNA to bone. However, recent reports suggested that long-term treatment of bisphosphonates-induced OB apoptosis (Orriss, Key, Colston, & Arnett, 2009), and can cause osteonecrosis of the jaw (Favus, 2007). Although the direct link between bisphosphonates and jaw osteonecrosis remains unclear, preventive treatment for jaw osteonecrosis is important during bisphosphonates therapy. An alternative delivery strategy is to conjugate the therapeutic compounds with tetracycline, an antibiotic with high affinity to calcium ions and thus have bone-targeting ability (Neale et al., 2009). The development of tetracycline and its derivative to conjugate with drug carrying vehicles could be an efficient strategy to deliver agents to the skeletal system without induction of jaw osteonecrosis (Dang et al., 2016).

Recent development of drug delivery models for the skeletal system has advanced to cell-specific targeting vehicles to avoid the toxic effects on nontargeted cells. The bone-forming surface is characterized by low crystallized hydroxyapatite and amorphous calcium phosphate compared to the high crystallized hydroxyapatite in bone-resorption surface (Dang et al., 2016). The differences of chemical and physical properties enable the design of OC- or OB-targeting vehicles. For example, the tripeptite asparate–serine–serine (DSS) binds to bone formation sites preferably than bone-resorption site in vivo (Zhang et al., 2012). On the other hand, the acid octapeptide with aspartic acid has high affinity to hydroxyapatite and accumulated at bone-resorption surfaces (Wang et al., 2007). Conjunction of the bone formation or bone-resorption selective peptide with dioleoyltrimethylammonium propane (DOTAP)-based cationic liposomes or HPMA nanoparticles successfully delivered siRNA, protein, or small molecule compounds to the desired region (Wang et al., 2007; Zhang et al., 2012).

A new developed technique, named as cell-SELEX (Gold, 1995), was able to screen and identify cell-specific molecules and target OB, OC, and even MSC, specifically (Dang et al., 2016). A random pool of 10^{13}–10^{16} of ssDNA or ssRNA is used to screen for the cell-specific targeting molecule. CH6 is an OB-specific aptamer screened from cell-SELEX system (Liang et al., 2015). Conjugation of CH6 with lipid nanoparticles was able to deliver siRNA to OB at cellular level in vivo. Antagomir-188 is another example of MSC-specific aptamer screened from cell-SELEX system (Li et al., 2015), which can specifically regulate miRNA-188 in bone marrow MSCs in vivo. Taken together, the advanced techniques enable the NF-κB-targeting therapy in the skeletal system at the cellular level, reduce the potential toxic effects to the immune system, and advance the translational application in inflammatory bone disorders.

7. CONCLUSION

Despite direct evidence from transgenic animal studies demonstrating the essential roles of NF-κB in the activation of OC and the suppression of OB differentiation, bone-remodeling processes are sensitive to the timing, duration, dose, and composition of upstream NF-κB stimulators (Fig. 4). Acute inflammation or low dose inflammatory cytokines are indispensable and beneficial for bone defect healing processes. Therefore, temporal modulation or delayed inhibition could optimize the therapeutic effects of targeting NF-κB pathway in bone diseases associated with acute inflammation. Alternatively, overwhelming inflammatory signals (e.g., infection) or unresolved chronic inflammation (e.g., periprosthetic osteolysis) alters the balance of bone remodeling toward osteolytic processes, thus effective inhibition of NF-κB signaling is required to mitigate these inflammation-associated bone diseases.

There are many modalities to target the canonical or noncanonical NF-κB signaling cascades. In many cases, the strategy of NF-κB inhibition focuses on suppressing IKK2 kinase activity or maintaining IκB protein stability. These approaches have proved to be effective in blocking the canonical pathway; however, they cannot modulate noncanonical signaling and consequent osteoclastogenesis (Ruocco et al., 2005). Although the strategic use of NF-κB decoy ODN could simultaneously target both canonical and noncanonical signaling, the stability of the decoy ODN in vivo, or the potential toxic effects induced by chemically modified ODN, remain an obstacle to their translational application

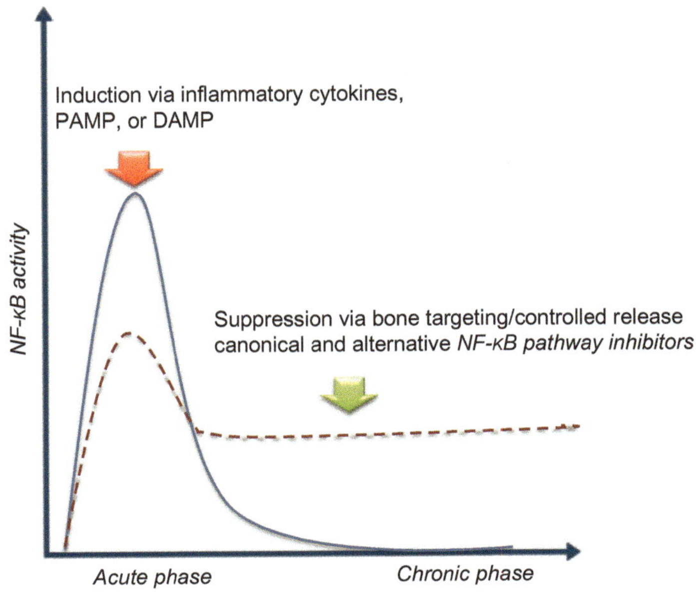

Fig. 4 Temporal modulation of NF-κB activity in inflammatory-associated bone diseases. *Blue line* represents optimal condition of NF-κB signaling for bone-healing/bone-remodeling process. *Red dash line* represents the impaired inflammatory response at acute phase and/or persistent response at chronic phase. The *arrows* indicate the optimal bone regeneration conditions by either induction of NF-κB signaling at acute phase of inflammation (*Red arrow*), or suppression at chronic phase with unresolved inflammation (*Green arrow*).

(Osako, Nakagami, & Morishita, 2012). Most likely, the best approach will be a combination of inhibitors that have specific activity, broad activity, and are synergistic, and provide a multistep inhibition of the inflammatory cascade. Enhanced understanding of situation-specific NF-κB activation and unique structural changes may also allow for the development of increasingly refined NF-κB target strategies.

The central roles of NF-κB as a master regulator of macrophages, OC, and OB make it possible to manipulate both the inflammation and bone-remodeling processes. Nevertheless, the critical functions of NF-κB in the immune system also raise concerns of toxicity with systemic treatment. Local delivery with controlled drug release systems is an efficient strategy for long-term manipulation of the NF-κB pathway with minimal side effects and can be applied to many bone diseases including fracture healing, periprosthetic osteolysis, osteoarthritis, etc. The discovery of bone-targeting drug delivery vehicles further advances the possibility to deliver siRNA, decoy ODN, or even small molecular inhibitors with minimal impacts on the patient's immune function.

Taken together, the in vitro and in vivo evidence strongly suggest that NF-κB-targeting therapy has great potential to restore impaired bone-remodeling processes. The natural differences in pathogenesis between acute and chronic inflammation suggest that the optimal timing, dose, and targeting strategy to modulate NF-κB pathways are critical in order to achieve the desired therapeutic effects. Temporal modulation of NF-κB activity with advanced drug delivery strategies could efficiently mitigate osteolytic processes and enhance bone formation in the patients with inflammation-associated bone diseases.

REFERENCES

AAOS (2013). http://www.aaos.org/Research/stats/patientstats.asp.

Abbas, S., Zhang, Y. H., Clohisy, J. C., & Abu-Amer, Y. (2003). Tumor necrosis factor-alpha inhibits pre-osteoblast differentiation through its type-1 receptor. *Cytokine, 22*(1–2), 33–41.

Abu-Amer, Y. (2013). NF-kappaB signaling and bone resorption. *Osteoporosis International, 24*(9), 2377–2386.

Alexander, K. A., Chang, M. K., Maylin, E. R., Kohler, T., Muller, R., Wu, A. C., et al. (2011). Osteal macrophages promote in vivo intramembranous bone healing in a mouse tibial injury model. *Journal of Bone and Mineral Research, 26*(7), 1517–1532.

Arruebo, M., Vilaboa, N., & Santamaria, J. (2010). Drug delivery from internally implanted biomedical devices used in traumatology and in orthopedic surgery. *Expert Opinion on Drug Delivery*, 7(5), 589–603.

Balmayor, E. R. (2015). Targeted delivery as key for the success of small osteoinductive molecules. *Advanced Drug Delivery Reviews, 94*, 13–27.

Beg, A. A., Sha, W. C., Bronson, R. T., Ghosh, S., & Baltimore, D. (1995). Embryonic lethality and liver degeneration in mice lacking the RelA component of NF-kappa B. *Nature, 376*(6536), 167–170.

Bhandari, K. H., Newa, M., Chapman, J., & Doschak, M. R. (2012). Synthesis, characterization and evaluation of bone targeting salmon calcitonin analogs in normal and osteoporotic rats. *Journal of Controlled Release, 158*(1), 44–52.

Bianchi, M. E. (2007). DAMPs, PAMPs and alarmins: All we need to know about danger. *Journal of Leukocyte Biology, 81*(1), 1–5.

Bielinska, A., Shivdasani, R. A., Zhang, L. Q., & Nabel, G. J. (1990). Regulation of gene expression with double-stranded phosphorothioate oligonucleotides. *Science, 250*(4983), 997–1000.

Bohuslav, J., Kravchenko, V. V., Parry, G. C., Erlich, J. H., Gerondakis, S., Mackman, N., et al. (1998). Regulation of an essential innate immune response by the p50 subunit of NF-kappaB. *The Journal of Clinical Investigation, 102*(9), 1645–1652.

Boston_University. http://www.bu.edu/nf-kb/gene-resources/target-genes/.

Boyce, B. F., Yao, Z., & Xing, L. (2010). Functions of nuclear factor kappaB in bone. *Annals of the New York Academy of Sciences, 1192*, 367–375.

Boyle, W. J., Simonet, W. S., & Lacey, D. L. (2003). Osteoclast differentiation and activation. *Nature, 423*(6937), 337–342.

Briolay, A., Lencel, P., Bessueille, L., Caverzasio, J., Buchet, R., & Magne, D. (2013). Autocrine stimulation of osteoblast activity by Wnt5a in response to TNF-alpha in human mesenchymal stem cells. *Biochemical and Biophysical Research Communications, 430*(3), 1072–1077.

Cavassani, K. A., Ishii, M., Wen, H., Schaller, M. A., Lincoln, P. M., Lukacs, N. W., et al. (2008). TLR3 is an endogenous sensor of tissue necrosis during acute inflammatory events. *Journal of Experimental Medicine, 205*(11), 2609–2621.

Chan, J. K., Glass, G. E., Ersek, A., Freidin, A., Williams, G. A., Gowers, K., et al. (2015). Low-dose TNF augments fracture healing in normal and osteoporotic bone by up-regulating the innate immune response. *EMBO Molecular Medicine*, 7(5), 547–561.

Chang, J., Liu, F., Lee, M., Wu, B., Ting, K., Zara, J. N., et al. (2013). NF-kappaB inhibits osteogenic differentiation of mesenchymal stem cells by promoting beta-catenin degradation. *Proceedings of the National Academy of Sciences of the United States of America, 110*(23), 9469–9474.

Chang, M. K., Raggatt, L. J., Alexander, K. A., Kuliwaba, J. S., Fazzalari, N. L., Schroder, K., et al. (2008). Osteal tissue macrophages are intercalated throughout human and mouse bone lining tissues and regulate osteoblast function in vitro and in vivo. *Journal of Immunology, 181*(2), 1232–1244.

Chang, J., Wang, Z., Tang, E., Fan, Z., McCauley, L., Franceschi, R., et al. (2009). Inhibition of osteoblastic bone formation by nuclear factor-kappaB. *Nature Medicine, 15*(6), 682–689.

Chen, Z. J. (2005). Ubiquitin signalling in the NF-kappaB pathway. *Nature Cell Biology*, 7(8), 758–765.

Chen, F. E., & Ghosh, G. (1999). Regulation of DNA binding by Rel/NF-kappaB transcription factors: Structural views. *Oncogene, 18*(49), 6845–6852.

Chen, G. Y., & Nunez, G. (2010). Sterile inflammation: Sensing and reacting to damage. *Nature Reviews. Immunology, 10*(12), 826–837.

Chiu, R., Ma, T., Smith, R. L., & Goodman, S. B. (2009). Ultrahigh molecular weight polyethylene wear debris inhibits osteoprogenitor proliferation and differentiation in vitro. *Journal of Biomedical Materials Research. Part A, 89*(1), 242–247.

Cho, H. H., Shin, K. K., Kim, Y. J., Song, J. S., Kim, J. M., Bae, Y. C., et al. (2010). NF-kappaB activation stimulates osteogenic differentiation of mesenchymal stem cells derived from human adipose tissue by increasing TAZ expression. *Journal of Cellular Physiology, 223*(1), 168–177.

Cho, S. W., Soki, F. N., Koh, A. J., Eber, M. R., Entezami, P., Park, S. I., et al. (2014). Osteal macrophages support physiologic skeletal remodeling and anabolic actions of parathyroid hormone in bone. *Proceedings of the National Academy of Sciences of the United States of America, 111*(4), 1545–1550.

Clinicaltrials.gov. (2011). https://clinicaltrials.gov/ct2/show/NCT01113333?term=IKK+inhibitor&rank=1.

Clinicaltrials.gov. (2012). https://clinicaltrials.gov/show/NCT01358669.

Clohisy, J. C., Hirayama, T., Frazier, E., Han, S. K., & Abu-Amer, Y. (2004). NF-kB signaling blockade abolishes implant particle-induced osteoclastogenesis. *Journal of Orthopaedic Research, 22*(1), 13–20.

Cobelli, N., Scharf, B., Crisi, G. M., Hardin, J., & Santambrogio, L. (2011). Mediators of the inflammatory response to joint replacement devices. *Nature Reviews. Rheumatology*, 7(10), 600–608.

Cole, L. E., Vargo-Gogola, T., & Roeder, R. K. (2016). Targeted delivery to bone and mineral deposits using bisphosphonate ligands. *Advanced Drug Delivery Reviews, 99*(Pt. A), 12–27.

Cordova, L. A., Stresing, V., Gobin, B., Rosset, P., Passuti, N., Gouin, F., et al. (2014). Orthopaedic implant failure: Aseptic implant loosening—The contribution and future challenges of mouse models in translational research. *Clinical Science, 127*(5), 277–293.

Courtois, G., & Gilmore, T. D. (2006). Mutations in the NF-kappaB signaling pathway: Implications for human disease. *Oncogene*, *25*(51), 6831–6843.
Craik, D. J., Fairlie, D. P., Liras, S., & Price, D. (2013). The future of peptide-based drugs. *Chemical Biology & Drug Design*, *81*(1), 136–147.
Croes, M., Oner, F. C., Kruyt, M. C., Blokhuis, T. J., Bastian, O., Dhert, W. J., et al. (2015). Proinflammatory mediators enhance the osteogenesis of human mesenchymal stem cells after lineage commitment. *PloS One*, *10*(7), e0132781.
Dang, L., Liu, J., Li, F., Wang, L., Li, D., Guo, B., et al. (2016). Targeted delivery systems for molecular therapy in skeletal disorders. *International Journal of Molecular Sciences*, *17*(3), 428.
De Rosa, G., & La Rotonda, M. I. (2009). Nano and microtechnologies for the delivery of oligonucleotides with gene silencing properties. *Molecules*, *14*(8), 2801–2823.
De Stefano, D. (2011). Oligonucleotides decoy to NF-kappaB: Becoming a reality? *Discovery Medicine*, *12*(63), 97–105.
de Villiers, M. M., Otto, D. P., Strydom, S. J., & Lvov, Y. M. (2011). Introduction to nanocoatings produced by layer-by-layer (LbL) self-assembly. *Advanced Drug Delivery Reviews*, *63*(9), 701–715.
DelaRosa, O., & Lombardo, E. (2010). Modulation of adult mesenchymal stem cells activity by toll-like receptors: Implications on therapeutic potential. *Mediators of Inflammation*, *2010*, 865601.
Ding, J., Ghali, O., Lencel, P., Broux, O., Chauveau, C., Devedjian, J. C., et al. (2009). TNF-alpha and IL-1beta inhibit RUNX2 and collagen expression but increase alkaline phosphatase activity and mineralization in human mesenchymal stem cells. *Life Sciences*, *84*(15–16), 499–504.
Dobbs, M. B., Buckwalter, J., & Saltzman, C. (1999). Osteoporosis: The increasing role of the orthopaedist. *The Iowa Orthopaedic Journal*, *19*, 43–52.
Eigenbrod, T., Park, J. H., Harder, J., Iwakura, Y., & Nunez, G. (2008). Cutting edge: Critical role for mesothelial cells in necrosis-induced inflammation through the recognition of IL-1 alpha released from dying cells. *Journal of Immunology*, *181*(12), 8194–8198.
Favus, M. J. (2007). Diabetes and the risk of osteonecrosis of the jaw. *The Journal of Clinical Endocrinology and Metabolism*, *92*(3), 817–818.
Franceschi, C., Bonafe, M., Valensin, S., Olivieri, F., De Luca, M., Ottaviani, E., et al. (2000). Inflamm-aging. An evolutionary perspective on immunosenescence. *Annals of the New York Academy of Sciences*, *908*, 244–254.
Franzoso, G., Carlson, L., Xing, L., Poljak, L., Shores, E. W., Brown, K. D., et al. (1997). Requirement for NF-kappaB in osteoclast and B-cell development. *Genes & Development*, *11*(24), 3482–3496.
Fujisaki, J., Tokunaga, Y., Takahashi, T., Kimura, S., Shimojo, F., & Hata, T. (1997). Osteotropic Drug Delivery System (ODDS) based on bisphosphonic prodrug. V. Biological disposition and targeting characteristics of osteotropic estradiol. *Biological & Pharmaceutical Bulletin*, *20*(11), 1183–1187.
Garcia-Pineres, A. J., Lindenmeyer, M. T., & Merfort, I. (2004). Role of cysteine residues of p65/NF-kappaB on the inhibition by the sesquiterpene lactone parthenolide and N-ethyl maleimide, and on its transactivating potential. *Life Sciences*, *75*(7), 841–856.
Gerondakis, S., Strasser, A., Metcalf, D., Grigoriadis, G., Scheerlinck, J. Y., & Grumont, R. J. (1996). Rel-deficient T cells exhibit defects in production of interleukin 3 and granulocyte-macrophage colony-stimulating factor. *Proceedings of the National Academy of Sciences of the United States of America*, *93*(8), 3405–3409.
Ghosh, S., May, M. J., & Kopp, E. B. (1998). NF-kappa B and Rel proteins: Evolutionarily conserved mediators of immune responses. *Annual Review of Immunology*, *16*, 225–260.
Gilbert, L. C., Rubin, J., & Nanes, M. S. (2005). The p55 TNF receptor mediates TNF inhibition of osteoblast differentiation independently of apoptosis. *American Journal of Physiology—Endocrinology & Metabolism*, *288*(5), E1011–E1018.

Gilmore, T. D. (2006). Introduction to NF-kappaB: Players, pathways, perspectives. *Oncogene, 25*(51), 6680–6684.

Gilmore, T. D., & Garbati, M. R. (2011). Inhibition of NF-kappaB signaling as a strategy in disease therapy. *Current Topics in Microbiology and Immunology, 349*, 245–263.

Gilmore, T. D., & Herscovitch, M. (2006). Inhibitors of NF-kappaB signaling: 785 and counting. *Oncogene, 25*(51), 6887–6899.

Glass, G. E., Chan, J. K., Freidin, A., Feldmann, M., Horwood, N. J., & Nanchahal, J. (2011). TNF-alpha promotes fracture repair by augmenting the recruitment and differentiation of muscle-derived stromal cells. *Proceedings of the National Academy of Sciences of the United States of America, 108*(4), 1585–1590.

Gold, L. (1995). The SELEX process: A surprising source of therapeutic and diagnostic compounds. *The Harvey Lectures, 91*, 47–57.

Goodman, S. B., Yao, Z., Keeney, M., & Yang, F. (2013). The future of biologic coatings for orthopaedic implants. *Biomaterials, 34*(13), 3174–3183.

Grigoriadis, A. E., Wang, Z. Q., Cecchini, M. G., Hofstetter, W., Felix, R., Fleisch, H. A., et al. (1994). c-Fos: A key regulator of osteoclast-macrophage lineage determination and bone remodeling. *Science, 266*(5184), 443–448.

Guihard, P., Danger, Y., Brounais, B., David, E., Brion, R., Delecrin, J., et al. (2012). Induction of osteogenesis in mesenchymal stem cells by activated monocytes/macrophages depends on oncostatin M signaling. *Stem Cells, 30*(4), 762–772.

Haleem-Smith, H., Argintar, E., Bush, C., Hampton, D., Postma, W. F., Chen, F. H., et al. (2012). Biological responses of human mesenchymal stem cells to titanium wear debris particles. *Journal of Orthopaedic Research, 30*(6), 853–863.

Hayden, M. S., & Ghosh, S. (2011). NF-kappaB in immunobiology. *Cell Research, 21*(2), 223–244.

Hayden, M. S., & Ghosh, S. (2014). Regulation of NF-kappaB by TNF family cytokines. *Seminars in Immunology, 26*(3), 253–266.

Heino, T. J., Hentunen, T. A., & Vaananen, H. K. (2002). Osteocytes inhibit osteoclastic bone resorption through transforming growth factor-beta: Enhancement by estrogen. *Journal of Cellular Biochemistry, 85*(1), 185–197.

Hess, K., Ushmorov, A., Fiedler, J., Brenner, R. E., & Wirth, T. (2009). TNFalpha promotes osteogenic differentiation of human mesenchymal stem cells by triggering the NF-kappaB signaling pathway. *Bone, 45*(2), 367–376.

Hirabayashi, H., Takahashi, T., Fujisaki, J., Masunaga, T., Sato, S., Hiroi, J., et al. (2001). Bone-specific delivery and sustained release of diclofenac, a non-steroidal anti-inflammatory drug, via bisphosphonic prodrug based on the Osteotropic Drug Delivery System (ODDS). *Journal of Controlled Release, 70*(1–2), 183–191.

Houghton, T. J., Tanaka, K. S. E., Kang, T., Dietrich, E., Lafontaine, Y., Delorme, D., et al. (2008). Linking bisphosphonates to the free amino groups in fluoroquinolones: Preparation of osteotropic prodrugs for the prevention of osteomyelitis. *Journal of Medicinal Chemistry, 51*(21), 6955–6969.

Huxford, T., Hoffmann, A., & Ghosh, G. (2011). Understanding the logic of IkappaB: NF-kappaB regulation in structural terms. *Current Topics in Microbiology and Immunology, 349*, 1–24.

Huxford, T., Huang, D. B., Malek, S., & Ghosh, G. (1998). The crystal structure of the IkappaBalpha/NF-kappaB complex reveals mechanisms of NF-kappaB inactivation. *Cell, 95*(6), 759–770.

Hwa Cho, H., Bae, Y. C., & Jung, J. S. (2006). Role of toll-like receptors on human adipose-derived stromal cells. *Stem Cells, 24*(12), 2744–2752.

Imaeda, A. B., Watanabe, A., Sohail, M. A., Mahmood, S., Mohamadnejad, M., Sutterwala, F. S., et al. (2009). Acetaminophen-induced hepatotoxicity in mice is dependent on Tlr9 and the Nalp3 inflammasome. *The Journal of Clinical Investigation, 119*(2), 305–314.

Iotsova, V., Caamano, J., Loy, J., Yang, Y., Lewin, A., & Bravo, R. (1997). Osteopetrosis in mice lacking NF-kappaB1 and NF-kappaB2. *Nature Medicine*, *3*(11), 1285–1289.
Itoh, K., Udagawa, N., Kobayashi, K., Suda, K., Li, X., Takami, M., et al. (2003). Lipopolysaccharide promotes the survival of osteoclasts via Toll-like receptor 4, but cytokine production of osteoclasts in response to lipopolysaccharide is different from that of macrophages. *Journal of Immunology*, *170*(7), 3688–3695.
Jacobs, M. D., & Harrison, S. C. (1998). Structure of an IkappaBalpha/NF-kappaB complex. *Cell*, *95*(6), 749–758.
Jagannath, S., Kyle, R. A., Palumbo, A., Siegel, D. S., Cunningham, S., & Berenson, J. (2010). The current status and future of multiple myeloma in the clinic. *Clinical Lymphoma, Myeloma & Leukemia*, *10*(1), 28–43.
Karin, M., & Ben-Neriah, Y. (2000). Phosphorylation meets ubiquitination: The control of NF-[kappa]B activity. *Annual Review of Immunology*, *18*, 621–663.
Karin, M., Yamamoto, Y., & Wang, Q. M. (2004). The IKK NF-kappa B system: A treasure trove for drug development. *Nature Reviews. Drug Discovery*, *3*(1), 17–26.
Karnes, J. M., Daffner, S. D., & Watkins, C. M. (2015). Multiple roles of tumor necrosis factor-alpha in fracture healing. *Bone*, *78*, 87–93.
Kawai, T., & Akira, S. (2010). The role of pattern-recognition receptors in innate immunity: Update on Toll-like receptors. *Nature Immunology*, *11*(5), 373–384.
Kawai, T., & Akira, S. (2011). Toll-like receptors and their crosstalk with other innate receptors in infection and immunity. *Immunity*, *34*(5), 637–650.
Keeney, M., Waters, H., Barcay, K., Jiang, X., Yao, Z., Pajarinen, J., et al. (2013). Mutant MCP-1 protein delivery from layer-by-layer coatings on orthopedic implants to modulate inflammatory response. *Biomaterials*, *34*(38), 10287–10295.
Kim, N., Kadono, Y., Takami, M., Lee, J., Lee, S. H., Okada, F., et al. (2005). Osteoclast differentiation independent of the TRANCE-RANK-TRAF6 axis. *Journal of Experimental Medicine*, *202*(5), 589–595.
Kim, J. H., & Kim, N. (2014). Regulation of NFATc1 in osteoclast differentiation. *Journal of Bone Metabolism*, *21*(4), 233–241.
Klement, J. F., Rice, N. R., Car, B. D., Abbondanzo, S. J., Powers, G. D., Bhatt, P. H., et al. (1996). IkappaBalpha deficiency results in a sustained NF-kappaB response and severe widespread dermatitis in mice. *Molecular and Cellular Biology*, *16*(5), 2341–2349.
Klenner, T., Valenzuela-Paz, P., Keppler, B. K., Angres, G., Scherf, H. R., Wingen, F., et al. (1990). Improvement of antitumor drug therapeutic index by toxicity reduction cisplatin-linked phosphonates in the treatment of the transplantable osteosarcoma in vitro and in vivo. *Cancer Treatment Reviews*, *17*(2), 253–259.
Kobayashi, K., Takahashi, N., Jimi, E., Udagawa, N., Takami, M., Kotake, S., et al. (2000). Tumor necrosis factor alpha stimulates osteoclast differentiation by a mechanism independent of the ODF/RANKL-RANK interaction. *Journal of Experimental Medicine*, *191*(2), 275–286.
Kono, H., & Rock, K. L. (2008). How dying cells alert the immune system to danger. *Nature Reviews. Immunology*, *8*(4), 279–289.
Kumar, H., Kawai, T., & Akira, S. (2009). Toll-like receptors and innate immunity. *Biochemical and Biophysical Research Communications*, *388*(4), 621–625.
Kwok, B. H., Koh, B., Ndubuisi, M. I., Elofsson, M., & Crews, C. M. (2001). The anti-inflammatory natural product parthenolide from the medicinal herb Feverfew directly binds to and inhibits IkappaB kinase. *Chemistry & Biology*, *8*(8), 759–766.
Lacey, D. L., Timms, E., Tan, H. L., Kelley, M. J., Dunstan, C. R., Burgess, T., et al. (1998). Osteoprotegerin ligand is a cytokine that regulates osteoclast differentiation and activation. *Cell*, *93*(2), 165–176.
Lam, J., Takeshita, S., Barker, J. E., Kanagawa, O., Ross, F. P., & Teitelbaum, S. L. (2000). TNF-alpha induces osteoclastogenesis by direct stimulation of macrophages exposed to permissive levels of RANK ligand. *The Journal of Clinical Investigation*, *106*(12), 1481–1488.

Lawrence, T. (2009). The nuclear factor NF-kappaB pathway in inflammation. *Cold Spring Harbor Perspectives in Biology*, *1*(6), a001651.
Lencel, P., Delplace, S., Pilet, P., Leterme, D., Miellot, F., Sourice, S., et al. (2011). Cell-specific effects of TNF-alpha and IL-1beta on alkaline phosphatase: Implication for syndesmophyte formation and vascular calcification. *Laboratory Investigation*, *91*(10), 1434–1442.
Letoha, T., Somlai, C., Takacs, T., Szabolcs, A., Jarmay, K., Rakonczay, Z., Jr., et al. (2005). A nuclear import inhibitory peptide ameliorates the severity of cholecystokinin-induced acute pancreatitis. *World Journal of Gastroenterology*, *11*(7), 990–999.
Li, C. J., Cheng, P., Liang, M. K., Chen, Y. S., Lu, Q., Wang, J. Y., et al. (2015). MicroRNA-188 regulates age-related switch between osteoblast and adipocyte differentiation. *The Journal of Clinical Investigation*, *125*(4), 1509–1522.
Li, Y., Li, A., Strait, K., Zhang, H., Nanes, M. S., & Weitzmann, M. N. (2007). Endogenous TNFalpha lowers maximum peak bone mass and inhibits osteoblastic Smad activation through NF-kappaB. *Journal of Bone and Mineral Research*, *22*(5), 646–655.
Liang, C., Guo, B., Wu, H., Shao, N., Li, D., Liu, J., et al. (2015). Aptamer-functionalized lipid nanoparticles targeting osteoblasts as a novel RNA interference-based bone anabolic strategy. *Nature Medicine*, *21*(3), 288–294.
Lin, T. H., Gibon, E., Loi, F., Pajarinen, J., Cordova, L. A., Nabeshima, A., et al. (2016). Decreased osteogenesis in mesenchymal stem cells derived from the aged mouse is associated with enhanced NF-kappaB activity. *Journal of Orthopaedic Research*, https://www.ncbi.nlm.nih.gov/pubmed/27105133, Epub ahead of print.
Lin, T. H., Pajarinen, J., Sato, T., Loi, F., Fan, C., Cordova, L. A., et al. (2016). NF-kappaB decoy oligodeoxynucleotide mitigates wear particle-associated bone loss in the murine continuous infusion model. *Acta Biomaterialia*, *41*, 273–281.
Lin, T. H., Sato, T., Barcay, K. R., Waters, H., Loi, F., Zhang, R., et al. (2015). NF-kappaB decoy oligodeoxynucleotide enhanced osteogenesis in mesenchymal stem cells exposed to polyethylene particle. *Tissue Engineering. Part A*, *21*(5–6), 875–883.
Lin, T. H., Tamaki, Y., Pajarinen, J., Waters, H. A., Woo, D. K., Yao, Z., et al. (2014). Chronic inflammation in biomaterial-induced periprosthetic osteolysis: NF-kappaB as a therapeutic target. *Acta Biomaterialia*, *10*(1), 1–10.
Lin, T. H., Yao, Z., Sato, T., Keeney, M., Li, C., Pajarinen, J., et al. (2014). Suppression of wear-particle-induced pro-inflammatory cytokine and chemokine production in macrophages via NF-kappaB decoy oligodeoxynucleotide: A preliminary report. *Acta Biomaterialia*, *10*(8), 3747–3755.
Lin, Y. Z., Yao, S. Y., Veach, R. A., Torgerson, T. R., & Hawiger, J. (1995). Inhibition of nuclear translocation of transcription factor NF-kappa B by a synthetic peptide containing a cell membrane-permeable motif and nuclear localization sequence. *The Journal of Biological Chemistry*, *270*(24), 14255–14258.
Lo, K. W. H., Ashe, K. M., Kan, H. M., & Laurencin, C. T. (2012). The role of small molecules in musculoskeletal regeneration. *Regenerative Medicine*, 7(4), 535–549.
Loi, F., Cordova, L. A., Pajarinen, J., Lin, T. H., Yao, Z., & Goodman, S. B. (2016). Inflammation, fracture and bone repair. *Bone*, *86*, 119–130.
Lombardo, E., DelaRosa, O., Mancheno-Corvo, P., Menta, R., Ramirez, C., & Buscher, D. (2009). Toll-like receptor-mediated signaling in human adipose-derived stem cells: Implications for immunogenicity and immunosuppressive potential. *Tissue Engineering. Part A*, *15*(7), 1579–1589.
Lu, X., Gilbert, L., He, X., Rubin, J., & Nanes, M. S. (2006). Transcriptional regulation of the osterix (Osx, Sp7) promoter by tumor necrosis factor identifies disparate effects of mitogen-activated protein kinase and NF kappa B pathways. *The Journal of Biological Chemistry*, *281*(10), 6297–6306.
Lu, Z., Wang, G., Dunstan, C. R., Chen, Y., Lu, W. Y., Davies, B., et al. (2013). Activation and promotion of adipose stem cells by tumour necrosis factor-alpha preconditioning for bone regeneration. *Journal of Cellular Physiology*, *228*(8), 1737–1744.

Lu, Z., Wang, G., Dunstan, C. R., & Zreiqat, H. (2012). Short-term exposure to tumor necrosis factor-alpha enables human osteoblasts to direct adipose tissue-derived mesenchymal stem cells into osteogenic differentiation. *Stem Cells and Development, 21*(13), 2420–2429.

Maia, F. R., Bidarra, S. J., Granja, P. L., & Barrias, C. C. (2013). Functionalization of biomaterials with small osteoinductive moieties. *Acta Biomaterialia, 9*(11), 8773–8789.

Mantovani, A., Biswas, S. K., Galdiero, M. R., Sica, A., & Locati, M. (2013). Macrophage plasticity and polarization in tissue repair and remodelling. *The Journal of Pathology, 229*(2), 176–185.

Mao, H. Q., Roy, K., Troung-Le, V. L., Janes, K. A., Lin, K. Y., Wang, Y., et al. (2001). Chitosan-DNA nanoparticles as gene carriers: Synthesis, characterization and transfection efficiency. *Journal of Controlled Release, 70*(3), 399–421.

Marsell, R., & Einhorn, T. A. (2011). The biology of fracture healing. *Injury, 42*(6), 551–555.

Martensson, K., Chrysis, D., & Savendahl, L. (2004). Interleukin-1beta and TNF-alpha act in synergy to inhibit longitudinal growth in fetal rat metatarsal bones. *Journal of Bone and Mineral Research, 19*(11), 1805–1812.

Martinez, F. O., Sica, A., Mantovani, A., & Locati, M. (2008). Macrophage activation and polarization. *Frontiers in Bioscience, 13*, 453–461.

Matzinger, P. (2002). The danger model: A renewed sense of self. *Science, 296*(5566), 301–305.

Miller, K., Eldar-Boock, A., Polyak, D., Segal, E., Benayoun, L., Shaked, Y., et al. (2011). Antiangiogenic antitumor activity of HPMA copolymer-paclitaxel-alendronate conjugate on breast cancer bone metastasis mouse model. *Molecular Pharmaceutics, 8*(4), 1052–1062.

Min, J., Braatz, R. D., & Hammond, P. T. (2014). Tunable staged release of therapeutics from layer-by-layer coatings with clay interlayer barrier. *Biomaterials, 35*(8), 2507–2517.

Miyake, T., Ihara, S., Miyake, T., Tsukada, Y., Watanabe, H., Matsuda, H., et al. (2014). Prevention of neointimal formation after angioplasty using nuclear factor-kappaB decoy oligodeoxynucleotide-coated balloon catheter in rabbit model. *Circulation. Cardiovascular Interventions*, 7(6), 787–796.

Mo, I. F., Yip, K. H., Chan, W. K., Law, H. K., Lau, Y. L., & Chan, G. C. (2008). Prolonged exposure to bacterial toxins downregulated expression of toll-like receptors in mesenchymal stromal cell-derived osteoprogenitors. *BMC Cell Biology, 9*, 52.

Morioka, M., Kamizono, A., Takikawa, H., Mori, A., Ueno, H., Kadowaki, S.-i., et al. (2010). Design, synthesis, and biological evaluation of novel estradiol–bisphosphonate conjugates as bone-specific estrogens. *Bioorganic & Medicinal Chemistry, 18*(3), 1143–1148.

Mountziaris, P. M., Dennis Lehman, E., Mountziaris, I., Sing, D. C., Kasper, F. K., & Mikos, A. G. (2013). Effect of temporally patterned TNF-alpha delivery on in vitro osteogenic differentiation of mesenchymal stem cells cultured on biodegradable polymer scaffolds. *Journal of Biomaterials Science Polymer Edition, 24*(15), 1794–1813.

Mountziaris, P. M., Tzouanas, S. N., & Mikos, A. G. (2010). Dose effect of tumor necrosis factor-alpha on in vitro osteogenic differentiation of mesenchymal stem cells on biodegradable polymeric microfiber scaffolds. *Biomaterials, 31*(7), 1666–1675.

Neale, J. R., Richter, N. B., Merten, K. E., Taylor, K. G., Singh, S., Waite, L. C., et al. (2009). Bone selective effect of an estradiol conjugate with a novel tetracycline-derived bone-targeting agent. *Bioorganic & Medicinal Chemistry Letters, 19*(3), 680–683.

Nguyen, N. D., Ahlborg, H. G., Center, J. R., Eisman, J. A., & Nguyen, T. V. (2007). Residual lifetime risk of fractures in women and men. *Journal of Bone and Mineral Research, 22*(6), 781–788.

Novack, D. V. (2011). Role of NF-kappaB in the skeleton. *Cell Research, 21*(1), 169–182.

Novack, D. V., Yin, L., Hagen-Stapleton, A., Schreiber, R. D., Goeddel, D. V., Ross, F. P., et al. (2003). The IkappaB function of NF-kappaB2 p100 controls stimulated osteoclastogenesis. *Journal of Experimental Medicine, 198*(5), 771–781.

Orriss, I. R., Key, M. L., Colston, K. W., & Arnett, T. R. (2009). Inhibition of osteoblast function in vitro by aminobisphosphonates. *Journal of Cellular Biochemistry, 106*(1), 109–118.

Osako, M. K., Nakagami, H., & Morishita, R. (2012). Modification of decoy oligodeoxynucleotides to achieve the stability and therapeutic efficacy. *Current Topics in Medicinal Chemistry, 12*(15), 1603–1607.

Osta, B., Benedetti, G., & Miossec, P. (2014). Classical and paradoxical effects of TNF-alpha on bone homeostasis. *Frontiers in Immunology, 5*(48), 48.

Panteli, M., Pountos, I., Jones, E., & Giannoudis, P. V. (2015). Biological and molecular profile of fracture non-union tissue: Current insights. *Journal of Cellular and Molecular Medicine, 19*(4), 685–713.

Panyam, J., Zhou, W. Z., Prabha, S., Sahoo, S. K., & Labhasetwar, V. (2002). Rapid endo-lysosomal escape of poly(DL-lactide-co-glycolide) nanoparticles: Implications for drug and gene delivery. *FASEB Journal, 16*(10), 1217–1226.

Pazianas, M., & Abrahamsen, B. (2011). Safety of bisphosphonates. *Bone, 49*(1), 103–110.

Pearl, J. I., Ma, T., Irani, A. R., Huang, Z., Robinson, W. H., Smith, R. L., et al. (2011). Role of the Toll-like receptor pathway in the recognition of orthopedic implant wear-debris particles. *Biomaterials, 32*(24), 5535–5542.

Pioletti, D. P., & Kottelat, A. (2004). The influence of wear particles in the expression of osteoclastogenesis factors by osteoblasts. *Biomaterials, 25*(27), 5803–5808.

Purdue, P. E., Koulouvaris, P., Potter, H. G., Nestor, B. J., & Sculco, T. P. (2007). The cellular and molecular biology of periprosthetic osteolysis. *Clinical Orthopaedics and Related Research, 454*, 251–261.

Quintana, F. J., & Cohen, I. R. (2005). Heat shock proteins as endogenous adjuvants in sterile and septic inflammation. *Journal of Immunology, 175*(5), 2777–2782.

Rad, S. M., Langroudi, L., Kouhkan, F., Yazdani, L., Koupaee, A. N., Asgharpour, S., et al. (2015). Transcription factor decoy: A pre-transcriptional approach for gene down-regulation purpose in cancer. *Tumour Biology, 36*(7), 4871–4881.

Raggatt, L. J., Wullschleger, M. E., Alexander, K. A., Wu, A. C., Millard, S. M., Kaur, S., et al. (2014). Fracture healing via periosteal callus formation requires macrophages for both initiation and progression of early endochondral ossification. *American Journal of Pathology, 184*(12), 3192–3204.

Raphel, J., Holodniy, M., Goodman, S. B., & Heilshorn, S. C. (2016). Multifunctional coatings to simultaneously promote osseointegration and prevent infection of orthopaedic implants. *Biomaterials, 84*, 301–314.

Reinholz, M. M., Zinnen, S. P., Dueck, A. C., Dingli, D., Reinholz, G. G., Jonart, L. A., et al. (2010). A promising approach for treatment of tumor-induced bone diseases: Utilizing bisphosphonate derivatives of nucleoside antimetabolites. *Bone, 47*(1), 12–22.

Rosenfeld, M. E., Prichard, L., Shiojiri, N., & Fausto, N. (2000). Prevention of hepatic apoptosis and embryonic lethality in RelA/TNFR-1 double knockout mice. *American Journal of Pathology, 156*(3), 997–1007.

Rudolph, D., Yeh, W. C., Wakeham, A., Rudolph, B., Nallainathan, D., Potter, J., et al. (2000). Severe liver degeneration and lack of NF-kappaB activation in NEMO/IKKgamma-deficient mice. *Genes & Development, 14*(7), 854–862.

Ruocco, M. G., Maeda, S., Park, J. M., Lawrence, T., Hsu, L. C., Cao, Y., et al. (2005). I{kappa}B kinase (IKK){beta}, but not IKK{alpha}, is a critical mediator of osteoclast survival and is required for inflammation-induced bone loss. *Journal of Experimental Medicine, 201*(10), 1677–1687.

Russell, R. G. G. (2011). Bisphosphonates: The first 40 years. *Bone, 49*(1), 2–19.

Salminen, A., Huuskonen, J., Ojala, J., Kauppinen, A., Kaarniranta, K., & Suuronen, T. (2008). Activation of innate immunity system during aging: NF-kB signaling is the molecular culprit of inflamm-aging. *Ageing Research Reviews,* 7(2), 83–105.

Scheidereit, C. (2006). IkappaB kinase complexes: Gateways to NF-kappaB activation and transcription. *Oncogene*, *25*(51), 6685–6705.
Serhan, C. N., & Savill, J. (2005). Resolution of inflammation: The beginning programs the end. *Nature Immunology*, *6*(12), 1191–1197.
Simonet, W. S., Lacey, D. L., Dunstan, C. R., Kelley, M., Chang, M. S., Luthy, R., et al. (1997). Osteoprotegerin: A novel secreted protein involved in the regulation of bone density. *Cell*, *89*(2), 309–319.
Soysa, N. S., & Alles, N. (2009). NF-kappaB functions in osteoclasts. *Biochemical and Biophysical Research Communications*, *378*(1), 1–5.
Suda, T., Takahashi, N., Udagawa, N., Jimi, E., Gillespie, M. T., & Martin, T. J. (1999). Modulation of osteoclast differentiation and function by the new members of the tumor necrosis factor receptor and ligand families. *Endocrine Reviews*, *20*(3), 345–357.
Sun, S. C. (2011). Non-canonical NF-kappaB signaling pathway. *Cell Research*, *21*(1), 71–85.
Tahara, K., Samura, S., Tsuji, K., Yamamoto, H., Tsukada, Y., Bando, Y., et al. (2011). Oral nuclear factor-kappaB decoy oligonucleotides delivery system with chitosan modified poly(D,L-lactide-co-glycolide) nanospheres for inflammatory bowel disease. *Biomaterials*, *32*(3), 870–878.
Takami, M., Kim, N., Rho, J., & Choi, Y. (2002). Stimulation by toll-like receptors inhibits osteoclast differentiation. *Journal of Immunology*, *169*(3), 1516–1523.
Takayanagi, H., Kim, S., Koga, T., Nishina, H., Isshiki, M., Yoshida, H., et al. (2002). Induction and activation of the transcription factor NFATc1 (NFAT2) integrate RANKL signaling in terminal differentiation of osteoclasts. *Developmental Cell*, *3*(6), 889–901.
Takeuchi, O., & Akira, S. (2010). Pattern recognition receptors and inflammation. *Cell*, *140*(6), 805–820.
Taylor, P. C., & Feldmann, M. (2009). Anti-TNF biologic agents: Still the therapy of choice for rheumatoid arthritis. *Nature Reviews. Rheumatology*, *5*(10), 578–582.
Thamake, S. I., Raut, S. L., Gryczynski, Z., Ranjan, A. P., & Vishwanatha, J. K. (2012). Alendronate coated poly-lactic-co-glycolic acid (PLGA) nanoparticles for active targeting of metastatic breast cancer. *Biomaterials*, *33*(29), 7164–7173.
Theoleyre, S., Wittrant, Y., Tat, S. K., Fortun, Y., Redini, F., & Heymann, D. (2004). The molecular triad OPG/RANK/RANKL: Involvement in the orchestration of pathophysiological bone remodeling. *Cytokine & Growth Factor Reviews*, *15*(6), 457–475.
Torchilin, V. P. (2000). Drug targeting. *European Journal of Pharmaceutical Sciences*, *11*(Suppl. 2), S81–S91.
Torgerson, T. R., Colosia, A. D., Donahue, J. P., Lin, Y. Z., & Hawiger, J. (1998). Regulation of NF-kappa B, AP-1, NFAT, and STAT1 nuclear import in T lymphocytes by noninvasive delivery of peptide carrying the nuclear localization sequence of NF-kappa B p50. *Journal of Immunology*, *161*(11), 6084–6092.
Torres Martin de Rosales, R., Finucane, C., Mather, S. J., & Blower, P. J. (2009). Bifunctional bisphosphonate complexes for the diagnosis and therapy of bone metastases. *Chemical Communications*, *32*, 4847–4849.
Vaira, S., Alhawagri, M., Anwisye, I., Kitaura, H., Faccio, R., & Novack, D. V. (2008). RelA/p65 promotes osteoclast differentiation by blocking a RANKL-induced apoptotic JNK pathway in mice. *The Journal of Clinical Investigation*, *118*(6), 2088–2097.
Vaira, S., Johnson, T., Hirbe, A. C., Alhawagri, M., Anwisye, I., Sammut, B., et al. (2008). RelB is the NF-kappaB subunit downstream of NIK responsible for osteoclast differentiation. *Proceedings of the National Academy of Sciences of the United States of America*, *105*(10), 3897–3902.
van Bezooijen, R. L., Roelen, B. A., Visser, A., van der Wee-Pals, L., de Wilt, E., Karperien, M., et al. (2004). Sclerostin is an osteocyte-expressed negative regulator of bone formation, but not a classical BMP antagonist. *Journal of Experimental Medicine*, *199*(6), 805–814.

Wahl, C., Liptay, S., Adler, G., & Schmid, R. M. (1998). Sulfasalazine: A potent and specific inhibitor of nuclear factor kappa B. *The Journal of Clinical Investigation, 101*(5), 1163–1174.

Wang, D., Miller, S. C., Shlyakhtenko, L. S., Portillo, A. M., Liu, X. M., Papangkorn, K., et al. (2007). Osteotropic peptide that differentiates functional domains of the skeleton. *Bioconjugate Chemistry, 18*(5), 1375–1378.

Wang, M. L., Nesti, L. J., Tuli, R., Lazatin, J., Danielson, K. G., Sharkey, P. F., et al. (2002). Titanium particles suppress expression of osteoblastic phenotype in human mesenchymal stem cells. *Journal of Orthopaedic Research, 20*(6), 1175–1184.

Wei, S., Kitaura, H., Zhou, P., Ross, F. P., & Teitelbaum, S. L. (2005). IL-1 mediates TNF-induced osteoclastogenesis. *The Journal of Clinical Investigation, 115*(2), 282–290.

Whiteside, S. T., & Israel, A. (1997). I kappa B proteins: Structure, function and regulation. *Seminars in Cancer Biology, 8*(2), 75–82.

Widler, L., Jahnke, W., & Green, J. R. (2012). The chemistry of bisphosphonates: From antiscaling agents to clinical therapeutics. *Anti-Cancer Agents of Medicinal Chemistry, 12*(2), 95–101.

Wu, J. Y., Scadden, D. T., & Kronenberg, H. M. (2009). Role of the osteoblast lineage in the bone marrow hematopoietic niches. *Journal of Bone and Mineral Research, 24*(5), 759–764.

Xu, J., Wu, H. F., Ang, E. S., Yip, K., Woloszyn, M., Zheng, M. H., et al. (2009). NF-kappaB modulators in osteolytic bone diseases. *Cytokine & Growth Factor Reviews, 20*(1), 7–17.

Yao, Z., Xing, L., & Boyce, B. F. (2009). NF-kappaB p100 limits TNF-induced bone resorption in mice by a TRAF3-dependent mechanism. *The Journal of Clinical Investigation, 119*(10), 3024–3034.

Yu, B., Chang, J., Liu, Y., Li, J., Kevork, K., Al-Hezaimi, K., et al. (2014). Wnt4 signaling prevents skeletal aging and inflammation by inhibiting nuclear factor-kappaB. *Nature Medicine, 20*(9), 1009–1017.

Zhang, G., Guo, B., Wu, H., Tang, T., Zhang, B. T., Zheng, L., et al. (2012). A delivery system targeting bone formation surfaces to facilitate RNAi-based anabolic therapy. *Nature Medicine, 18*(2), 307–314.

Zhao, Z., Hou, X., Yin, X., Li, Y., Duan, R., Boyce, B. F., et al. (2015). TNF induction of NF-kappaB RelB enhances RANKL-induced osteoclastogenesis by promoting inflammatory macrophage differentiation but also limits it through suppression of NFATc1 expression. *PloS One, 10*(8), e0135728.

Zhao, L., Huang, J., Zhang, H., Wang, Y., Matesic, L. E., Takahata, M., et al. (2011). Tumor necrosis factor inhibits mesenchymal stem cell differentiation into osteoblasts via the ubiquitin E3 ligase Wwp1. *Stem Cells, 29*(10), 1601–1610.

Zwerina, J., Redlich, K., Polzer, K., Joosten, L., Kronke, G., Distler, J., et al. (2007). TNF-induced structural joint damage is mediated by IL-1. *Proceedings of the National Academy of Sciences of the United States of America, 104*(28), 11742–11747.

CHAPTER FIVE

The Functional Stability of FOXP3 and RORγt in Treg and Th17 and Their Therapeutic Applications

J. Ren*, B. Li*,†,1

*Key Laboratory of Molecular Virology and Immunology, CAS Center for Excellence in Molecular Cell Science, Unit of Molecular Immunology, Institut Pasteur of Shanghai, Shanghai Institutes for Biological Sciences, Chinese Academy of Sciences, Universities and Colleges Admissions Service, Shanghai, PR China
†Shanghai Institute of Immunology, Shanghai JiaoTong University School of Medicine, Shanghai, PR China
[1]Corresponding author: e-mail address: binli@sibs.ac.cn

Contents

Abstract

The balance of $CD4^+CD25^+FOXP3^+$ regulatory T cells (Tregs) and effector T cells plays a key role in maintaining immune homeostasis, while the imbalance of them is related to many inflammatory diseases in both human and mice. Here we discuss about the plasticity of Tregs and Th17 cells, and the related human diseases resulted from the imbalance of them. Further, we will focus on the mechanisms regulating the plasticity

Advances in Protein Chemistry and Structural Biology, Volume 107
ISSN 1876-1623
http://dx.doi.org/10.1016/bs.apcsb.2016.10.002

between Tregs and Th17 cells and the potential therapeutic strategies by targeting regulators of the expression and activity of FOXP3 and RORγt or regulators of Treg/Th17 balance in autoimmune diseases, allergy, infection, and cancer.

1. INTRODUCTION

1.1 $CD4^+CD25^+FOXP3^+$ Tregs

Maintaining immune homeostasis needs precise control that allows the immune system to effectively eliminate excessive inflammation caused by pathogens and environmental insults, and meanwhile tolerates self-antigens. Regulatory T cells play a pivotal role in maintaining immunological self-tolerance and excessive immune responses by actively suppressing both self-reactive and foreign antigen-reactive lymphocytes. Forkhead box P3 (FOXP3) was first identified as a master regulator in Treg development by Sakaguchi, Rudensky, Ramsdell, and their colleagues (Fontenot, Gavin, & Rudensky, 2003; Hori, Nomura, & Sakaguchi, 2003; Khattri, Cox, Yasayko, & Ramsdell, 2003). Point mutation of FOXP3 can lead to lethal X-linked syndrome "Immune Dysregulation, Polyendocrinopathy, Enteropathy, X-Linked Syndrome" (IPEX) in human and the scurfy phenotype in mice, which associate with a lot of multiorgan autoimmune disease such as type I diabetes, inflammatory bowel disease (IBD) (Bennett et al., 2001). Therefore, $CD4^+CD25^+FOXP3^+$ Tregs are considered as a $CD4^+$ T cell subset with suppressive activity. Actually, apart from Tregs, there are other T cells that harbor immunosuppressive activity including IL-10- and TGF-β-secreting $CD4^+$ T cells including Tr1 cells (Chen, Kuchroo, Inobe, Hafler, & Weiner, 1994), TGF-β-producing Th3 cells (Sakaguchi, Yamaguchi, Nomura, & Ono, 2008), and also $CD8^+$, $CD4^-CD8^-$, and γ/δ T cells (Shevach et al., 2006).

In this review, we will focus on $CD4^+CD25^+FOXP3^+$ Tregs, which can directly develop from thymus called natural regulatory T cells (nTreg) or generated in periphery from naïve $CD4^+$ T cells called induced regulatory T cells (iTreg) (Sakaguchi et al., 2008). Of note, both nTreg and iTreg need transcription factor FOXP3 to develop and maintain the suppression function; interestingly, nTregs are already functionally mature due to antigen-primed in thymus in contrast to other thymus-produced naïve T cells (Itoh et al., 1999; Sakaguchi, Takahashi, & Nishizuka, 1982). Evidence suggests that nTregs develop in the thymus requires a strong signal via TCR (Koonpaew, Shen, Flowers, & Zhang, 2006), they have TCRs with higher affinity for thymic MHC/self-peptide ligands and would be committed to the Treg cell lineage during thymic T cell selection. Besides, a great number of T cell accessory molecules and their ligands such as CD28, CD40, CD11a/CD18 (LFA1), or CD80 and CD86 (B7) and so on also contribute to the development of nTregs (Sakaguchi, 2005). Also, medullary thymic epithelial cells (mTECs) and dendritic cells are also involved in generation of nTregs. For example, ectopically expression of tissue-specific antigens (TSAs) by mTECs may contribute to generation of TSA-specific nTregs (Aschenbrenner et al., 2007). For iTregs, TGF-β, retinoic acid, and IL-2 are the extracellular signaling molecules required for theirs generation of in the periphery (Benson, Pino-Lagos, Rosemblatt, & Noelle, 2007; Coombes et al., 2007; Sun et al., 2007).

Mechanisms by which Treg-mediated immune suppression include secreting inhibitory cytokines, cytolysis, disrupting metabolism, targeting dendritic cell (Vignali, Collison, & Workman, 2008), and by secreting miRNA-containing exosomes (Okoye et al., 2014). Notably, Treg is not a terminally differentiated cell subset, it exhibits its plasticity by acquiring phenotypes of other helper T cells and becomes T-helper-like Tregs including Th1-like T-bet$^+$IFN-γ$^+$ Tregs, Th2-like IRF4$^+$GATA3$^+$IL-4$^+$ Tregs, Tfh-like CXCR5$^+$Bcl-6$^+$ Tregs, and Th17-like RORγt$^+$IL-17$^+$ Tregs (Chung et al., 2011; Koch et al., 2009; Krishnamoorthy et al., 2012; Tartar et al., 2010). Supporting this, accumulating evidence shows that Tregs can transdifferentiate into proinflammatory cytokine-producing T-helper-like cells under inflammation (Beriou et al., 2009; Du, Zhao, Yan, & Li, 2014; Duarte, Zelenay, Bergman, Martins, & Demengeot, 2009; Hovhannisyan, Treatman, Littman, & Mayer, 2011; Pesenacker, Broady, & Levings, 2015; Takahashi & Yoshimura, 2014; Voo et al., 2009). However, it is still not fully understood whether the T-helper-like Tregs retain their immunosuppressive activity or not (Beriou et al., 2009; Bovenschen et al., 2011; Koenen et al., 2008; Voo et al., 2009). Besides, Treg may also lose FOXP3 expression and becomes "exTreg" that no longer exerts suppression (Zhou, Bailey-Bucktrout, Jeker, & Bluestone, 2009).

1.2 Th17

In a long time, Th1/Th2 paradigm has taken a leading role in explaining the mechanism underlying the development of multiple autoimmune diseases; however, there are some studies that could not explained by Th1/Th2 responses, such as lack of STAT1 and IL-12 receptor β2 in animals lead to increased susceptibility to experimental autoimmune encephalomyelitis (EAE) (Bettelli et al., 2004; Zhang et al., 2003). IL-17-producing Th17 cells exhibit distinct development pattern compared with that in Th1 and Th2 cells, of which the most important feature is that RORγt directs the developmental program of Th17 cells (Harrington et al., 2005; Ivanov et al., 2006; Park et al., 2005).

Upon activation in vitro, naïve $CD4^+$ T cells differentiated into Th17 under the presence of IL-6 plus TGF-β (Bettelli et al., 2006; Veldhoen, Hocking, Atkins, Locksley, & Stockinger, 2006). Notably, the phenotypic feature of Th17 cells is the production of IL-17 including IL-17A, IL-17B, IL-17C, IL-17D, IL-17E, and IL-17F, which play a fundamental role in Th17-mediated immune response either protecting host from extracellular pathogen infection or mediating autoimmune diseases (Iizuka et al., 2014; Raffatellu et al., 2008).

IL-17-producing Th17 is identified as effector T cells to exacerbate tissue inflammation through triggering other proinflammatory cytokines and chemokines in a great number of autoimmune diseases including rheumatoid arthritis (RA), psoriasis, systemic lupus erythematosus, multiple sclerosis (MS), and inflammatory colitis (Korn, Bettelli, Oukka, & Kuchroo, 2009). Actually, not all IL-17-producing Th17 cells are pathogenic; for instance, regulatory Th17 cells with immune suppressive phenotype features have been described in small intestine (SI) (Korn et al., 2009). Mechanistically, Th17 cells need IL-23 receptor (IL-23R) to become pathogenic (Bettelli et al., 2006; Huang, Wang, & Chi, 2012), the transcriptional mechanism is recently described by Meyer Zu Horste et al., that is, canonical Notch signaling mediator RBPJ binds to *IL23r* promoter to induce IL-23R expression, and on the other hand also constrains antiinflammatory IL-10 production by Th17 cells (Meyer Zu Horste et al., 2016).

2. REGULATION OF FOXP3 FUNCTION IN Tregs

2.1 Transcriptional Regulation of FOXP3 Expression

It has been reported that a number of factors can influence the expression of FOXP3 including conserved noncoding DNA sequence (CNS) at the foxp3 gene locus, and transcription factors such as NF-κB, NFAT, SMAD3, STAT5, and so on.

Many studies have demonstrated that *Foxp3* locus comprises three proximal intronic CNS elements, CNS1, CNS2, and CNS3, which plays a role in the expression of FOXP3 both in iTregs and mature nTreg cells in a

TGF-β-induced manner, and contributes to their generation in the periphery and thymus, respectively (Kim et al., 2007; Tone et al., 2008; Zheng et al., 2010). Using transgenic mice with increased or decreased NF-κB activity, Long et al. showed that TCR-induced NF-κB pathway upregulates FOXP3 expression (Long, Park, Strickland, Hayden, & Ghosh, 2009). NFAT family contains five members including NFAT1, NFAT2, NFAT3, NFAT4, and NFAT5, among which NFAT1 and NFAT2 together with SMAD contribute to FOXP3 expression via binding to promoter and enhancer region of FOXP3 (Mantel et al., 2006; Ruan et al., 2009; Tone et al., 2008). Additionally, TGF-β plus TCR stimulation triggers naïve $CD4^{+}CD25^{-}FOXP3^{-}$ T cells to differentiate into $FOXP3^{+}$ Tregs, via a SMAD-dependent pathway in mice (Bettelli et al., 2006; Chen et al., 2003; Luo et al., 2007). In contrast, the function of TGF-β signaling on nTregs development remains elucidated. Marie et al. showed that TGF-β1 is required for maintaining FOXP3 expression and regulatory function of peripheral $CD4^{+}CD25^{+}$ Tregs, but is not required for their thymic development (Marie, Letterio, Gavin, & Rudensky, 2005). However, evidence from Liu et al. suggests that TGF-β is critical for nTreg development in the thymus (Liu et al., 2008). RUNX proteins are crucial for differentiation and proliferation of various cells. Indeed, previous evidence found that complexes of the transcription factors RUNX and CBFβ play a pivotal role in inducing and maintaining FOXP3 expression in nTregs and iTregs (Kitoh et al., 2009; Rudra et al., 2009).

2.2 Posttranslational Regulation of FOXP3

FOXP3 can be regulated by various posttranscriptional modifications including ubiquitinylation, acetylation, and phosphorylation, which can influence both its stability and function.

2.2.1 Ubiquitinylation

FOXP3 can be regulated by ubiquitinylation and deubiquitinylation (Chen, Barbi, et al., 2013; Chen, Wu, Pier, Zhao, & Shen, 2013; Dang et al., 2011; van Loosdregt et al., 2013, 2010). A FOXP3 E3 ligase STUB1 have recently been identified by Chen et al., it targets FOXP3 to degradation by promoting K48-linked polyubiquitinylation of FOXP3 in a heat shock protein 70-dependent manner. Furthermore, ectopic expression of STUB1 impaired Treg-mediated suppression in vitro and in vivo, and endowed Tregs a Th1-like phenotype (Chen, Wu, et al., 2013). Conversely, USP7 has been characterized as deubiquitinylase of FOXP3 in Tregs. Enforced expression of USP7 can reduce the polyubiquitinylation of FOXP3 and increase the FOXP3 expression. In contrast, knockdown of USP7 impaired the suppressive activity of Tregs both in vitro and in vivo (van Loosdregt et al., 2013). Besides, USP21 has been found to deubiquitinylase and stabilize GATA3 in Tregs, thus contributing to limiting inflammation by Tregs (Zhang et al., 2013). Based on the role STUB1 and USP7 play in regulating the stability of FOXP3, targeting STUB1 or USP7 may be beneficial in inflammation conditions and cancer, respectively.

2.2.2 Acetylation

It has now been well established that FOXP3 acetylation is associated with DNA-binding activity and FOXP3 protein stability. There are two histone acetyltransferases (HATs) involving in FOXP3 acetylation, Tat-interacting protein (TIP) 60 (Li et al., 2007; van Loosdregt et al., 2010), and p300 (van Loosdregt et al., 2010). Functionally, Treg-specific deletion of TIP60 or p300 showed impaired immunosuppressive function of Tregs (Xiao, Nagai, et al., 2014). Inversely, FOXP3 can be deacetylated by SIRT1 (Beier et al., 2011, 2012; van Loosdregt et al., 2011, 2010), HDAC6 (Beier et al., 2012; de Zoeten et al., 2011), HDAC7 (Li et al., 2007), and HDAC9 (Beier et al., 2012; de Zoeten, Wang, Sai, Dillmann, & Hancock, 2010). All of them could affect the suppressive activity of Tregs, among which SIRT1 and HDAC6 could also affect the protein stability of FOXP3. TGF-β and IL-6 are the upstream signals regulating the acetylation of FOXP3. Li et al. found that TGF-β could increase the acetylated level of FOXP3 and promote the recruitment of FOXP3 to IL-2 promoter. Inversely, IL-6 treatment reverses this effect, suggesting the inflammatory signals downregulate Treg function partially through regulating FOXP3 acetylation. Further, histone deacetylation inhibitor (HDACi) treatment abolished the decreased acetylation of FOXP3 by TGF-β and IL-6, suggesting TGF-β and IL-6 may through upregulating HDACs to deacetylate FOXP3 (Samanta et al., 2008). Therefore, the HDACi can be a potential drug for promoting Treg function. Actually, HDACi therapy in various animal disease models including arthritis, IBD, and graft-vs-host disease showed effective therapeutic effects (Saouaf et al., 2009; Tao et al., 2007). Moreover, in human Tregs and RA patients, HDACi also proved to be effective (Akimova et al., 2010; Grabiec et al., 2010). Due to the important role of acetylation in Treg function, both HATs and HDACs are potential targets to enhance or reduce Treg suppressive activity in various diseases.

2.2.3 *Phosphorylation*

Nie et al. found that FOXP3 can be phosphorylated at serine 418 site upon TCR stimulation, which is decreased by TNF-α treatment, functionally, decreased phosphorylation at serine 418 site leads to impaired binding of IL-2 promotor by FOXP3, and deficient suppressive activity of Tregs (Nie et al., 2013). Besides, phosphorylation of FOXP3 can also regulate the stability of itself. Morawski et al. showed that mutation of the four CDK phosphosites to alanine residues reduced FOXP3 degradation, and thus stabilized the protein level of FOXP3 (Morawski, Mehra, Chen, Bhatti, & Wells, 2013). How FOXP3 phosphorylation results in increased protein stability is an open question, but it is possible that phosphorylation of FOXP3 can modulate its polyubiquitinylation, and lead to "phosphorylation-dependent ubiquitinylation." Recently, Li et al. showed that PIM1 could mediate phosphorylation of human FOXP3 at serine 422 site and could negatively regulate the DNA-binding activity of FOXP3 (Li, Lin, et al., 2014). Moreover, kaempferol has been found to reduce PIM1-mediated FOXP3 phosphorylation at serine 422 site, and further to enhance the function of Tregs. Notably, kaempferol treatment ameliorated the pathological symptoms in a collagen-induced arthritis rat model (Lin et al., 2015). Later, PIM2, another member of PIM family kinases, has also been found to be able to phosphorylate FOXP3, which correlates to the lineage stability and suppression function of Tregs (Deng et al., 2015). Therefore, the kinases of FOXP3, especially those negative regulators, could be a therapeutic target for inflammatory-associated diseases, here, kaempferol act as an example by targeting PIM1. Actually, compared with ubiquitinylation and acetylation, phosphorylation of FOXP3 is less well studied, so more studies for identifying the kinases and phosphatases need to be done in future to deepen our understandings of phosphorylation regulation in Tregs.

3. REGULATION OF RORγt FUNCTION IN Th17

3.1 Regulation of RORγt Stability in Th17

RORγt is an isoform of RORγ that belongs to retinoid acid-related orphan receptor (ROR) subgroup, it is best known as the mast regulator of Th17 cells (Ivanov et al., 2006) and the third group of innate lymphoid cells (ILC3s) (Spits et al., 2013; Spits & Di Santo, 2011), and more broadly the transcription factor controlling IL-17 production in other cells (Ivanov, Zhou, & Littman, 2007; Korn et al., 2009). The expression of RORγt is restricted to immune tissues and cells (Hirose, Smith, & Jetten, 1994; Jetten, 2009), developmentally, RORγt is induced at the transition stage of thymic T cells from double negative to the double positive state and is required for thymocyte survival and lymphoid organogenesis (Kurebayashi et al., 2000; Sun et al., 2000).

Because ubiquitinylation and acetylation will compete for the same lysine residues, acetylation can protect proteins from ubiquitin-mediated proteasomal degradation. Thus, the regulation of RORγt acetylation can also regulate the stability of RORγt. RORγt has been reported to be acetylated by p300 and reciprocally deacetylated by SIRT1 and HDAC1 (Lim et al., 2015; Wu et al., 2015).

Ubiquitinylation and deubiquitinylation is an inverse process that can regulate protein stability dynamically. K48-linked polyubiquitinylation can trigger the degradation of the target proteins. Several E3 ligases have been identified to regulate RORγt activity including TRAF5, ITCH, and UBR5, among which TRAF5 promotes RORγt function by K63-linked polyubiquitinylation of RORγt (Wang, Yang, et al., 2015), whereas ITCH and UBR5 targets RORγt for proteasome degradation (Kathania et al., 2016; Rutz et al., 2015). Besides E3 ligases, several deubiquitinases have also been identified to directly interact with and stabilize RORγt including USP17 and USP4 (Han et al., 2014; Yang, Qu, et al., 2015). Both USP17 and USP4 reduce K48-linked polyubiquitinylation of RORγt. Moreover, it is also found that TGF-β plus IL-6 which are important signals for Th17 differentiation could enhance the deubiquitinylation mediated by USP4, suggesting USP4 may play a role in Th17 development. Another deubiquitinase DUBA (OTUD5) has also been shown to regulate RORγt stability, but it do not directly interact with RORγt (Rutz et al., 2015).

3.2 Functional Regulation of RORγt and Th17 Cells by Posttranslational Modifications

3.2.1 Ubiquitinylation

Degradation-associated ubiquitinylation has been discussed earlier, so here we just focus on the functional regulation of RORγt by ubiquitinylation. TRAF5 mediate K63-linked polyubiquitinylation of RORγt (Wang, Yang, et al., 2015), likely to modulate the transcriptional activity of RORγt, evidenced by reduced production of IL-17A and IL-17F in TRAF5 down-regulated human Th17 cells. Deubiquitinases have also been found to regulate the function of Th17 cells. The deubiquitinase USP25 targets TRAF5 and TRAF6 to indirectly regulate the transcriptional activity of RORγt

(Zhong et al., 2012). Another deubiquitinase USP18 has been found to regulate Th17 differentiation by deubiquitinylating TAK1-TAB1 complex (Liu et al., 2013).

3.2.2 Acetylation

Acetylation of transcription factors could affect their degradation, subcellular localization, and DNA-binding activity. The regulation of RORγt by p300 might affect the stability and more likely the DNA-binding activity of RORγt (Wu et al., 2015). Naïve $CD4^+$ T cells transduced with a construct mimicking constitutively acetylated RORγt have a deficiency in Th17 polarization, suggesting a role acetylation plays in Th17 differentiation (Ivanov et al., 2007). Genetic ablation or pharmacological inhibition of SIRT1 that can deacetylate RORγt ameliorates the severity of EAE (Ivanov et al., 2007), suggesting it may play a role in regulating the transcriptional activity of RORγt in Th17 cells.

3.2.3 Phosphorylation

Phosphorylation has shown to be able to decrease the ligand-binding affinity (Anbalagan, Huderson, Murphy, & Rowan, 2012; Faus & Haendler, 2006; Rochette-Egly, 2003). Considering the ligand-binding feature of RORγt, phosphorylation could possibly regulate the ligand-binding affinity of RORγt with its ligands. However, there are just several phosphorylation sites of RORγt that have been detected (Lim et al., 2015), and no related kinases or phosphatases have been identified.

3.3 Cofactors of RORγt

The interactions of RORγt with different cofactors lead to totally different results in target gene expression; hence, a comprehensive understanding of the cofactor landscape will provide important information for precisely targeting RORγt activity.

First, coactivator complexes can promote transcription via epigenetic changes, whereas corepressors can recruit deacetylation for exhibiting its function (Glass & Rosenfeld, 2000; Perissi & Rosenfeld, 2005; Xu, 2005). Previous studies have identified a variety of coactivators of RORγt including NCOA1 (SRC-1), NCOA2 (TIF2 or GRIP1), PGC-1α, p300, and cholesterol biosynthetic pathway (CBP); and corepressors including NCOR1, NCOR2, RIP140, and NIX1 to form complexes with ROR proteins (Atkins et al., 1999; Gold et al., 2003; Greiner et al., 2000; Harding, Atkins, Jaffe, Seo, & Lazar, 1997; Harris, Lau, Chen, & Muscat, 2002;

Jetten, 2004; Jetten & Joo, 2006; Johnson, Lovett, Hirsch, Xia, & Chen, 2004; Kurebayashi et al., 2004; Lau, Bailey, Dowhan, & Muscat, 1999; Lau, Nixon, Parton, & Muscat, 2004; Littman et al., 1999; Xie, Sadim, & Sun, 2005).

Transcription factors are also involved in regulation of RORγt. Dong and colleagues established that FOXP3 can inhibit RORγt transcriptional activity during Th17 differentiation (Du, Huang, Zhou, & Ziegler, 2008). Later, the transcription factor RUNX1, IkappaBzeta, and hypoxia-inducible factor 1 (HIF-1α) were identified to be positive regulators for Th17 cell development (Dang et al., 2011; Okamoto et al., 2010; Zhang, Meng, & Strober, 2008). Later, Ciofani et al. have delineated a regulatory network in Th17 cells to specify the Th17 cell lineage (Ciofani et al., 2012). More recently, Jain et al. revealed that Blimp-1 is an important downstream transcription factor of IL-23 signaling, and it together with RORγt regulates several Th17 signature genes (Jain et al., 2016).

3.4 RORγt Ligands

Though RORγt has its full name as retinoic acid-related orphan receptor gamma t, it is possible that the endogenous ligands exist to regulate its activity. Santori et al. show that sterols produced in CBP are necessary and sufficient to promote RORγt transcriptional activity, to be more precisely, the RORγt ligand is downstream of lanosterol and upstream of zymosterol. Moreover, manipulation of CBP could modulate Th17 cell differentiation and lymph node development (Santori et al., 2015). Consistently, another group also found sterol synthesis products could regulate RORγt activity and thus affect Th17 differentiation (Hu et al., 2015). Supporting these studies, Wang et al. identified CD5-like (CD5L) protein as key switch of Th17 functional state (pathogenic or nonpathogenic) through regulating polyunsaturated fatty acids/saturated fatty acids balance and RORγt ligand availability (Wang, Yosef, et al., 2015). Therefore, the endogenous ligand links lipid metabolism to RORγt activity, opening a door for lipid metabolism to regulate proinflammatory conditions by modulating RORγt target gene expression.

Besides the endogenous ligand, many ligands acting as potent agonists, inverse agonists, and antagonists have been synthesized as potential drugs to target RORγt activity through recruiting coactivator or corepressor to modulate RORγt downstream gene expression. Some of the compounds identified have been tested in mouse disease models including MS, psoriasis, or arthritis and showed promising therapeutic effects on the diseases (Chang, Lyda, Kamenecka, & Griffin, 2014; Huh et al., 2011; Skepner et al., 2014; Solt et al., 2011; Xiao, Yosef, et al., 2014; Xu et al., 2011). VTP-43742, an inverse agonist for RORγt, is used for treating psoriasis in phase II clinical trials currently. Ouyang and Rutz have recently written a comprehensive review about the endogenous and synthetic ligands of RORγt (Rutz, Eidenschenk, Kiefer, & Ouyang, 2016).

4. PLASTICITY OF Treg AND Th17 IN HEALTH AND DISEASES

Th17-like Tregs are involved in several diseases including psoriasis, IBD, periodontitis, arthritis, MS, cancer, allergy, and infections; here, we will discuss about the phenotypes and roles of these Th17-like Tregs in these diseases.

Psoriasis is a chronic inflammatory disease of skin where Th17/IL-23 axis and the inflammatory mediator IL-17A are critical for maintaining the inflammation (Golden, McCormick, & Ward, 2013). Sugiyama et al. demonstrated that both in blood and in psoriatic lesional skin of psoriatic, Tregs were dysfunctional showing reduced suppression function and therefore lead to hyperproliferation of the pathogenic effector T cells in skin (Sugiyama et al., 2005). These Tregs in psoriatic were later found to be Th17-like Tregs that showed capacity of producing IL-17A and a memory phenotype by expressing CD45RO (Bovenschen et al., 2011; Sanchez Rodriguez et al., 2014; Voo et al., 2009; Zhang, Yang, Cheng, Hui, & Gao, 2010). Functionally, these Tregs showed enhanced tendency to differentiate into IL-17A-producing cells ex vivo, accompanied by increased RORγt levels and loss of FOXP3 (Bovenschen et al., 2011). Therefore, the Th17-like Tregs in psoriatic have impaired suppression function on the one hand and exert their pathogenic role through producing inflammatory mediator IL-17A on the other hand.

Crohn's disease and ulcerative colitis are the principal types of IBD. Tregs play a protective role in maintaining tolerance of commensals and dietary antigens in the intestine (Arpaia et al., 2013; Cebula et al., 2013; Cong, Feng, Fujihashi, Schoeb, & Elson, 2009), and break of the tolerance can lead to IBD. It has been shown that Th17-like Tregs increased in IBD patients compared with healthy controls (Hovhannisyan et al., 2011; Kryczek et al., 2011; Li & Boussiotis, 2013; Ueno et al., 2013). Functionally, these Tregs retain their immunosuppressive function (Beriou et al., 2009;

Kryczek et al., 2011), though one study showed that these Tregs have impaired suppression function (Ueno et al., 2013).

Break of tolerance of oral commensals could lead to the onset of periodontitis. Recent study determined the association between Th17-like Tregs and the pathogenesis of periodontitis; however, the role that Th17-like Tregs play in pathogenesis of periodontitis has not been fully understood (Okui, Aoki, Ito, Honda, & Yamazaki, 2012).

Th17 cells play a pathological role in various kinds of arthritis. Th17-like Tregs showed a higher frequency in RA patients compared with healthy controls (Afzali et al., 2013). These $FOXP3^+$ Tregs defined by CD161 are capable of producing IL-17A, IFN-γ, IL-2, and expressing RORγt and T-bet (Pesenacker et al., 2013; Pesenacker & Wedderburn, 2013). Importantly, these Tregs have suppressive functions comparable to Tregs from healthy subjects.

Besides Th1 cells, recent studies showed evidences that Th17 cells also play a crucial role in the pathogenesis of MS (Cua et al., 2003; El-Behi et al., 2011; Kleinewietfeld et al., 2013; Lee, Menezes, Umesaki, & Mazmanian, 2011; Nylander & Hafler, 2012). Whereas Th1-like Tregs contribute to the pathogenesis of MS, Th17-like Tregs inversely correlate with the inflammation, suggesting they may act as controllers of Th17-mediated inflammation, so the exact role of Th17-like Tregs in MS patients remains to be further investigated.

Ye et al. first described that tumor-infiltrating Th17 cells from colon and ovarian cancers have the capacity to differentiate into IFN-γ^+ and $FOXP3^+$ T cells with potent suppression function (Ye et al., 2011). Consistently, Kryczek et al. also found that IL-17^+Foxp3^+CD4^+ T cells from colon carcinoma have effective suppression function but can also produce inflammatory cytokines in the colitic tissues (Kryczek et al., 2011). These studies demonstrate that Th17-like Tregs may link chronic inflammation to cancer development by restraining local T cell immunity as well as producing inflammatory cytokines.

RORγt^+FOXP3^+IL-17A^+ Th17-like Tregs also occur in allergic rhinitis patients, especially those with polyps. However, whether these Tregs contribute to pathogenesis of the allergy is not clear.

In oropharyngeal candidiasis (OPC) Th17 cell infection model, Th17 cells are considered protective in antifungi immunity, what is more, Tregs are also regarded as protective cells that can promote fungal clearance and recovery from infection. Mechanistically, Tregs can promote Th17 response by promoting Th17 expansion as well as IL-17A production, and moreover, Tregs themselves can acquire Th17 characteristics and become IL-17A-producing Th17-like Tregs and therefore contribute to fungal clearance (Bhaskaran, Cohen, Zhang, Weinberg, & Pandiyan, 2015; Bhaskaran, Weinberg, & Pandiyan, 2015; Hernandez-Santos et al., 2013; Pandiyan et al., 2011; Whibley & Gaffen, 2014).

5. PLASTICITY OF Treg AND Th17 CELL LINEAGES

5.1 Plasticity of Treg and Th17 Cell Lineages

Treg and Th17 cells have proved to be regulated reciprocally during differentiation. Dong and his colleagues showed that in mice iTregs can be reprogrammed to Th17 cells in vitro and nTregs can be reprogrammed to Th17 cells in vitro and in vivo in the presence of inflammatory cytokines. Interestingly, in the presence of all-*trans* retinoic acid, iTregs were resistant to redifferentiation toward Th17 cells (Yang et al., 2008). By detecting suppressive activity of IL-17^+ and IL-17^- nTregs from in vitro IL-6 cultured condition, they found IL-17^- nTreg inhibit the proliferation of naïve T cells as nonmanipulated nTregs, whereas IL-17^+ nTreg have greatly reduced suppression function.

In consistent, IL-17A-producing FOXP3^+ Tregs were found in human peripheral blood and lymphoid tissues, a little different with that in mice, these Tregs retain potent suppressive activity (Voo et al., 2009). Also, it has been proved that human FOXP3^+ Tregs can differentiate to IL-17A-producing cells in the presence of lineage-specific polarizing factors in vitro (Koenen et al., 2008; Valmori, Raffin, Raimbaud, & Ayyoub, 2010). It is worth noting that human thymus dose not contain IL-17A-producing Tregs, indicating these IL-17A-producing Tregs are generated in the periphery. In the same year, Beriou et al. found that there were IL-17-producing Tregs with HLA-DR^- phenotype; however, this study showed that the suppressive activity of these Tregs is inversely correlated with IL-17A production, that is, the conversion of non-IL-17A-producing Tregs to IL-17A-producing Tregs dampened the suppression function of Tregs (Beriou et al., 2009). These results were consistent with that in mice revealed by Dong and his colleagues, but inconsistent with the study by Voo et al.

Actually, more and more studies have reported IL-17A-producing Tregs in various human diseases including psoriasis, IBD, periodontitis, arthritis, MS, cancer, allergy, and infections. Whether the suppression functions of these IL-17A-producing Tregs are impaired remain varying in different diseases. In psoriatic, it is considered that IL-17A-producing Tregs have impaired suppressive activity (Bovenschen et al., 2011; Sanchez Rodriguez et al., 2014; Voo et al., 2009; Zhang et al., 2010), in IBD patients, one study showed that the Th17-like Tregs have defective suppression function (Ueno et al., 2013), while others reported that they retain their suppression function (Beriou et al., 2009; Kryczek et al., 2011). A recent study reported that $FOXP3^{+}ROR\gamma t^{+}$ T cells represent a stable cell lineage in vivo by adoptively transferring these cells to a T cell transfer colitis model, and epigenetic study confirmed that these cells contain Treg-specific epigenetic signatures, more importantly, these $FOXP3^{+}ROR\gamma t^{+}$ T cells showed enhanced suppression function compared with $FOXP3^{+}ROR\gamma t^{-}$ Tregs, indicating that RORγ expression in Tregs endows Tregs with optimal suppressive activity (Yang et al., 2016). In periodontitis patients, the suppression function of Th17-like Tregs was not detected. In MS and RA patients, the Th17-like Tregs retain their suppressive activity (Afzali et al., 2013; Cua et al., 2003; El-Behi et al., 2011; Kleinewietfeld et al., 2013; Lee et al., 2011; Nylander & Hafler, 2012; Pesenacker et al., 2013; Pesenacker & Wedderburn, 2013), also, in colon and ovarian cancers, the Th17-derived Tregs and Th17-like Tregs retain their potent suppressive function (Kryczek et al., 2011; Ye et al., 2011). In OPC infection model, instead of suppressing Th17 cells, Th17-like Tregs are considered as helper of Th17 cells by promoting Th17 response and becoming IL-17A-producing Th17-like Tregs to contribute to fungal clearance (Bhaskaran, Cohen, et al., 2015; Bhaskaran, Weinberg, et al., 2015; Hernandez-Santos et al., 2013; Pandiyan et al., 2011; Whibley & Gaffen, 2014).

Besides Tregs, Th17 cells display high degree of plasticity. Here we just focus on Th17 cells transdifferentiate to Tregs. Ye et al. first described that tumor-infiltrating Th17 cells have the capacity to transdifferentiate into $IFN\text{-}\gamma^{+}$ and $FOXP3^{+}$ T cells with potent suppressive function in vitro (Ye et al., 2011). Further study by Carbo et al. demonstrated that Th17 cells could transdifferentiate to Tregs in vivo in an adoptive transfer induced mouse colitis model (Carbo et al., 2013). Later, Obermajer et al. showed evidence for a direct conversion of Th17 to Tregs when transferring mesenchymal stem cells to induce long-term tolerance of allogeneic heart grafts in mice (Obermajer et al., 2014).

5.2 Regulation of the Plasticity of Treg and Th17 Cell Lineages

Various signals, factors, epigenetic modifications, metabolic pathways, and microbiota have been shown to regulate the plasticity between Tregs and Th17 cells (Fig. 1).

5.2.1 Regulation of Treg/Th17 Plasticity by Extracellular Signals

Dong and his colleagues found that in mice FOXP3 antagonizes the function of RORγt and RORα; however, IL-6 could prevent this suppressive effect by FOXP3; moreover, IL-6 in combination with IL-1 could induce reprogramming of FOXP3$^+$ Tregs, among which RORγt, RORα and STAT3 were required for IL-17 production in Tregs. Reciprocally, STAT3 but not RORγt or RORα downregulates FOXP3 expression. They also found that TGF-β receptor signaling was required for FOXP3 as well as IL-17 induction, whereas SMAD4 was only needed for FOXP3 induction. In summary, their study revealed molecular antagonism responsible for Treg and Th17 differentiation program, demonstrating the underlying molecular mechanism for plasticity of Treg and Th17 cells (Yang et al., 2008).

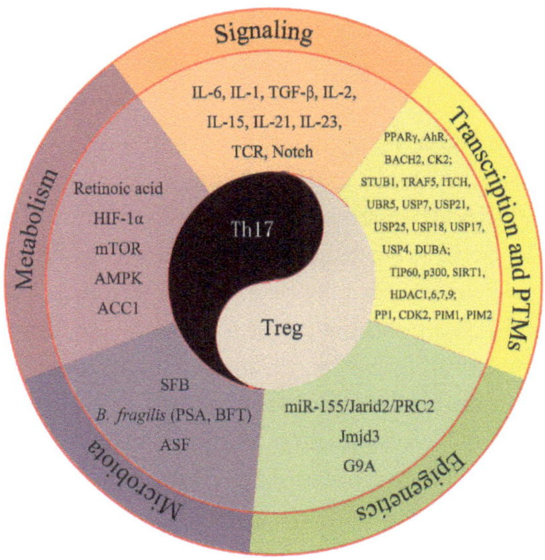

Fig. 1 Regulation of the plasticity of Treg and Th17 cell lineages. Understanding the functional stability and plasticity of Tregs and Th17 cells under inflammatory stimuli and the related diseases will provide helpful instructions for Treg-based immunotherapy. Regulators of Tregs and Th17 cells represent intriguing targets to delicately manipulate immune response and avoid side effects.

Koenen et al. demonstrated that human $CD25^{high}Foxp3^+$ Tregs differentiated into IL-17A-producing cells depends on IL-2 and IL-15, and is boosted by IL-1β, IL-21, and IL-23 (Koenen et al., 2008). Later, another group demonstrated that human $CD4^+FOXP3^+CCR6^-$ Tregs differentiated to IL-17A-producing cells in the presence of TCR stimulation and IL-2, IL-1β, IL-21, IL-23, as well as human serum (Voo et al., 2009). Consistently, Ye et al. found that TCR stimulation is required for IFN-γ and FOXP3 expression during the conversion from Th17 to Th17-like Tregs, and this process is accompanied by the demethylation of the Treg-specific demethylated region (TSDR) within the FOXP3 CpG island in expanded Th17 cells, suggesting epigenetic regulation occurred during the trans-differentiation (Ye et al., 2011). A recent study focused on IL-15 and investigated its role in modulating Treg/Th17 balance in IBD by affecting the expression of FOXP3 and RORγt, suggesting IL-15 treatment might be a potential therapeutic scheme in Treg/Th17 imbalance-based inflammatory diseases (Tosiek, Fiette, El Daker, Eberl, & Freitas, 2016).

Though TGF-β is required for iTreg and Th17 differentiation, it can inhibit Treg differentiation in the presence of other inflammatory cytokines like IL-6 and IL-21 (Korn et al., 2009). What is more, TGF-β can inhibit IL-17A production by Tregs in the presence of strong TCR signals, IL-6 and IL-1β (Beriou et al., 2009). However, another study found that human $FOXP3^+$ naïve Tregs ($CD4^+CD45RA^+CD25^+CD127^{low}$) can differentiate to Th17 cells under IL-2 and IL-1β, and this is increased by IL-23 and TGF-β (Valmori et al., 2010). A more recent study revealed that impaired TGF-β/BAMBI pathway is associated with increased Th17/Treg ratio in chronic obstructive pulmonary disease (COPD), suggesting the imbalance of Th17 and Tregs may be a new mechanism in COPD patients and BAMBI could be a potential target for the disease (Zhang et al., 2016).

Yu et al. demonstrated that Notch signaling could also modulate Th17/Treg balance in immune thrombocytopenia (ITP) patients, and DAPT, an effective inhibitor of Notch signaling, could inhibit Th17 differentiation in ITP patients as well as in vitro polarized Th17 cells, indicating inactivation of Notch signaling in ITP patients might be a potential immunotherapeutic strategy (Yu et al., 2015).

5.2.2 Regulation of Treg/Th17 Plasticity by Transcription Factors and Other Proteins

PPARγ could regulate Th17 and Treg plasticity by enhancing FOXP3 expression and decreasing RORγ and IL-17A expression. Pioglitazone, an agonist of PPARγ, belongs to thiazolidinedione class of antidiabetic drugs, could promote Th17 cells to redifferentiate to Tregs in vitro and in an adoptive transfer induced colitis model (Carbo et al., 2013).

Quintana et al. revealed that aryl hydrocarbon receptor (AHR) could regulate Treg and Th17 differentiation in a ligand-specific manner. AHR activation by 2,3,7,8-tetrachlorodibenzo-*p*-dioxin (TCDD) promotes Treg development and remitted EAE; nevertheless, AHR activation by another agonist 6-formylindolo[3,2-*b*]carbazole (FICZ) suppressed Treg differentiation but promote Th17 development and aggravated EAE (Quintana et al., 2008). Consistently, another study using FICZ to activate AHR also showed enhanced Th17 proliferation and function and worsened EAE (Veldhoen et al., 2008). Therefore, based on the ligand regulation nature of AHR, it can be an intriguingly target to regulate Th17 and Treg balance.

Roychoudhuri et al. identified BACH2 as potent regulator to promote Treg-mediated immunoregulatory capacity while inhibiting differentiation of multiple effector $CD4^+$ T cells including Th1, Th2, and Th17, thus targeting this protein can be a potential strategy for cancer immunotherapy (Roychoudhuri et al., 2013).

Ulges et al. identified protein kinase CK2 (CK2) as a regulator of Th17/Treg balance, in an EAE model, they found pharmacological inhibition or genetic deletion of CK2 reduced the severity of EAE. Mechanistically, inhibition of CK2 resulted in decreased STAT3 phosphorylation and the significantly reduced expression of IL-23 receptor, IL-17, and GM-CSF (Ulges et al., 2016). This study provided that CK2 can be a potential target for treatment of Th17-driven autoimmune diseases.

5.2.3 Regulation of Treg/Th17 Plasticity by Epigenetic Modifications

Genomewide analysis of epigenetic features in naïve $CD4^+$ T cells, T-helper 1 (Th1), Th2, Th17, nTreg, and iTreg cells displayed a comprehensive epigenetic map to us. The analysis of H3K4me3 and H3K27me3 at signature cytokine genes and transcription factor genes revealed that the epigenetic modifications at signature cytokine genes matched the conventional expectations, while those at signature master regulator genes showed a nonexclusive feature, suggesting plasticity of T-helper cells (Wei et al., 2009). Therefore, epigenetic modifications at cytokine gene locus and transcription factor locus could be the underlying mechanism directing T-helper cell plasticity. Supporting this notion, Allan et al. found that SUV39H1, a histone methylase involved in the trimethylation of histone H3 on lysine 9 (H3K9me3), could regulate Th2 lineage stability, the results showed that Th2 cells lacking of SUV39H1 have a propensity to differentiate to Th1 cells in a Th1 polarization condition (Allan et al., 2012).

There are also many studies in epigenetic regulation in Tregs and Th17 cells. For example, DNA demethylase TET proteins (Wang et al., 2013; Yang, Xu, et al., 2015), histone/protein acetyltransferases CBP, P300, TIP60 (Li et al., 2007; Liu et al., 2014), histone/protein deacetylase 3, 7, 9 (Li et al., 2007; Wang, Liu, et al., 2015), and chromatin-modifying enzyme EZH2 (Arvey et al., 2014; DuPage et al., 2015) are crucial for maintaining Treg stability. Note that p300 has also been reported to be involved in ROR protein complexes to regulate RORγt function (Wu et al., 2015). A recent study by Muljo and colleagues found that deficiency of miR-155 lead to both impaired development of Th17 cells and Tregs partially through the inability to control a DNA-binding protein named Jarid2 that recruits the Polycomb Repressive Complex 2 (PRC2) to chromatin, leading to increased H3K27 histone methylation and the reduced production of Th17-associated cytokines and FOXP3 (Escobar et al., 2014). Besides, Jmjd3, a histone demethylase, has been found to act as an epigenetic factor to regulate $CD4^+$ T cell differentiation and plasticity through modulating histone methylation and target gene expression. Deletion of Jmjd3 promotes the development of Th2 and Th17 cells in the colon and SI, and restrains the conversion of Th2, Th17, and Treg cells to Th1 cells (Li, Zou, et al., 2014). Similarly, a histone lysine methyltransferase G9A has been found to regulate the differentiation of Treg and Th17 differentiation by mediating H3K9me2 and therefore regulating the chromatin accessibility and increase TGF-β1 sensitivity. Deletion of G9A in a T cell transfer model of colitis resulted in largely reduced severity of the disease, indicating G9A is a potential drug target for treating intestinal inflammation (Antignano et al., 2014). In summary, epigenetic modifications confine gene expression to regulate T-helper cell plasticity and the enzymes responsible for epigenetic modifications could be attractive targets for regulating Treg/Th17 plasticity.

5.2.4 Regulation of Treg/Th17 Plasticity by Cell Metabolism

Mucida et al. identified a metabolite of vitamin A retinoic acid as a critical regulator of TGF-β-dependent immune response, which inhibits IL-6-driven Th17 differentiation while promotes the differentiation of Tregs (Mucida et al., 2007). This study showed that a common metabolite could regulate the balance of Tregs and Th17 cells.

HIF-1 is a key metabolic sensor, which is also a regulator of Treg and Th17 differentiation. Dang et al. found that HIF-1α on the one hand can enhance Th17 development through transcriptional activation of RORγt and IL-17A. On the other hand, HIF-1α suppresses Treg development by targeting FOXP3 for proteasomal degradation under both normoxic and hypoxic circumstances, what is more, HIF-1α-deficient mice showed resistance for EAE induction that associated with reduced Th17 cells and increased Tregs (Dang et al., 2011). Another study revealed HIF-1α and mammalian target of rapamycin (mTOR) contributed to Th17/Treg lineage choice by promoting glycolytic activity in T cells; moreover, lack of HIF-1α leads to protection of mice from autoimmune neuroinflammation that is associated with reduced Th17 and enhanced Treg differentiation (Shi et al., 2011). Thus, blocking mTOR and HIF-1α could be effective ways to promote Tregs while inhibit Th17 development (Delgoffe et al., 2009; Shi et al., 2011).

On the other hand, AMP-activated kinase (AMPK) is critical for Tregs by promoting fatty acid oxidation and inhibiting mTOR-dependent pathways (Michalek et al., 2011; Pollizzi & Powell, 2014). A recent study found that the direct activator of AMPK 5-aminoimidazole-4-carboxamide ribonucleotide (AICAR) could modulate Treg/Th17 axis through regulating fatty acid oxidation. By using a murine model of in vivo Treg cell induction by IL-2 complexes, they found AICAR could induce Tregs in vivo (Gualdoni et al., 2016). Therefore, AMPK and mTOR represent two major metabolic pathways influencing the cell fate of T cells, and hence, targeting the two pathways becomes new attractive approach for manipulating T cell fate and function.

Further, Sparwasser and colleagues investigated the role of glycolytic-lipogenic metabolic pathway for regulating Tregs and Th17 cells, they found Th17 cells but not Tergs are dependent on acetyl-CoA carboxylase 1 (ACC1)-mediated de novo fatty acid synthesis; moreover, inhibition of ACC1 by soraphen A attenuates Th17-mediated EAE in mice, providing that ACC1 or fatty acid synthesis pathway could be exploited as strategies for manipulating metabolism in Tregs and Th17 cells and further influence their cell fates (Berod, Friedrich, Nandan, & Freitag, 2014; O'Sullivan & Pearce, 2014).

These findings highlight metabolic cues and pathways in regulating T cell fate determination, suggesting targeting metabolic pathways could be new strategies for treating diseases involving Treg and effector T cell plasticity.

5.2.5 Regulation of Treg/Th17 Plasticity by Microbiota

Microbiota has recently been found a new factor to modulate Treg and Th17 differentiation. Segmented filamentous bacteria (SFB) were reported to induce Th17 cells in the SI and outside SI during autoimmune responses (Gaboriau-Routhiau et al., 2009; Ivanov et al., 2009; Lee et al., 2011; Wu et al., 2010). Actually, Treg-specific deletion of MyD88 leads to impaired IgA responses and overgrowth of SFB and increased of Th17 cells, suggesting Tregs control Th17 cells partially through promoting IgA responses and therefore restraining SFB (Wang, Charbonnier, et al., 2015).

Polysaccharide A (PSA), expressed by *Bacteroides fragilis*, could promote IL-10-producing FOXP3^{+} Tregs formation from CD4^{+} T cells through TLR-2 and suppress Th17 responses (Round et al., 2011); therefore, PSA could be a therapeutic strategy to promote immune tolerance in the intestine. Supporting this, in an experimental colitis mouse model, PSA has proved to be able to prevent and even cured the disease (Round & Mazmanian, 2010). However, another component secreted by *B. fragilis* named *B. fragilis* toxin (BFT) is involved in IBD and Th17 response induced colon cancer (Rabizadeh et al., 2007; Rhee et al., 2009; Wu et al., 2009). Thus, *B. fragilis* plays a dual role in Treg and Th17 development depending on the different metabolites produced by it. Besides, *Clostridia* can also induce Tregs in the gut (Atarashi et al., 2011, 2013).

Similarly, colonization of a standardized microbial cocktail altered Schaedler flora (ASF) representing normal microbiota of mice leads to de novo induction of Tregs and decreased of Th1 and Th17 immunity (Geuking et al., 2011).

Mechanistically, several studies indicate that microbiota-derived short-chain fatty acids are involved in T cell regulation in the gut (Arpaia et al., 2013; Furusawa et al., 2013; Smith et al., 2013). Dietary factors like high-fat diet can lead to impaired Th17 development and therefore lead to reduced inflammation (Garidou et al., 2015). In summary, these findings in microbiota and microbiota-derived metabolites or dietary factors provide new strategies by manipulating our bacteria to modulate T cell immune responses.

6. CONCLUDING REMARKS

Human Tregs are promising to be used in clinic, therefore understanding the functional stability and plasticity of Tregs under inflammatory stimuli and the related diseases will provide helpful instructions for

Treg-based immunotherapy. In this review, we focused on the plasticity between Tregs and Th17 cells and the mechanisms regulating the plasticity. As the plasticity between Tregs and Th17 cells is closely related to autoimmune diseases, allergy, infection, and cancer, regulators of Treg/Th17 plasticity represent intriguing targets to delicately manipulate immune response and avoid side effects.

ACKNOWLEDGMENTS

Our research is supported by National Basic Research Program of China Grants 2014CB541803 and 2014CB541903; NSFC 31370863, 81330072. B.L. is a recipient of Shanghai "Rising Star" program 10QA1407900, CAS "100-talent" program and National Science Foundation for Distinguished Young Scholars 31525008. We thank members of Li laboratory for helpful discussion.

REFERENCES

Afzali, B., Mitchell, P. J., Edozie, F. C., Povoleri, G. A., Dowson, S. E., Demandt, L., ... Lombardi, G. (2013). CD161 expression characterizes a subpopulation of human regulatory T cells that produces IL-17 in a STAT3-dependent manner. *European Journal of Immunology*, *43*(8), 2043–2054. http://dx.doi.org/10.1002/eji.201243296.

Akimova, T., Ge, G., Golovina, T., Mikheeva, T., Wang, L., Riley, J. L., & Hancock, W. W. (2010). Histone/protein deacetylase inhibitors increase suppressive functions of human FOXP3+ Tregs. *Clinical Immunology*, *136*(3), 348–363. http://dx.doi.org/10.1016/j.clim.2010.04.018.

Allan, R. S., Zueva, E., Cammas, F., Schreiber, H. A., Masson, V., Belz, G. T., ... Amigorena, S. (2012). An epigenetic silencing pathway controlling T helper 2 cell lineage commitment. *Nature*, *487*(7406), 249–253. http://dx.doi.org/10.1038/nature11173.

Anbalagan, M., Huderson, B., Murphy, L., & Rowan, B. G. (2012). Post-translational modifications of nuclear receptors and human disease. *Nuclear Receptor Signaling*, *10*, e001. http://dx.doi.org/10.1621/nrs.10001.

Antignano, F., Burrows, K., Hughes, M. R., Han, J. M., Kron, K. J., Penrod, N. M., ... Zaph, C. (2014). Methyltransferase G9A regulates T cell differentiation during murine intestinal inflammation. *The Journal of Clinical Investigation*, *124*(5), 1945–1955. http://dx.doi.org/10.1172/jci69592.

Arpaia, N., Campbell, C., Fan, X., Dikiy, S., van der Veeken, J., deRoos, P., ... Rudensky, A. Y. (2013). Metabolites produced by commensal bacteria promote peripheral regulatory T-cell generation. *Nature*, *504*(7480), 451–455. http://dx.doi.org/10.1038/nature12726.

Arvey, A., van der Veeken, J., Samstein, R. M., Feng, Y., Stamatoyannopoulos, J. A., & Rudensky, A. Y. (2014). Inflammation-induced repression of chromatin bound by the transcription factor Foxp3 in regulatory T cells. *Nature Immunology*, *15*(6), 580–587. http://dx.doi.org/10.1038/ni.2868.

Aschenbrenner, K., D'Cruz, L. M., Vollmann, E. H., Hinterberger, M., Emmerich, J., Swee, L. K., ... Klein, L. (2007). Selection of Foxp3+ regulatory T cells specific for self antigen expressed and presented by Aire+medullary thymic epithelial cells. *Nature Immunology*, *8*(4), 351–358. http://dx.doi.org/10.1038/ni1444.

Atarashi, K., Tanoue, T., Oshima, K., Suda, W., Nagano, Y., Nishikawa, H., ... Honda, K. (2013). Treg induction by a rationally selected mixture of Clostridia strains from the human microbiota. *Nature*, *500*(7461), 232–236. http://dx.doi.org/10.1038/nature12331.

Atarashi, K., Tanoue, T., Shima, T., Imaoka, A., Kuwahara, T., Momose, Y., … Honda, K. (2011). Induction of colonic regulatory T cells by indigenous Clostridium species. *Science*, *331*(6015), 337–341. http://dx.doi.org/10.1126/science.1198469.

Atkins, G. B., Hu, X., Guenther, M. G., Rachez, C., Freedman, L. P., & Lazar, M. A. (1999). Coactivators for the orphan nuclear receptor RORalpha. *Molecular Endocrinology*, *13*(9), 1550–1557. http://dx.doi.org/10.1210/mend.13.9.0343.

Beier, U. H., Wang, L., Bhatti, T. R., Liu, Y., Han, R., Ge, G., & Hancock, W. W. (2011). Sirtuin-1 targeting promotes Foxp3+ T-regulatory cell function and prolongs allograft survival. *Molecular and Cellular Biology*, *31*(5), 1022–1029. http://dx.doi.org/10.1128/mcb.01206-10.

Beier, U. H., Wang, L., Han, R., Akimova, T., Liu, Y., & Hancock, W. W. (2012). Histone deacetylases 6 and 9 and sirtuin-1 control Foxp3+ regulatory T cell function through shared and isoform-specific mechanisms. *Science Signaling*, *5*(229), ra45. http://dx.doi.org/10.1126/scisignal.2002873.

Bennett, C. L., Christie, J., Ramsdell, F., Brunkow, M. E., Ferguson, P. J., Whitesell, L., … Ochs, H. D. (2001). The immune dysregulation, polyendocrinopathy, enteropathy, X-linked syndrome (IPEX) is caused by mutations of FOXP3. *Nature Genetics*, *27*(1), 20–21. http://dx.doi.org/10.1038/83713.

Benson, M. J., Pino-Lagos, K., Rosemblatt, M., & Noelle, R. J. (2007). All-trans retinoic acid mediates enhanced T reg cell growth, differentiation, and gut homing in the face of high levels of co-stimulation. *The Journal of Experimental Medicine*, *204*(8), 1765–1774. http://dx.doi.org/10.1084/jem.20070719.

Beriou, G., Costantino, C. M., Ashley, C. W., Yang, L., Kuchroo, V. K., Baecher-Allan, C., & Hafler, D. A. (2009). IL-17-producing human peripheral regulatory T cells retain suppressive function. *Blood*, *113*(18), 4240–4249. http://dx.doi.org/10.1182/blood-2008-10-183251.

Berod, L., Friedrich, C., Nandan, A., & Freitag, J. (2014). De novo fatty acid synthesis controls the fate between regulatory T and T helper 17 cells. *Nature Medicine*, *20*(11), 1327–1333. http://dx.doi.org/10.1038/nm.3704.

Bettelli, E., Carrier, Y., Gao, W., Korn, T., Strom, T. B., Oukka, M., … Kuchroo, V. K. (2006). Reciprocal developmental pathways for the generation of pathogenic effector TH17 and regulatory T cells. *Nature*, *441*(7090), 235–238. http://dx.doi.org/10.1038/nature04753.

Bettelli, E., Sullivan, B., Szabo, S. J., Sobel, R. A., Glimcher, L. H., & Kuchroo, V. K. (2004). Loss of T-bet, but not STAT1, prevents the development of experimental autoimmune encephalomyelitis. *The Journal of Experimental Medicine*, *200*(1), 79–87. http://dx.doi.org/10.1084/jem.20031819.

Bhaskaran, N., Cohen, S., Zhang, Y., Weinberg, A., & Pandiyan, P. (2015). TLR-2 signaling promotes IL-17A production in CD4+CD25+Foxp3+ regulatory cells during oropharyngeal candidiasis. *Pathogens (Basel, Switzerland)*, *4*(1), 90–110. http://dx.doi.org/10.3390/pathogens4010090.

Bhaskaran, N., Weinberg, A., & Pandiyan, P. (2015 Feb 18). Th17 inflammation model of oropharyngeal candidiasis in immunodeficient mice. *Journal of Visualized Experiments*, *96*. http://dx.doi.org/10.3791/52538.

Bovenschen, H. J., van de Kerkhof, P. C., van Erp, P. E., Woestenenk, R., Joosten, I., & Koenen, H. J. (2011). Foxp3+ regulatory T cells of psoriasis patients easily differentiate into IL-17A-producing cells and are found in lesional skin. *The Journal of Investigative Dermatology*, *131*(9), 1853–1860. http://dx.doi.org/10.1038/jid.2011.139.

Carbo, A., Hontecillas, R., Kronsteiner, B., Viladomiu, M., Pedragosa, M., Lu, P., … Bassaganya-Riera, J. (2013). Systems modeling of molecular mechanisms controlling cytokine-driven CD4+ T cell differentiation and phenotype plasticity. *PLoS Computational Biology*, *9*(4), e1003027. http://dx.doi.org/10.1371/journal.pcbi.1003027.

Cebula, A., Seweryn, M., Rempala, G. A., Pabla, S. S., McIndoe, R. A., Denning, T. L., ... Ignatowicz, L. (2013). Thymus-derived regulatory T cells contribute to tolerance to commensal microbiota. *Nature*, *497*(7448), 258–262. http://dx.doi.org/10.1038/nature12079.

Chang, M. R., Lyda, B., Kamenecka, T. M., & Griffin, P. R. (2014). Pharmacologic repression of retinoic acid receptor-related orphan nuclear receptor gamma is therapeutic in the collagen-induced arthritis experimental model. *Arthritis & Rhematology*, *66*(3), 579–588. http://dx.doi.org/10.1002/art.38272.

Chen, Z., Barbi, J., Bu, S., Yang, H. Y., Li, Z., Gao, Y., ... Li, B. (2013). The ubiquitin ligase Stub1 negatively modulates regulatory T cell suppressive activity by promoting degradation of the transcription factor Foxp3. *Immunity*, *39*(2), 272–285. http://dx.doi.org/10.1016/j.immuni.2013.08.006.

Chen, W., Jin, W., Hardegen, N., Lei, K. J., Li, L., Marinos, N., ... Wahl, S. M. (2003). Conversion of peripheral CD4+CD25− naive T cells to CD4+CD25+ regulatory T cells by TGF-beta induction of transcription factor Foxp3. *The Journal of Experimental Medicine*, *198*(12), 1875–1886. http://dx.doi.org/10.1084/jem.20030152.

Chen, Y., Kuchroo, V. K., Inobe, J., Hafler, D. A., & Weiner, H. L. (1994). Regulatory T cell clones induced by oral tolerance: Suppression of autoimmune encephalomyelitis. *Science*, *265*(5176), 1237–1240.

Chen, L., Wu, J., Pier, E., Zhao, Y., & Shen, Z. (2013). mTORC2-PKBalpha/Akt1 Serine 473 phosphorylation axis is essential for regulation of FOXP3 Stability by chemokine CCL3 in psoriasis. *The Journal of Investigative Dermatology*, *133*(2), 418–428. http://dx.doi.org/10.1038/jid.2012.333.

Chung, Y., Tanaka, S., Chu, F., Nurieva, R. I., Martinez, G. J., Rawal, S., ... Dong, C. (2011). Follicular regulatory T cells expressing Foxp3 and Bcl-6 suppress germinal center reactions. *Nature Medicine*, *17*(8), 983–988. http://dx.doi.org/10.1038/nm.2426.

Ciofani, M., Madar, A., Galan, C., Sellars, M., Mace, K., Pauli, F., ... Littman, D. R. (2012). A validated regulatory network for Th17 cell specification. *Cell*, *151*(2), 289–303. http://dx.doi.org/10.1016/j.cell.2012.09.016.

Cong, Y., Feng, T., Fujihashi, K., Schoeb, T. R., & Elson, C. O. (2009). A dominant, coordinated T regulatory cell-IgA response to the intestinal microbiota. *Proceedings of the National Academy of Sciences of the United States of America*, *106*(46), 19256–19261. http://dx.doi.org/10.1073/pnas.0812681106.

Coombes, J. L., Siddiqui, K. R., Arancibia-Carcamo, C. V., Hall, J., Sun, C. M., Belkaid, Y., & Powrie, F. (2007). A functionally specialized population of mucosal CD103+ DCs induces Foxp3+ regulatory T cells via a TGF-beta and retinoic acid-dependent mechanism. *The Journal of Experimental Medicine*, *204*(8), 1757–1764. http://dx.doi.org/10.1084/jem.20070590.

Cua, D. J., Sherlock, J., Chen, Y., Murphy, C. A., Joyce, B., Seymour, B., ... Sedgwick, J. D. (2003). Interleukin-23 rather than interleukin-12 is the critical cytokine for autoimmune inflammation of the brain. *Nature*, *421*(6924), 744–748. http://dx.doi.org/10.1038/nature01355.

Dang, E. V., Barbi, J., Yang, H. Y., Jinasena, D., Yu, H., Zheng, Y., ... Pan, F. (2011). Control of T(H)17/T(reg) balance by hypoxia-inducible factor 1. *Cell*, *146*(5), 772–784. http://dx.doi.org/10.1016/j.cell.2011.07.033.

de Zoeten, E. F., Wang, L., Butler, K., Beier, U. H., Akimova, T., Sai, H., ... Hancock, W. W. (2011). Histone deacetylase 6 and heat shock protein 90 control the functions of Foxp3(+) T-regulatory cells. *Molecular and Cellular Biology*, *31*(10), 2066–2078. http://dx.doi.org/10.1128/mcb.05155-11.

de Zoeten, E. F., Wang, L., Sai, H., Dillmann, W. H., & Hancock, W. W. (2010). Inhibition of HDAC9 increases T regulatory cell function and prevents colitis in mice. *Gastroenterology*, *138*(2), 583–594. http://dx.doi.org/10.1053/j.gastro.2009.10.037.

Delgoffe, G. M., Kole, T. P., Zheng, Y., Zarek, P. E., Matthews, K. L., Xiao, B., … Powell, J. D. (2009). The mTOR kinase differentially regulates effector and regulatory T cell lineage commitment. *Immunity*, *30*(6), 832–844. http://dx.doi.org/10.1016/j.immuni.2009.04.014.

Deng, G., Nagai, Y., Xiao, Y., Li, Z., Dai, S., Ohtani, T., … Greene, M. I. (2015). Pim-2 kinase influences regulatory T cell function and stability by mediating Foxp3 protein N-terminal phosphorylation. *The Journal of Biological Chemistry*, *290*(33), 20211–20220. http://dx.doi.org/10.1074/jbc.M115.638221.

Du, J., Huang, C., Zhou, B., & Ziegler, S. F. (2008). Isoform-specific inhibition of ROR alpha-mediated transcriptional activation by human FOXP3. *Journal of Immunology*, *180*(7), 4785–4792.

Du, R., Zhao, H., Yan, F., & Li, H. (2014). IL-17+Foxp3+ T cells: An intermediate differentiation stage between Th17 cells and regulatory T cells. *Journal of Leukocyte Biology*, *96*(1), 39–48. http://dx.doi.org/10.1189/jlb.1RU0114-010RR.

Duarte, J. H., Zelenay, S., Bergman, M. L., Martins, A. C., & Demengeot, J. (2009). Natural Treg cells spontaneously differentiate into pathogenic helper cells in lymphopenic conditions. *European Journal of Immunology*, *39*(4), 948–955. http://dx.doi.org/10.1002/eji.200839196.

DuPage, M., Chopra, G., Quiros, J., Rosenthal, W. L., Morar, M. M., Holohan, D., … Bluestone, J. A. (2015). The chromatin-modifying enzyme Ezh2 is critical for the maintenance of regulatory T cell identity after activation. *Immunity*, *42*(2), 227–238. http://dx.doi.org/10.1016/j.immuni.2015.01.007.

El-Behi, M., Ciric, B., Dai, H., Yan, Y., Cullimore, M., Safavi, F., … Rostami, A. (2011). The encephalitogenicity of T(H)17 cells is dependent on IL-1- and IL-23-induced production of the cytokine GM-CSF. *Nature Immunology*, *12*(6), 568–575. http://dx.doi.org/10.1038/ni.2031.

Escobar, T. M., Kanellopoulou, C., Kugler, D. G., Kilaru, G., Nguyen, C. K., Nagarajan, V., … Muljo, S. A. (2014). miR-155 activates cytokine gene expression in Th17 cells by regulating the DNA-binding protein Jarid2 to relieve polycomb-mediated repression. *Immunity*, *40*(6), 865–879. http://dx.doi.org/10.1016/j.immuni.2014.03.014.

Faus, H., & Haendler, B. (2006). Post-translational modifications of steroid receptors. *Biomedicine & Pharmacotherapy*, *60*(9), 520–528. http://dx.doi.org/10.1016/j.biopha.2006.07.082.

Fontenot, J. D., Gavin, M. A., & Rudensky, A. Y. (2003). Foxp3 programs the development and function of CD4(+)CD25(+) regulatory T cells. *Nature Immunology*, *4*(4), 330–336. http://dx.doi.org/10.1038/ni904.

Furusawa, Y., Obata, Y., Fukuda, S., Endo, T. A., Nakato, G., Takahashi, D., … Ohno, H. (2013). Commensal microbe-derived butyrate induces the differentiation of colonic regulatory T cells. *Nature*, *504*(7480), 446–450. http://dx.doi.org/10.1038/nature12721.

Gaboriau-Routhiau, V., Rakotobe, S., Lecuyer, E., Mulder, I., Lan, A., Bridonneau, C., … Cerf-Bensussan, N. (2009). The key role of segmented filamentous bacteria in the coordinated maturation of gut helper T cell responses. *Immunity*, *31*(4), 677–689. http://dx.doi.org/10.1016/j.immuni.2009.08.020.

Garidou, L., Pomie, C., Klopp, P., Waget, A., Charpentier, J., Aloulou, M., … Burcelin, R. (2015). The gut microbiota regulates intestinal CD4 T cells expressing RORgammat and controls metabolic disease. *Cell Metabolism*, *22*(1), 100–112. http://dx.doi.org/10.1016/j.cmet.2015.06.001.

Geuking, M. B., Cahenzli, J., Lawson, M. A., Ng, D. C., Slack, E., Hapfelmeier, S., … Macpherson, A. J. (2011). Intestinal bacterial colonization induces mutualistic regulatory T cell responses. *Immunity*, *34*(5), 794–806. http://dx.doi.org/10.1016/j.immuni.2011.03.021.

Glass, C. K., & Rosenfeld, M. G. (2000). The coregulator exchange in transcriptional functions of nuclear receptors. *Genes & Development*, *14*(2), 121–141.

Gold, D. A., Baek, S. H., Schork, N. J., Rose, D. W., Larsen, D. D., Sachs, B. D., … Hamilton, B. A. (2003). RORalpha coordinates reciprocal signaling in cerebellar development through sonic hedgehog and calcium-dependent pathways. *Neuron, 40*(6), 1119–1131.

Golden, J. B., McCormick, T. S., & Ward, N. L. (2013). IL-17 in psoriasis: Implications for therapy and cardiovascular co-morbidities. *Cytokine, 62*(2), 195–201. http://dx.doi.org/10.1016/j.cyto.2013.03.013.

Grabiec, A. M., Krausz, S., de Jager, W., Burakowski, T., Groot, D., Sanders, M. E., … Reedquist, K. A. (2010). Histone deacetylase inhibitors suppress inflammatory activation of rheumatoid arthritis patient synovial macrophages and tissue. *Journal of Immunology, 184*(5), 2718–2728. http://dx.doi.org/10.4049/jimmunol.0901467.

Greiner, E. F., Kirfel, J., Greschik, H., Huang, D., Becker, P., Kapfhammer, J. P., & Schule, R. (2000). Differential ligand-dependent protein-protein interactions between nuclear receptors and a neuronal-specific cofactor. *Proceedings of the National Academy of Sciences of the United States of America, 97*(13), 7160–7165.

Gualdoni, G. A., Mayer, K. A., Goschl, L., Boucheron, N., Ellmeier, W., & Zlabinger, G. J. (2016). The AMP analog AICAR modulates the Treg/Th17 axis through enhancement of fatty acid oxidation. *FASEB Journal: Official Publication of the Federation Of American Societies for Experimental Biology, 30*(11), 3800–3809. http://dx.doi.org/10.1096/fj.201600522R.

Han, L., Yang, J., Wang, X., Wu, Q., Yin, S., Li, Z., … Li, B. (2014). The E3 deubiquitinase USP17 is a positive regulator of retinoic acid-related orphan nuclear receptor gammat (RORgammat) in Th17 cells. *The Journal of Biological Chemistry, 289*(37), 25546–25555. http://dx.doi.org/10.1074/jbc.M114.565291.

Harding, H. P., Atkins, G. B., Jaffe, A. B., Seo, W. J., & Lazar, M. A. (1997). Transcriptional activation and repression by RORalpha, an orphan nuclear receptor required for cerebellar development. *Molecular Endocrinology, 11*(11), 1737–1746. http://dx.doi.org/10.1210/mend.11.11.0002.

Harrington, L. E., Hatton, R. D., Mangan, P. R., Turner, H., Murphy, T. L., Murphy, K. M., & Weaver, C. T. (2005). Interleukin 17-producing CD4+ effector T cells develop via a lineage distinct from the T helper type 1 and 2 lineages. *Nature Immunology, 6*(11), 1123–1132. http://dx.doi.org/10.1038/ni1254.

Harris, J. M., Lau, P., Chen, S. L., & Muscat, G. E. (2002). Characterization of the retinoid orphan-related receptor-alpha coactivator binding interface: A structural basis for ligand-independent transcription. *Molecular Endocrinology, 16*(5), 998–1012. http://dx.doi.org/10.1210/mend.16.5.0837.

Hernandez-Santos, N., Huppler, A. R., Peterson, A. C., Khader, S. A., McKenna, K. C., & Gaffen, S. L. (2013). Th17 cells confer long-term adaptive immunity to oral mucosal Candida albicans infections. *Mucosal Immunology, 6*(5), 900–910. http://dx.doi.org/10.1038/mi.2012.128.

Hirose, T., Smith, R. J., & Jetten, A. M. (1994). ROR gamma: The third member of ROR/RZR orphan receptor subfamily that is highly expressed in skeletal muscle. *Biochemical and Biophysical Research Communications, 205*(3), 1976–1983. http://dx.doi.org/10.1006/bbrc.1994.2902.

Hori, S., Nomura, T., & Sakaguchi, S. (2003). Control of regulatory T cell development by the transcription factor Foxp3. *Science, 299*(5609), 1057–1061. http://dx.doi.org/10.1126/science.1079490.

Hovhannisyan, Z., Treatman, J., Littman, D. R., & Mayer, L. (2011). Characterization of interleukin-17-producing regulatory T cells in inflamed intestinal mucosa from patients with inflammatory bowel diseases. *Gastroenterology, 140*(3), 957–965. http://dx.doi.org/10.1053/j.gastro.2010.12.002.

Hu, X., Wang, Y., Hao, L. Y., Liu, X., Lesch, C. A., Sanchez, B. M., … Glick, G. D. (2015). Sterol metabolism controls T(H)17 differentiation by generating endogenous ROR-gamma agonists. *Nature Chemical Biology*, *11*(2), 141–147. http://dx.doi.org/10.1038/nchembio.1714.

Huang, G., Wang, Y., & Chi, H. (2012). Regulation of TH17 cell differentiation by innate immune signals. *Cellular & Molecular Immunology*, *9*(4), 287–295. http://dx.doi.org/10.1038/cmi.2012.10.

Huh, J. R., Leung, M. W., Huang, P., Ryan, D. A., Krout, M. R., Malapaka, R. R., … Littman, D. R. (2011). Digoxin and its derivatives suppress TH17 cell differentiation by antagonizing RORgammat activity. *Nature*, *472*(7344), 486–490. http://dx.doi.org/10.1038/nature09978.

Iizuka, M., Wakasa, Y., Tsuboi, H., Asashima, H., Hirota, T., Kondo, Y., … Sumida, T. (2014). Suppression of collagen-induced arthritis by oral administration of transgenic rice seeds expressing altered peptide ligands of type II collagen. *Plant Biotechnology Journal*, *12*(8), 1143–1152. http://dx.doi.org/10.1111/pbi.12223.

Itoh, M., Takahashi, T., Sakaguchi, N., Kuniyasu, Y., Shimizu, J., Otsuka, F., & Sakaguchi, S. (1999). Thymus and autoimmunity: Production of CD25+CD4+ naturally anergic and suppressive T cells as a key function of the thymus in maintaining immunologic self-tolerance. *Journal of Immunology*, *162*(9), 5317–5326.

Ivanov, I. I., Atarashi, K., Manel, N., Brodie, E. L., Shima, T., Karaoz, U., … Littman, D. R. (2009). Induction of intestinal Th17 cells by segmented filamentous bacteria. *Cell*, *139*(3), 485–498. http://dx.doi.org/10.1016/j.cell.2009.09.033.

Ivanov, I. I., McKenzie, B. S., Zhou, L., Tadokoro, C. E., Lepelley, A., Lafaille, J. J., … Littman, D. R. (2006). The orphan nuclear receptor RORgammat directs the differentiation program of proinflammatory IL-17+ T helper cells. *Cell*, *126*(6), 1121–1133. http://dx.doi.org/10.1016/j.cell.2006.07.035.

Ivanov, I. I., Zhou, L., & Littman, D. R. (2007). Transcriptional regulation of Th17 cell differentiation. *Seminars in Immunology*, *19*(6), 409–417. http://dx.doi.org/10.1016/j.smim.2007.10.011.

Jain, R., Chen, Y., Kanno, Y., Joyce-Shaikh, B., Vahedi, G., Hirahara, K., … Cua, D. J. (2016). Interleukin-23-induced transcription factor blimp-1 promotes pathogenicity of T helper 17 cells. *Immunity*, *44*(1), 131–142. http://dx.doi.org/10.1016/j.immuni.2015.11.009.

Jetten, A. M. (2004). Recent advances in the mechanisms of action and physiological functions of the retinoid-related orphan receptors (RORs). *Current Drug Targets Inflammation and Allergy*, *3*(4), 395–412.

Jetten, A. M. (2009). Retinoid-related orphan receptors (RORs): Critical roles in development, immunity, circadian rhythm, and cellular metabolism. *Nuclear Receptor Signaling*, *7*, e003. http://dx.doi.org/10.1621/nrs.07003.

Jetten, A. M., & Joo, J. H. (2006). Retinoid-related orphan receptors (RORs): Roles in cellular differentiation and development. *Advances in Developmental Biology*, *16*, 313–355. http://dx.doi.org/10.1016/s1574-3349(06)16010-x.

Johnson, D. R., Lovett, J. M., Hirsch, M., Xia, F., & Chen, J. D. (2004). NuRD complex component Mi-2beta binds to and represses RORgamma-mediated transcriptional activation. *Biochemical and Biophysical Research Communications*, *318*(3), 714–718. http://dx.doi.org/10.1016/j.bbrc.2004.04.087.

Kathania, M., Khare, P., Zeng, M., Cantarel, B., Zhang, H., & Ueno, H. (2016). Itch inhibits IL-17-mediated colon inflammation and tumorigenesis by ROR-gammat ubiquitination. *Nature Immunology*, *17*(8), 997–1004. http://dx.doi.org/10.1038/ni.3488.

Khattri, R., Cox, T., Yasayko, S. A., & Ramsdell, F. (2003). An essential role for Scurfin in CD4+CD25+ T regulatory cells. *Nature Immunology*, *4*(4), 337–342. http://dx.doi.org/10.1038/ni909.

Kim, J. Y., Kim, H. J., Hurt, E. M., Chen, X., Howard, O. M., & Farrar, W. L. (2007). Functional and genomic analyses of FOXP3-transduced Jurkat-T cells as regulatory T (Treg)-like cells. *Biochemical and Biophysical Research Communications*, *362*(1), 44–50. http://dx.doi.org/10.1016/j.bbrc.2007.07.187.

Kitoh, A., Ono, M., Naoe, Y., Ohkura, N., Yamaguchi, T., Yaguchi, H., … Sakaguchi, S. (2009). Indispensable role of the Runx1-Cbfbeta transcription complex for in vivo-suppressive function of FoxP3+ regulatory T cells. *Immunity*, *31*(4), 609–620. http://dx.doi.org/10.1016/j.immuni.2009.09.003.

Kleinewietfeld, M., Manzel, A., Titze, J., Kvakan, H., Yosef, N., Linker, R. A., … Hafler, D. A. (2013). Sodium chloride drives autoimmune disease by the induction of pathogenic TH17 cells. *Nature*, *496*(7446), 518–522. http://dx.doi.org/10.1038/nature11868.

Koch, M. A., Tucker-Heard, G., Perdue, N. R., Killebrew, J. R., Urdahl, K. B., & Campbell, D. J. (2009). The transcription factor T-bet controls regulatory T cell homeostasis and function during type 1 inflammation. *Nature Immunology*, *10*(6), 595–602. http://dx.doi.org/10.1038/ni.1731.

Koenen, H. J., Smeets, R. L., Vink, P. M., van Rijssen, E., Boots, A. M., & Joosten, I. (2008). Human CD25highFoxp3pos regulatory T cells differentiate into IL-17-producing cells. *Blood*, *112*(6), 2340–2352. http://dx.doi.org/10.1182/blood-2008-01-133967.

Koonpaew, S., Shen, S., Flowers, L., & Zhang, W. (2006). LAT-mediated signaling in CD4+CD25+ regulatory T cell development. *The Journal of Experimental Medicine*, *203*(1), 119–129. http://dx.doi.org/10.1084/jem.20050903.

Korn, T., Bettelli, E., Oukka, M., & Kuchroo, V. K. (2009). IL-17 and Th17 cells. *Annual Review of Immunology*, *27*, 485–517. http://dx.doi.org/10.1146/annurev.immunol.021908.132710.

Krishnamoorthy, N., Khare, A., Oriss, T. B., Raundhal, M., Morse, C., Yarlagadda, M., … Ray, P. (2012). Early infection with respiratory syncytial virus impairs regulatory T cell function and increases susceptibility to allergic asthma. *Nature Medicine*, *18*(10), 1525–1530. http://dx.doi.org/10.1038/nm.2896.

Kryczek, I., Wu, K., Zhao, E., Wei, S., Vatan, L., Szeliga, W., … Zou, W. (2011). IL-17+ regulatory T cells in the microenvironments of chronic inflammation and cancer. *Journal of Immunology*, *186*(7), 4388–4395. http://dx.doi.org/10.4049/jimmunol.1003251.

Kurebayashi, S., Nakajima, T., Kim, S. C., Chang, C. Y., McDonnell, D. P., Renaud, J. P., & Jetten, A. M. (2004). Selective LXXLL peptides antagonize transcriptional activation by the retinoid-related orphan receptor RORgamma. *Biochemical and Biophysical Research Communications*, *315*(4), 919–927. http://dx.doi.org/10.1016/j.bbrc.2004.01.131.

Kurebayashi, S., Ueda, E., Sakaue, M., Patel, D. D., Medvedev, A., Zhang, F., & Jetten, A. M. (2000). Retinoid-related orphan receptor gamma (RORgamma) is essential for lymphoid organogenesis and controls apoptosis during thymopoiesis. *Proceedings of the National Academy of Sciences of the United States of America*, *97*(18), 10132–10137.

Lau, P., Bailey, P., Dowhan, D. H., & Muscat, G. E. (1999). Exogenous expression of a dominant negative RORalpha1 vector in muscle cells impairs differentiation: RORalpha1 directly interacts with p300 and myoD. *Nucleic Acids Research*, *27*(2), 411–420.

Lau, P., Nixon, S. J., Parton, R. G., & Muscat, G. E. (2004). RORalpha regulates the expression of genes involved in lipid homeostasis in skeletal muscle cells: Caveolin-3 and CPT-1 are direct targets of ROR. *The Journal of Biological Chemistry*, *279*(35), 36828–36840. http://dx.doi.org/10.1074/jbc.M404927200.

Lee, Y. K., Menezes, J. S., Umesaki, Y., & Mazmanian, S. K. (2011). Proinflammatory T-cell responses to gut microbiota promote experimental autoimmune encephalomyelitis. *Proceedings of the National Academy of Sciences of the United States of America*, *108*(Suppl. 1), 4615–4622. http://dx.doi.org/10.1073/pnas.1000082107.

Li, L., & Boussiotis, V. A. (2013). The role of IL-17-producing Foxp3+ CD4+ T cells in inflammatory bowel disease and colon cancer. *Clinical Immunology, 148*(2), 246–253. http://dx.doi.org/10.1016/j.clim.2013.05.003.

Li, Z., Lin, F., Zhuo, C., Deng, G., Chen, Z., Yin, S., … Li, B. (2014). PIM1 kinase phosphorylates the human transcription factor FOXP3 at serine 422 to negatively regulate its activity under inflammation. *The Journal of Biological Chemistry, 289*(39), 26872–26881. http://dx.doi.org/10.1074/jbc.M114.586651.

Li, B., Samanta, A., Song, X., Iacono, K. T., Bembas, K., Tao, R., … Greene, M. I. (2007). FOXP3 interactions with histone acetyltransferase and class II histone deacetylases are required for repression. *Proceedings of the National Academy of Sciences of the United States of America, 104*(11), 4571–4576. http://dx.doi.org/10.1073/pnas.0700298104.

Li, Q., Zou, J., Wang, M., Ding, X., Chepelev, I., Zhou, X., … Wang, R. F. (2014). Critical role of histone demethylase Jmjd3 in the regulation of CD4+ T-cell differentiation. *Nature Communications, 5*, 5780. http://dx.doi.org/10.1038/ncomms6780.

Lim, H. W., Kang, S. G., Ryu, J. K., Schilling, B., Fei, M., Lee, I. S., … Verdin, E. (2015a). SIRT1 deacetylates RORgammat and enhances Th17 cell generation. *The Journal of Experimental Medicine, 212*(5), 607–617. http://dx.doi.org/10.1084/jem.20132378.

Lin, F., Luo, X., Tsun, A., Li, Z., Li, D., & Li, B. (2015). Kaempferol enhances the suppressive function of Treg cells by inhibiting FOXP3 phosphorylation. *International Immunopharmacology, 28*(2), 859–865. http://dx.doi.org/10.1016/j.intimp.2015.03.044.

Littman, D. R., Sun, Z., Unutmaz, D., Sunshine, M. J., Petrie, H. T., & Zou, Y. R. (1999). Role of the nuclear hormone receptor ROR gamma in transcriptional regulation, thymocyte survival, and lymphoid organogenesis. *Cold Spring Harbor Symposia on Quantitative Biology, 64*, 373–381.

Liu, X., Li, H., Zhong, B., Blonska, M., Gorjestani, S., Yan, M., … Dong, C. (2013). USP18 inhibits NF-kappaB and NFAT activation during Th17 differentiation by deubiquitinating the TAK1-TAB1 complex. *The Journal of Experimental Medicine, 210*(8), 1575–1590. http://dx.doi.org/10.1084/jem.20122327.

Liu, Y., Wang, L., Han, R., Beier, U. H., Akimova, T., Bhatti, T., … Hancock, W. W. (2014). Two histone/protein acetyltransferases, CBP and p300, are indispensable for Foxp3+ T-regulatory cell development and function. *Molecular and Cellular Biology, 34*(21), 3993–4007. http://dx.doi.org/10.1128/mcb.00919-14.

Liu, Y., Zhang, P., Li, J., Kulkarni, A. B., Perruche, S., & Chen, W. (2008). A critical function for TGF-beta signaling in the development of natural CD4+CD25+Foxp3+ regulatory T cells. *Nature Immunology, 9*(6), 632–640. http://dx.doi.org/10.1038/ni.1607.

Long, M., Park, S. G., Strickland, I., Hayden, M. S., & Ghosh, S. (2009). Nuclear factor-kappaB modulates regulatory T cell development by directly regulating expression of Foxp3 transcription factor. *Immunity, 31*(6), 921–931. http://dx.doi.org/10.1016/j.immuni.2009.09.022.

Luo, X., Tarbell, K. V., Yang, H., Pothoven, K., Bailey, S. L., Ding, R., … Suthanthiran, M. (2007). Dendritic cells with TGF-beta1 differentiate naive CD4+CD25− T cells into islet-protective Foxp3+ regulatory T cells. *Proceedings of the National Academy of Sciences of the United States of America, 104*(8), 2821–2826. http://dx.doi.org/10.1073/pnas.0611646104.

Mantel, P. Y., Ouaked, N., Ruckert, B., Karagiannidis, C., Welz, R., Blaser, K., & Schmidt-Weber, C. B. (2006). Molecular mechanisms underlying FOXP3 induction in human T cells. *Journal of Immunology, 176*(6), 3593–3602.

Marie, J. C., Letterio, J. J., Gavin, M., & Rudensky, A. Y. (2005). TGF-beta1 maintains suppressor function and Foxp3 expression in CD4+CD25+ regulatory T cells. *The Journal of Experimental Medicine, 201*(7), 1061–1067. http://dx.doi.org/10.1084/jem.20042276.

Meyer Zu Horste, G., Wu, C., Wang, C., Cong, L., Pawlak, M., Lee, Y., ... Kuchroo, V. K. (2016). RBPJ controls development of pathogenic Th17 cells by regulating IL-23 receptor expression. *Cell Reports*, *16*(2), 392–404. http://dx.doi.org/10.1016/j.celrep.2016.05.088.

Michalek, R. D., Gerriets, V. A., Jacobs, S. R., Macintyre, A. N., MacIver, N. J., Mason, E. F., ... Rathmell, J. C. (2011). Cutting edge: Distinct glycolytic and lipid oxidative metabolic programs are essential for effector and regulatory CD4+ T cell subsets. *Journal of Immunology*, *186*(6), 3299–3303. http://dx.doi.org/10.4049/jimmunol.1003613.

Morawski, P. A., Mehra, P., Chen, C., Bhatti, T., & Wells, A. D. (2013). Foxp3 protein stability is regulated by cyclin-dependent kinase 2. *The Journal of Biological Chemistry*, *288*(34), 24494–24502. http://dx.doi.org/10.1074/jbc.M113.467704.

Mucida, D., Park, Y., Kim, G., Turovskaya, O., Scott, I., Kronenberg, M., & Cheroutre, H. (2007). Reciprocal TH17 and regulatory T cell differentiation mediated by retinoic acid. *Science*, *317*(5835), 256–260. http://dx.doi.org/10.1126/science.1145697.

Nie, H., Zheng, Y., Li, R., Guo, T. B., He, D., Fang, L., ... Zhang, J. Z. (2013). Phosphorylation of FOXP3 controls regulatory T cell function and is inhibited by TNF-alpha in rheumatoid arthritis. *Nature Medicine*, *19*(3), 322–328. http://dx.doi.org/10.1038/nm.3085.

Nylander, A., & Hafler, D. A. (2012). Multiple sclerosis. *The Journal of Clinical Investigation*, *122*(4), 1180–1188. http://dx.doi.org/10.1172/jci58649.

Obermajer, N., Popp, F. C., Soeder, Y., Haarer, J., Geissler, E. K., Schlitt, H. J., & Dahlke, M. H. (2014). Conversion of Th17 into IL-17A(neg) regulatory T cells: A novel mechanism in prolonged allograft survival promoted by mesenchymal stem cell-supported minimized immunosuppressive therapy. *Journal of Immunology*, *193*(10), 4988–4999. http://dx.doi.org/10.4049/jimmunol.1401776.

Okamoto, K., Iwai, Y., Oh-Hora, M., Yamamoto, M., Morio, T., Aoki, K., ... Takayanagi, H. (2010). IkappaBzeta regulates T(H)17 development by cooperating with ROR nuclear receptors. *Nature*, *464*(7293), 1381–1385. http://dx.doi.org/10.1038/nature08922.

Okoye, I. S., Coomes, S. M., Pelly, V. S., Czieso, S., Papayannopoulos, V., Tolmachova, T., ... Wilson, M. S. (2014). MicroRNA-containing T-regulatory-cell-derived exosomes suppress pathogenic T helper 1 cells. *Immunity*, *41*(1), 89–103. http://dx.doi.org/10.1016/j.immuni.2014.05.019.

Okui, T., Aoki, Y., Ito, H., Honda, T., & Yamazaki, K. (2012). The presence of IL-17+/FOXP3+ double-positive cells in periodontitis. *Journal of Dental Research*, *91*(6), 574–579. http://dx.doi.org/10.1177/0022034512446341.

O'Sullivan, D., & Pearce, E. L. (2014). Fatty acid synthesis tips the TH17-Treg cell balance. *Nature Medicine*, *20*(11), 1235–1236. http://dx.doi.org/10.1038/nm.3744.

Pandiyan, P., Conti, H. R., Zheng, L., Peterson, A. C., Mathern, D. R., Hernandez-Santos, N., ... Lenardo, M. J. (2011). CD4(+)CD25(+)Foxp3(+) regulatory T cells promote Th17 cells in vitro and enhance host resistance in mouse Candida albicans Th17 cell infection model. *Immunity*, *34*(3), 422–434. http://dx.doi.org/10.1016/j.immuni.2011.03.002.

Park, H., Li, Z., Yang, X. O., Chang, S. H., Nurieva, R., Wang, Y. H., ... Dong, C. (2005). A distinct lineage of CD4 T cells regulates tissue inflammation by producing interleukin 17. *Nature Immunology*, *6*(11), 1133–1141. http://dx.doi.org/10.1038/ni1261.

Perissi, V., & Rosenfeld, M. G. (2005). Controlling nuclear receptors: The circular logic of cofactor cycles. *Nature Reviews. Molecular Cell Biology*, *6*(7), 542–554. http://dx.doi.org/10.1038/nrm1682.

Pesenacker, A. M., Bending, D., Ursu, S., Wu, Q., Nistala, K., & Wedderburn, L. R. (2013). CD161 defines the subset of FoxP3+ T cells capable of producing proinflammatory cytokines. *Blood*, *121*(14), 2647–2658. http://dx.doi.org/10.1182/blood-2012-08-443473.

Pesenacker, A. M., Broady, R., & Levings, M. K. (2015). Control of tissue-localized immune responses by human regulatory T cells. *European Journal of Immunology, 45*(2), 333–343. http://dx.doi.org/10.1002/eji.201344205.

Pesenacker, A. M., & Wedderburn, L. R. (2013). T regulatory cells in childhood arthritis—Novel insights. *Expert Reviews in Molecular Medicine. 15*, e13. http://dx.doi.org/10.1017/erm.2013.14.

Pollizzi, K. N., & Powell, J. D. (2014). Integrating canonical and metabolic signalling programmes in the regulation of T cell responses. *Nature Reviews. Immunology, 14*(7), 435–446. http://dx.doi.org/10.1038/nri3701.

Quintana, F. J., Basso, A. S., Iglesias, A. H., Korn, T., Farez, M. F., Bettelli, E., … Weiner, H. L. (2008). Control of T(reg) and T(H)17 cell differentiation by the aryl hydrocarbon receptor. *Nature, 453*(7191), 65–71. http://dx.doi.org/10.1038/nature06880.

Rabizadeh, S., Rhee, K. J., Wu, S., Huso, D., Gan, C. M., Golub, J. E., … Sears, C. L. (2007). Enterotoxigenic Bacteroides fragilis: A potential instigator of colitis. *Inflammatory Bowel Diseases, 13*(12), 1475–1483. http://dx.doi.org/10.1002/ibd.20265.

Raffatellu, M., Santos, R. L., Verhoeven, D. E., George, M. D., Wilson, R. P., Winter, S. E., … Baumler, A. J. (2008). Simian immunodeficiency virus-induced mucosal interleukin-17 deficiency promotes Salmonella dissemination from the gut. *Nature Medicine, 14*(4), 421–428. http://dx.doi.org/10.1038/nm1743.

Rhee, K. J., Wu, S., Wu, X., Huso, D. L., Karim, B., Franco, A. A., … Sears, C. L. (2009). Induction of persistent colitis by a human commensal, enterotoxigenic Bacteroides fragilis, in wild-type C57BL/6 mice. *Infection and Immunity, 77*(4), 1708–1718. http://dx.doi.org/10.1128/iai.00814-08.

Rochette-Egly, C. (2003). Nuclear receptors: Integration of multiple signalling pathways through phosphorylation. *Cellular Signalling, 15*(4), 355–366.

Round, J. L., Lee, S. M., Li, J., Tran, G., Jabri, B., Chatila, T. A., & Mazmanian, S. K. (2011). The Toll-like receptor 2 pathway establishes colonization by a commensal of the human microbiota. *Science, 332*(6032), 974–977. http://dx.doi.org/10.1126/science.1206095.

Round, J. L., & Mazmanian, S. K. (2010). Inducible Foxp3 + regulatory T-cell development by a commensal bacterium of the intestinal microbiota. *Proceedings of the National Academy of Sciences of the United States of America, 107*(27), 12204–12209. http://dx.doi.org/10.1073/pnas.0909122107.

Roychoudhuri, R., Hirahara, K., Mousavi, K., Clever, D., Klebanoff, C. A., Bonelli, M., … Restifo, N. P. (2013). BACH2 represses effector programs to stabilize T(reg)-mediated immune homeostasis. *Nature, 498*(7455), 506–510. http://dx.doi.org/10.1038/nature12199.

Ruan, Q., Kameswaran, V., Tone, Y., Li, L., Liou, H. C., Greene, M. I., … Chen, Y. H. (2009). Development of Foxp3(+) regulatory t cells is driven by the c-Rel enhanceosome. *Immunity, 31*(6), 932–940. http://dx.doi.org/10.1016/j.immuni.2009.10.006.

Rudra, D., Egawa, T., Chong, M. M., Treuting, P., Littman, D. R., & Rudensky, A. Y. (2009). Runx-CBFbeta complexes control expression of the transcription factor Foxp3 in regulatory T cells. *Nature Immunology, 10*(11), 1170–1177. http://dx.doi.org/10.1038/ni.1795.

Rutz, S., Eidenschenk, C., Kiefer, J. R., & Ouyang, W. (2016). Post-translational regulation of RORgammat-A therapeutic target for the modulation of interleukin-17-mediated responses in autoimmune diseases. *Cytokine & Growth Factor Reviews, 30*, 1–17. http://dx.doi.org/10.1016/j.cytogfr.2016.07.004.

Rutz, S., Kayagaki, N., Phung, Q. T., Eidenschenk, C., Noubade, R., Wang, X., … Dixit, V. M. (2015). Deubiquitinase DUBA is a post-translational brake on interleukin-17 production in T cells. *Nature, 518*(7539), 417–421. http://dx.doi.org/10.1038/nature13979.

Sakaguchi, S. (2005). Naturally arising Foxp3-expressing CD25+CD4+ regulatory T cells in immunological tolerance to self and non-self. *Nature Immunology*, *6*(4), 345–352. http://dx.doi.org/10.1038/ni1178.

Sakaguchi, S., Takahashi, T., & Nishizuka, Y. (1982). Study on cellular events in post-thymectomy autoimmune oophoritis in mice. II. Requirement of Lyt-1 cells in normal female mice for the prevention of oophoritis. *The Journal of Experimental Medicine*, *156*(6), 1577–1586.

Sakaguchi, S., Yamaguchi, T., Nomura, T., & Ono, M. (2008). Regulatory T cells and immune tolerance. *Cell*, *133*(5), 775–787. http://dx.doi.org/10.1016/j.cell.2008.05.009.

Samanta, A., Li, B., Song, X., Bembas, K., Zhang, G., Katsumata, M., … Greene, M. I. (2008). TGF-beta and IL-6 signals modulate chromatin binding and promoter occupancy by acetylated FOXP3. *Proceedings of the National Academy of Sciences of the United States of America*, *105*(37), 14023–14027. http://dx.doi.org/10.1073/pnas.0806726105.

Sanchez Rodriguez, R., Pauli, M. L., Neuhaus, I. M., Yu, S. S., Arron, S. T., Harris, H. W., … Rosenblum, M. D. (2014). Memory regulatory T cells reside in human skin. *The Journal of Clinical Investigation*, *124*(3), 1027–1036. http://dx.doi.org/10.1172/jci72932.

Santori, F. R., Huang, P., van de Pavert, S. A., Douglass, E. F., Jr., Leaver, D. J., Haubrich, B. A., … Littman, D. R. (2015). Identification of natural RORgamma ligands that regulate the development of lymphoid cells. *Cell Metabolism*, *21*(2), 286–297. http://dx.doi.org/10.1016/j.cmet.2015.01.004.

Saouaf, S. J., Li, B., Zhang, G., Shen, Y., Furuuchi, N., Hancock, W. W., & Greene, M. I. (2009). Deacetylase inhibition increases regulatory T cell function and decreases incidence and severity of collagen-induced arthritis. *Experimental and Molecular Pathology*, *87*(2), 99–104. http://dx.doi.org/10.1016/j.yexmp.2009.06.003.

Shevach, E. M., DiPaolo, R. A., Andersson, J., Zhao, D. M., Stephens, G. L., & Thornton, A. M. (2006). The lifestyle of naturally occurring CD4+ CD25+ Foxp3+ regulatory T cells. *Immunological Reviews*, *212*, 60–73. http://dx.doi.org/10.1111/j.0105-2896.2006.00415.x.

Shi, L. Z., Wang, R., Huang, G., Vogel, P., Neale, G., Green, D. R., & Chi, H. (2011). HIF1alpha-dependent glycolytic pathway orchestrates a metabolic checkpoint for the differentiation of TH17 and Treg cells. *The Journal of Experimental Medicine*, *208*(7), 1367–1376. http://dx.doi.org/10.1084/jem.20110278.

Skepner, J., Ramesh, R., Trocha, M., Schmidt, D., Baloglu, E., Lobera, M., … Yang, J. (2014). Pharmacologic inhibition of RORgammat regulates Th17 signature gene expression and suppresses cutaneous inflammation in vivo. *Journal of Immunology*, *192*(6), 2564–2575. http://dx.doi.org/10.4049/jimmunol.1302190.

Smith, P. M., Howitt, M. R., Panikov, N., Michaud, M., Gallini, C. A., Bohlooly, Y. M., … Garrett, W. S. (2013). The microbial metabolites, short-chain fatty acids, regulate colonic Treg cell homeostasis. *Science*, *341*(6145), 569–573. http://dx.doi.org/10.1126/science.1241165.

Solt, L. A., Kumar, N., Nuhant, P., Wang, Y., Lauer, J. L., Liu, J., … Burris, T. P. (2011). Suppression of TH17 differentiation and autoimmunity by a synthetic ROR ligand. *Nature*, *472*(7344), 491–494. http://dx.doi.org/10.1038/nature10075.

Spits, H., Artis, D., Colonna, M., Diefenbach, A., Di Santo, J. P., Eberl, G., … Vivier, E. (2013). Innate lymphoid cells—A proposal for uniform nomenclature. *Nature Reviews. Immunology*, *13*(2), 145–149. http://dx.doi.org/10.1038/nri3365.

Spits, H., & Di Santo, J. P. (2011). The expanding family of innate lymphoid cells: Regulators and effectors of immunity and tissue remodeling. *Nature Immunology*, *12*(1), 21–27. http://dx.doi.org/10.1038/ni.1962.

Sugiyama, H., Gyulai, R., Toichi, E., Garaczi, E., Shimada, S., Stevens, S. R., … Cooper, K. D. (2005). Dysfunctional blood and target tissue CD4+CD25 high regulatory T cells in psoriasis: Mechanism underlying unrestrained pathogenic effector T cell proliferation. *Journal of Immunology*, *174*(1), 164–173.

Sun, C. M., Hall, J. A., Blank, R. B., Bouladoux, N., Oukka, M., Mora, J. R., & Belkaid, Y. (2007). Small intestine lamina propria dendritic cells promote de novo generation of Foxp3 T reg cells via retinoic acid. *The Journal of Experimental Medicine*, *204*(8), 1775–1785. http://dx.doi.org/10.1084/jem.20070602.

Sun, Z., Unutmaz, D., Zou, Y. R., Sunshine, M. J., Pierani, A., Brenner-Morton, S., … Littman, D. R. (2000). Requirement for RORgamma in thymocyte survival and lymphoid organ development. *Science*, *288*(5475), 2369–2373.

Takahashi, R., & Yoshimura, A. (2014). SOCS1 and regulation of regulatory T cells plasticity. *Journal of Immunology Research*, *2014*, 943149. http://dx.doi.org/10.1155/2014/943149.

Tao, R., de Zoeten, E. F., Ozkaynak, E., Chen, C., Wang, L., Porrett, P. M., … Hancock, W. W. (2007). Deacetylase inhibition promotes the generation and function of regulatory T cells. *Nature Medicine*, *13*(11), 1299–1307. http://dx.doi.org/10.1038/nm1652.

Tartar, D. M., VanMorlan, A. M., Wan, X., Guloglu, F. B., Jain, R., Haymaker, C. L., … Zaghouani, H. (2010). FoxP3+RORgammat+T helper intermediates display suppressive function against autoimmune diabetes. *Journal of Immunology*, *184*(7), 3377–3385. http://dx.doi.org/10.4049/jimmunol.0903324.

Tone, Y., Furuuchi, K., Kojima, Y., Tykocinski, M. L., Greene, M. I., & Tone, M. (2008). Smad3 and NFAT cooperate to induce Foxp3 expression through its enhancer. *Nature Immunology*, *9*(2), 194–202. http://dx.doi.org/10.1038/ni1549.

Tosiek, M. J., Fiette, L., El Daker, S., Eberl, G., & Freitas, A. A. (2016). IL-15-dependent balance between Foxp3 and RORgammat expression impacts inflammatory bowel disease. *Nature Communications*, 7, 10888. http://dx.doi.org/10.1038/ncomms10888.

Ueno, A., Jijon, H., Chan, R., Ford, K., Hirota, C., Kaplan, G. G., … Ghosh, S. (2013). Increased prevalence of circulating novel IL-17 secreting Foxp3 expressing CD4+ T cells and defective suppressive function of circulating Foxp3+ regulatory cells support plasticity between Th17 and regulatory T cells in inflammatory bowel disease patients. *Inflammatory Bowel Diseases*, *19*(12), 2522–2534. http://dx.doi.org/10.1097/MIB.0b013e3182a85709.

Ulges, A., Witsch, E. J., Pramanik, G., Klein, M., Birkner, K., Buhler, U., … Bopp, T. (2016). Protein kinase CK2 governs the molecular decision between encephalitogenic TH17 cell and Treg cell development. *Proceedings of the National Academy of Sciences of the United States of America*, *113*, 10145–10150. http://dx.doi.org/10.1073/pnas.1523869113.

Valmori, D., Raffin, C., Raimbaud, I., & Ayyoub, M. (2010). Human RORgammat+ TH17 cells preferentially differentiate from naive FOXP3+Treg in the presence of lineage-specific polarizing factors. *Proceedings of the National Academy of Sciences of the United States of America*, *107*(45), 19402–19407. http://dx.doi.org/10.1073/pnas.1008247107.

van Loosdregt, J., Brunen, D., Fleskens, V., Pals, C. E., Lam, E. W., & Coffer, P. J. (2011). Rapid temporal control of Foxp3 protein degradation by sirtuin-1. *PloS One*. *6*(4). e19047. http://dx.doi.org/10.1371/journal.pone.0019047.

van Loosdregt, J., Fleskens, V., Fu, J., Brenkman, A. B., Bekker, C. P., Pals, C. E., … Coffer, P. J. (2013). Stabilization of the transcription factor Foxp3 by the deubiquitinase USP7 increases Treg-cell-suppressive capacity. *Immunity*, *39*(2), 259–271. http://dx.doi.org/10.1016/j.immuni.2013.05.018.

van Loosdregt, J., Vercoulen, Y., Guichelaar, T., Gent, Y. Y., Beekman, J. M., van Beekum, O., … Coffer, P. J. (2010). Regulation of Treg functionality by acetylation-mediated Foxp3 protein stabilization. *Blood*, *115*(5), 965–974. http://dx.doi.org/10.1182/blood-2009-02-207118.

Veldhoen, M., Hirota, K., Westendorf, A. M., Buer, J., Dumoutier, L., Renauld, J. C., & Stockinger, B. (2008). The aryl hydrocarbon receptor links TH17-cell-mediated autoimmunity to environmental toxins. *Nature*, *453*(7191), 106–109. http://dx.doi.org/10.1038/nature06881.

Veldhoen, M., Hocking, R. J., Atkins, C. J., Locksley, R. M., & Stockinger, B. (2006). TGFbeta in the context of an inflammatory cytokine milieu supports de novo differentiation of IL-17-producing T cells. *Immunity*, *24*(2), 179–189. http://dx.doi.org/10.1016/j.immuni.2006.01.001.

Vignali, D. A. A., Collison, L. W., & Workman, C. J. (2008). How regulatory T cells work. *Nature Reviews. Immunology*, *8*(7), 523–532. http://dx.doi.org/10.1038/nri2343.

Voo, K. S., Wang, Y. H., Santori, F. R., Boggiano, C., Wang, Y. H., Arima, K., … Liu, Y. J. (2009). Identification of IL-17-producing FOXP3+ regulatory T cells in humans. *Proceedings of the National Academy of Sciences of the United States of America*, *106*(12), 4793–4798. http://dx.doi.org/10.1073/pnas.0900408106.

Wang, S., Charbonnier, L. M., Noval Rivas, M., Georgiev, P., Li, N., Gerber, G., … Chatila, T. A. (2015). MyD88 adaptor-dependent microbial sensing by regulatory T cells promotes mucosal tolerance and enforces commensalism. *Immunity*, *43*(2), 289–303. http://dx.doi.org/10.1016/j.immuni.2015.06.014.

Wang, L., Liu, Y., Han, R., Beier, U. H., Bhatti, T. R., Akimova, T., … Hancock, W. W. (2015). FOXP3+ regulatory T cell development and function require histone/protein deacetylase 3. *The Journal of Clinical Investigation*, *125*(3), 1111–1123. http://dx.doi.org/10.1172/jci77088.

Wang, L., Liu, Y., Han, R., Beier, U. H., Thomas, R. M., Wells, A. D., & Hancock, W. W. (2013). Mbd2 promotes foxp3 demethylation and T-regulatory-cell function. *Molecular and Cellular Biology*, *33*(20), 4106–4115. http://dx.doi.org/10.1128/mcb.00144-13.

Wang, X., Yang, J., Han, L., Zhao, K., Wu, Q., Bao, L., … Li, B. (2015). TRAF5-mediated Lys-63-linked polyubiquitination plays an essential role in positive regulation of RORgammat in promoting IL-17A expression. *The Journal of Biological Chemistry*, *290*(48), 29086–29094. http://dx.doi.org/10.1074/jbc.M115.664573.

Wang, C., Yosef, N., Gaublomme, J., Wu, C., Lee, Y., Clish, C. B., … Kuchroo, V. K. (2015). CD5L/AIM regulates lipid biosynthesis and restrains Th17 cell pathogenicity. *Cell*, *163*(6), 1413–1427. http://dx.doi.org/10.1016/j.cell.2015.10.068.

Wei, G., Wei, L., Zhu, J., Zang, C., Hu-Li, J., Yao, Z., … Zhao, K. (2009). Global mapping of H3K4me3 and H3K27me3 reveals specificity and plasticity in lineage fate determination of differentiating CD4+ T cells. *Immunity*, *30*(1), 155–167. http://dx.doi.org/10.1016/j.immuni.2008.12.009.

Whibley, N., & Gaffen, S. L. (2014). Brothers in arms: Th17 and Treg responses in Candida albicans immunity. *PLoS Pathogens*. *10*(12). e1004456. http://dx.doi.org/10.1371/journal.ppat.1004456.

Wu, H. J., Ivanov, I. I., Darce, J., Hattori, K., Shima, T., Umesaki, Y., … Mathis, D. (2010). Gut-residing segmented filamentous bacteria drive autoimmune arthritis via T helper 17 cells. *Immunity*, *32*(6), 815–827. http://dx.doi.org/10.1016/j.immuni.2010.06.001.

Wu, Q., Nie, J., Gao, Y., Xu, P., Sun, Q., Yang, J., … Li, B. (2015). Reciprocal regulation of RORgammat acetylation and function by p300 and HDAC1. *Scientific Reports*, *5*, 16355. http://dx.doi.org/10.1038/srep16355.

Wu, S., Rhee, K. J., Albesiano, E., Rabizadeh, S., Wu, X., Yen, H. R., … Sears, C. L. (2009). A human colonic commensal promotes colon tumorigenesis via activation of T helper type 17T cell responses. *Nature Medicine*, *15*(9), 1016–1022. http://dx.doi.org/10.1038/nm.2015.

Xiao, Y., Nagai, Y., Deng, G., Ohtani, T., Zhu, Z., Zhou, Z., … Greene, M. I. (2014). Dynamic interactions between TIP60 and p300 regulate FOXP3 function through a structural switch defined by a single lysine on TIP60. *Cell Reports*, 7(5), 1471–1480. http://dx.doi.org/10.1016/j.celrep.2014.04.021.

Xiao, S., Yosef, N., Yang, J., Wang, Y., Zhou, L., Zhu, C., … Kuchroo, V. K. (2014). Small-molecule RORgammat antagonists inhibit T helper 17 cell transcriptional network by divergent mechanisms. *Immunity*, *40*(4), 477–489. http://dx.doi.org/10.1016/j.immuni.2014.04.004.

Xie, H., Sadim, M. S., & Sun, Z. (2005). RORgammat recruits steroid receptor coactivators to ensure thymocyte survival. *Journal of Immunology*, *175*(6), 3800–3809.

Xu, W. (2005). Nuclear receptor coactivators: The key to unlock chromatin. *Biochemistry and Cell Biology*, *83*(4), 418–428. http://dx.doi.org/10.1139/o05-057.

Xu, T., Wang, X., Zhong, B., Nurieva, R. I., Ding, S., & Dong, C. (2011). Ursolic acid suppresses interleukin-17 (IL-17) production by selectively antagonizing the function of RORgamma t protein. *The Journal of Biological Chemistry*, *286*(26), 22707–22710. http://dx.doi.org/10.1074/jbc.C111.250407.

Yang, B. H., Hagemann, S., Mamareli, P., Lauer, U., Hoffmann, U., Beckstette, M., … Lochner, M. (2016). Foxp3(+) T cells expressing RORgammat represent a stable regulatory T-cell effector lineage with enhanced suppressive capacity during intestinal inflammation. *Mucosal Immunology*, *9*(2), 444–457. http://dx.doi.org/10.1038/mi.2015.74.

Yang, X. O., Nurieva, R., Martinez, G. J., Kang, H. S., Chung, Y., Pappu, B. P., … Dong, C. (2008). Molecular antagonism and plasticity of regulatory and inflammatory T cell programs. *Immunity*, *29*(1), 44–56. http://dx.doi.org/10.1016/j.immuni.2008.05.007.

Yang, R., Qu, C., Zhou, Y., Konkel, J. E., Shi, S., Liu, Y., … Shi, S. (2015). Hydrogen sulfide promotes Tet1- and Tet2-mediated Foxp3 demethylation to drive regulatory T cell differentiation and maintain immune homeostasis. *Immunity*, *43*(2), 251–263. http://dx.doi.org/10.1016/j.immuni.2015.07.017.

Yang, J., Xu, P., Han, L., Guo, Z., Wang, X., Chen, Z., … Li, B. (2015). Cutting edge: Ubiquitin-specific protease 4 promotes Th17 cell function under inflammation by deubiquitinating and stabilizing RORgammat. *Journal of Immunology*, *194*(9), 4094–4097. http://dx.doi.org/10.4049/jimmunol.1401451.

Ye, J., Su, X., Hsueh, E. C., Zhang, Y., Koenig, J. M., Hoft, D. F., & Peng, G. (2011). Human tumor-infiltrating Th17 cells have the capacity to differentiate into IFN-gamma+ and FOXP3+ T cells with potent suppressive function. *European Journal of Immunology*, *41*(4), 936–951. http://dx.doi.org/10.1002/eji.201040682.

Yu, S., Liu, C., Li, L., Tian, T., Wang, M., Hu, Y., … Ma, D. (2015). Inactivation of Notch signaling reverses the Th17/Treg imbalance in cells from patients with immune thrombocytopenia. *Laboratory Investigation*, *95*(2), 157–167. http://dx.doi.org/10.1038/labinvest.2014.142.

Zhang, J. C., Chen, G., Chen, L., Meng, Z. J., Xiong, X. Z., Liu, H. J., … Sun, S. W. (2016). TGF-beta/BAMBI pathway dysfunction contributes to peripheral Th17/Treg imbalance in chronic obstructive pulmonary disease. *Scientific Reports*, *6*, 31911. http://dx.doi.org/10.1038/srep31911.

Zhang, J., Chen, C., Hou, X., Gao, Y., Lin, F., Yang, J., … Li, B. (2013). Identification of the E3 deubiquitinase ubiquitin-specific peptidase 21 (USP21) as a positive regulator of the transcription factor GATA3. *The Journal of Biological Chemistry*, *288*(13), 9373–9382. http://dx.doi.org/10.1074/jbc.M112.374744.

Zhang, G. X., Gran, B., Yu, S., Li, J., Siglienti, I., Chen, X., … Rostami, A. (2003). Induction of experimental autoimmune encephalomyelitis in IL-12 receptor-beta 2-deficient mice: IL-12 responsiveness is not required in the pathogenesis of inflammatory demyelination in the central nervous system. *Journal of Immunology*, *170*(4), 2153–2160.

Zhang, F., Meng, G., & Strober, W. (2008). Interactions among the transcription factors Runx1, RORgammat and Foxp3 regulate the differentiation of interleukin 17-producing T cells. *Nature Immunology*, *9*(11), 1297–1306. http://dx.doi.org/10.1038/ni.1663.

Zhang, L., Yang, X. Q., Cheng, J., Hui, R. S., & Gao, T. W. (2010). Increased Th17 cells are accompanied by FoxP3(+) Treg cell accumulation and correlated with psoriasis disease severity. *Clinical Immunology*, *135*(1), 108–117. http://dx.doi.org/10.1016/j.clim.2009.11.008.

Zheng, Y., Josefowicz, S., Chaudhry, A., Peng, X. P., Forbush, K., & Rudensky, A. Y. (2010). Role of conserved non-coding DNA elements in the Foxp3 gene in regulatory T-cell fate. *Nature*, *463*(7282), 808–812. http://dx.doi.org/10.1038/nature08750.

Zhong, B., Liu, X., Wang, X., Chang, S. H., Liu, X., Wang, A., … Dong, C. (2012). Negative regulation of IL-17-mediated signaling and inflammation by the ubiquitin-specific protease USP25. *Nature Immunology*, *13*(11), 1110–1117. http://dx.doi.org/10.1038/ni.2427.

Zhou, X., Bailey-Bucktrout, S., Jeker, L. T., & Bluestone, J. A. (2009). Plasticity of CD4(+) FoxP3(+) T cells. *Current Opinion in Immunology*, *21*(3), 281–285. http://dx.doi.org/10.1016/j.coi.2009.05.007.

CHAPTER SIX

The Complex Role of the ZNF224 Transcription Factor in Cancer

E. Cesaro*, G. Sodaro*, G. Montano†, M. Grosso*, A. Lupo‡, P. Costanzo*,[1]

*University of Naples Federico II, Naples, Italy
†BioMedical Center, Lund University, Lund, Sweden
‡University of Sannio, Benevento, Italy
[1]Corresponding author: e-mail address: paola.costanzo@unina.it

Contents

Abstract

ZNF224 is a member of the Kruppel-associated box zinc finger proteins (KRAB-ZFPs) family. It was originally identified as a transcriptional repressor involved in gene-specific silencing through the recruitment of the corepressor KAP1, chromatin-modifying activities, and the arginine methyltransferase PRMT5 on the promoter of its target genes.

Recent findings indicate that ZNF224 can behave both as a tumor suppressor or an oncogene in different human cancers. The transcriptional regulatory properties of ZNF224 in these systems appear to be complex and influenced by specific sets of interactors. ZNF224 can also act as a transcription cofactor for other DNA-binding proteins. A role for ZNF224 in transcriptional activation has also emerged.

Here, we review the state of the literature supporting both roles of ZNF224 in cancer. We also examine the functional activity of ZNF224 as a transcription factor and the influence of protein partners on its dual behavior.

Advances in Protein Chemistry and Structural Biology, Volume 107
ISSN 1876-1623
http://dx.doi.org/10.1016/bs.apcsb.2016.11.003

Increasing information on the mechanism through which ZNF224 can operate could lead to the identification of agents capable of modulating ZNF224 function, thus potentially paving the way to new therapeutic strategies for treatment of cancer.

1. INTRODUCTION

1.1 KRAB-ZFPs: A Protein Family Derived From Combinations of Ancient and Modern Structural and Functional Modules

Gene repression represents a powerful tool in controlling gene expression in order to prevent the metabolic dysregulation during the development or the differentiation program, especially in eukaryotes. Inhibition of basal transcription machinery formation constitutes the first mechanism through which downregulation of gene expression can be produced. This type of transcriptional repression can be obtained by direct disruption of functional interactions of a transcription factor with the basal transcription complex (Fig. 1A). Inactivation of the transcriptional function of an activator may represent the second mechanism of gene repression. In this case the repressor may affect the availability of an activator through different mechanisms (degradation, sequestration, covalent modification, localization, site competition, and protein–protein interaction) (Fig. 1B). In both these conditions the repression mechanisms require the specific binding of the repressor molecule to one or more components of the basal transcription machinery or to the transcriptional activator itself. A more recently described mechanism strictly depends on the presence of epigenetic modifications that can affect the chromatin structure close to the target genes (Fig. 1C; Gaston & Jayaraman, 2003; Payankaulam, Li, & Arnosti, 2010). Kruppel-associated box zinc finger proteins (KRAB-ZFPs) constitute one of the most widespread and numerous protein family of transcription repressors that cause gene-specific transcriptional repression by physical interaction with chromatin-remodeling factors (Huntley et al., 2006; Shannon, Hamilton, Gordon, Branscomb, & Stubbs, 2003).

Most of the KRAB-ZFPs have been generated during the recent evolutionary history from the expansion of clustered genes by subsequent duplication events. Indeed, in humans KRAB-ZFPs represent one-third of about 800 members of the large zinc finger protein family (Emerson & Thomas, 2010; Nowick, Hamilton, Zhang, & Stubbs, 2010).

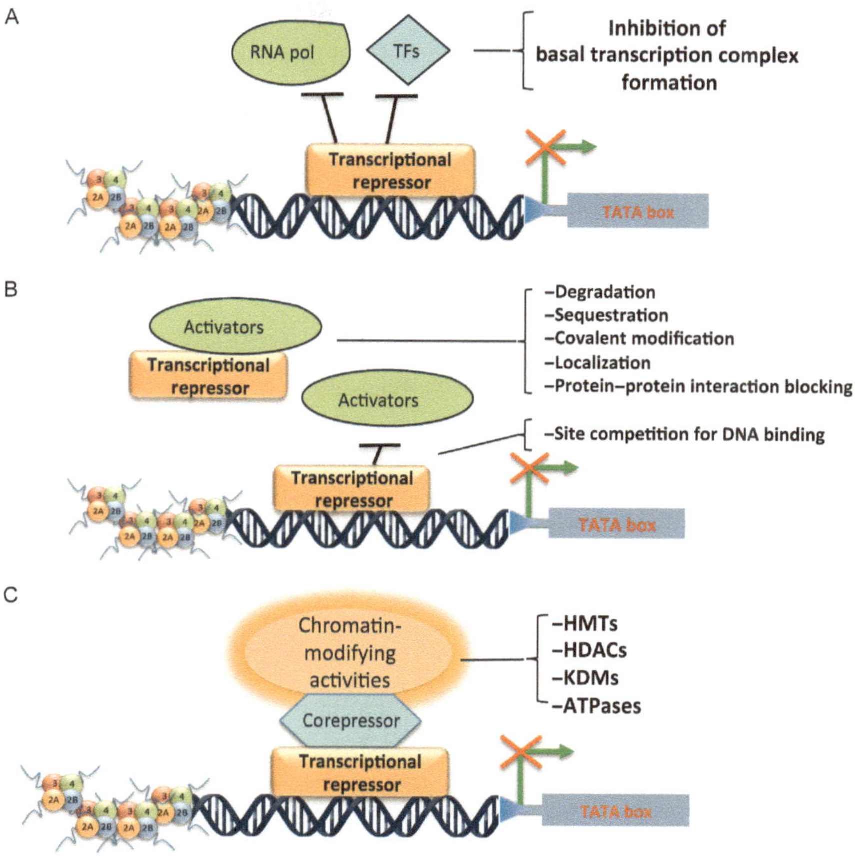

Fig. 1 Schematic representation of different mechanisms of transcriptional repression.

KRAB-ZFPs are characterized by the presence, in their NH2-terminus, of the potent KRAB repression module, while, at the COOH-terminus, they contain tandem repeats of C2H2 zinc finger DNA-binding domains. Repression activity of the KRAB-ZFPs is strictly dependent on the ability of the KRAB repression module to recruit the corepressor KRAB-associated protein 1 (KAP1) (Kim et al., 1996; Peng et al., 2000; Urrutia, 2003). Interaction between the KRAB module and KAP1 has been demonstrated to be crucial to display the maximal repression effect on target gene transcription, through the enrollment of chromatin modifiers, comprising the histone methyltransferase SETDB1, the CHD3/Mi2 component of the NuRD complex, and heterochromatin protein 1 (HP1) (Schultz, Ayyanathan, Negorev, Maul, & Rauscher, 2002; Schultz, Friedman, & Rauscher, 2001; Sripathy, Stevens, & Schultz, 2006).

Several studies have shown that the KRAB-ZFPs/KAP1 system plays a relevant role in embryonic differentiation and morphogenesis, in the establishment of genomic imprinting and pluripotency, and in self-renewal maintenance of embryonic stem cells (Hu et al., 2009; Messerschmidt et al., 2012; Quenneville et al., 2011; Seki et al., 2010; Shibata, Blauvelt, Liem, & García-García, 2011).

Furthermore, an increasing number of reports demonstrate that some KRAB-ZFPs are also associated with metabolism regulation, apoptosis, and cancer (Lupo et al., 2013).

Regulation of transcription, through KRAB-mediated repression mechanism, is also a defense mechanism against a variety of environment injuries. Recently, indeed, it has been proposed that the KRAB-ZFPs/KAP1 system could be involved in the epigenetic control of the adaptive response of mouse and human specialized immune cells. Particularly, the KRAB-ZFPs/KAP1-mediated repression mechanism seems to drive antibody maturation and diversification in activated B cells, as shown in different mouse models. Different studies carried out in KO mice demonstrated that KAP1 is also involved in controlling T cell-mediated response and tolerance. Finally, the regulatory function of KRAB-ZFPs/KAP1 has been confirmed also in human models where KAP1 plays a significant role in the maintenance of human immune tolerance (Santoni de Sio, 2014).

More recently, it has been speculated that the KRAB-ZFPs have evolved in mammals as an adaptive genomic surveillance system against mobile parasitic DNA (Wolf, Greenberg, & MacFarland, 2015).

Retroviral and nonretroviral DNA constitute the endogenous retroviruses including sequences derived from the infection by retroviruses and nonretroviral retrotransposons (LINE, SINEs, etc.). These interspersed sequences may be activated in specific conditions and, somehow, cause insurgence of cancer (Kassiotis, 2014; Katoh & Kurata, 2013; Lamprecht et al., 2010).

To repress these endogenous retroelements and protect the integrity of the genome, the tetrapod species developed a very powerful strategy based on the KRAB-ZFPs-mediated repression mechanism to induce a stable and silenced chromatin structure (Trono, 2016; Wolf et al., 2015).

Despite the considerable amount of data focused on the biochemical mechanisms of the KRAB-ZFPs/KAP1 system, several aspects concerning the roles of these protein complexes in adult tissues remain elusive. In addition, for many members of the KRAB-ZFP family so far identified it has not been possible to recognize a specific function, because their target genes are still unknown. However, it is expected that biochemical and functional characterization of a growing number of KRAB-ZFP proteins could contribute to clarify their physiological functions.

In this review, we report the most recent experimental data concerning ZNF224, a member of KRAB-ZFP proteins family, and discuss its functions in cancer.

Interestingly, ZNF224 gene is differentially expressed in human cells and tissues and, in accordance with the distinctive cellular environment, displays different essential functions in transcriptional or posttranscriptional gene regulation. In addition, physical interaction between ZNF224 and different protein partners in various cellular contexts may generate a differential response (repression or activation) toward specific sets of target genes. In this scenario, ZNF224 may assume the role of a typical tumor suppressor or oncogene, thus eliciting apoptosis or stimulating cell proliferation.

1.2 ZNF224: One Gene, Two Messenger RNAs, and Two Proteins With Differential Biological Functions

The gene encoding for ZNF224, a member of the KRAB-ZFP family, is mapped at the human gene locus 19q13.2 and has no functional mouse homologue. In fact, this human region is the product of duplications of a common ancestral gene resulting, therefore, in a gain of novel functions deriving from the appearance of new genes. On the contrary, analysis of the mouse genome revealed that a loss of these common parental sequences occurred after the human and mice lineage divergence. These data suggest that these new genes have conferred an evolutionary advantage to humans with respect to mice (Huntley et al., 2006; Lupo et al., 2013; Shannon et al., 2003).

Analysis of the ZNF224 genomic sequences and comparison with the related 2466-bp-long cDNA put in evidence the existence of six exonic regions generating the 707-aa-long protein. All together, exons I, II, and III encode for the 303-bp-long 5′ UTR, while the fusion of exons IV, V, and VI generates the complete coding region plus a short 3′ UTR of the mRNA.

It is interesting to note that exons IV and V encode for the KRAB-A and KRAB-b motifs, while numerous zinc finger modules are generated from the single large exon VI, as also found in other KRAB-ZFPs (Medugno et al., 2003). Identification of a second transcript, namely, ZNF255, supported the notion that a single gene coding for KRAB-ZFPs may give rise to different functional mRNAs through alternative splicing or different promoter usage (Medugno et al., 2007). With respect to the ZNF224 mRNA, the ZNF255 transcript contains the same 3′ UTR and a more extended 5′ UTR. Translation of this transcript produces a novel protein of 622 amino acids that spans the region between amino acids 86 and 707 of ZNF224, but lacks the KRAB domain and includes the 19 zinc finger domains. Therefore, the ZNF224 and ZNF255 proteins, originating from a single gene, differ by the presence or absence of the KRAB domain, respectively (Fig. 2). These results are in accordance with data obtained for other KRAB-containing ZFPs (Lupo, Izzo, & Costanzo, 2012).

While ZNF224, that contains the transcriptional repression domain KRAB-A, is known to repress gene transcription, the KRAB-less ZNF255 variant displays a reduced ability to inhibit gene transcription (Medugno et al., 2007, 2005).

Besides the different ability in repressing gene transcription, which is related to the presence or absence of the KRAB-A domain, ZNF224 and ZNF255 also have a distinctive cellular localization, as shown by immunofluorescent staining of the FLAG-tagged fusion proteins expressed in several cell types (Medugno et al., 2007). ZNF224 is largely found in the nucleus as expected for a transcription factor, whereas ZNF255 is localized in both cytoplasm and nucleus. In this latter subcellular compartment, ZNF255 displays a speckled distribution, a pattern commonly associated with nuclear splicing factors, in contrast to ZNF224 which shows a more diffuse nuclear distribution.

Intriguingly, ZNF255 was also found to be associated with the nucleolus; this finding, along with the nuclear splecked distribution of ZNF255, strongly indicates a role for this isoform in splicing and in ribosomal biosynthesis (Medugno et al., 2007).

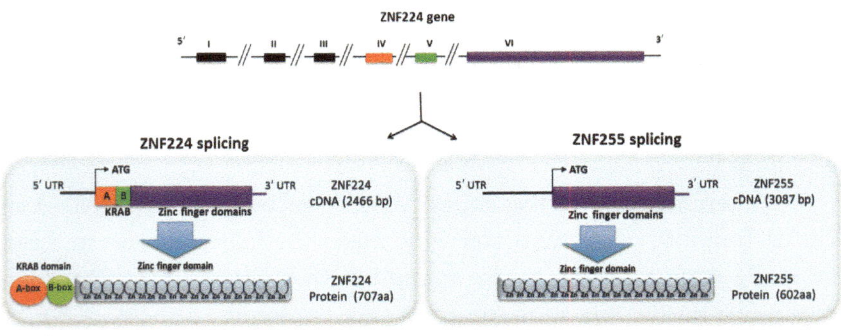

Fig. 2 Origin of the two different transcripts and proteins arising from the ZNF224 gene by alternative splicing.

Therefore, ZNF224 and ZNF255 display typical features of KRAB-ZF proteins that, although originating from a single gene, give rise to multiple isoforms with distinct subcellular localizations and relevant functional differences (Cote et al., 2001; Sun et al., 2005).

2. ZNF224 INTERACTS WITH DISTINCT PARTNERS TO ELICIT DIFFERENTIAL REGULATORY PATHWAYS

ZNF224 was originally identified as the transcriptional repressor of the human aldolase A gene. In particular, ZNF224 binds to a negative regulatory element (AldA-NRE), located in the distal promoter of the aldolase A gene, and suppresses the transcriptional activity of this promoter (Medugno et al., 2005). Subsequently, the mitochondrial carrier citrate gene (CIC), a gene involved in fatty acids metabolism, was identified as another target of transcriptional repression mediated by ZNF224 (Iacobazzi et al., 2009), thus indicating a more general role of ZNF224 in controlling metabolic pathways.

The repressive activity of ZNF224 requires the specific interaction of its KRAB-A domain with the KAP1 corepressor. Indeed, the RBCC motif of KAP1, including a Ring finger, two B-box zinc fingers, and a coiled-coil domain, binds with high affinity the KRAB repression module of KRAB-ZFPs (Friedman et al., 1996; Kim et al., 1996). Tethering KAP1 corepressor to ZNF224 elicits the recruitment of histone deacetylases, histone methylases, and other chromatin modifiers to the promoter regions of ZNF224 target genes through the C-terminus tandem PHD and the bromodomain of KAP1. In addition, the central region of KAP1 contains the binding domain of HP1, a specific protein partner involved in heterochromatin packaging. In this way, KAP1 is able to induce a facultative heterochromatin environment required for the mediated KRAB-ZFPs gene silencing (Lechner, Begg, Speicher, & Rauscher, 2000; Sripathy et al., 2006).

The study of the biochemical and functional features of ZNF224 KRAB-A/KAP1 interaction led us to identify the arginine methyltransferase PRMT5 as an additional component of the ZNF224 transcriptional repressor complex. Particularly, we demonstrated that methylation of arginine 3 of histone H4 (H4R3) by PRMT5 is necessary for ZNF224-mediated gene repression, thus highlighting a chromatin modification required for ZNF224 activity (Cesaro et al., 2009). Moreover, this finding establishes a link between histone arginine methylation and the KRAB-zinc finger-mediated transcriptional repression (Fig. 3).

In these years the relationship between chromatin structure and gene regulation by KRAB-ZFPs is becoming ever more evident. KRAB/KAP1-mediated epigenetic regulation, by modulating the dynamics of hetero- and euchromatin maintenance, plays important roles in different physiological and pathological conditions, such as the maintenance of genome integrity, the response to behavioral stress, the regulation of differentiation in hematopoiesis, the tumorigenesis, and the onset of neurodegenerative diseases (Jakobsson et al., 2008; Santoni de Sio et al., 2012; Shin et al., 2011; Zheng et al., 2000).

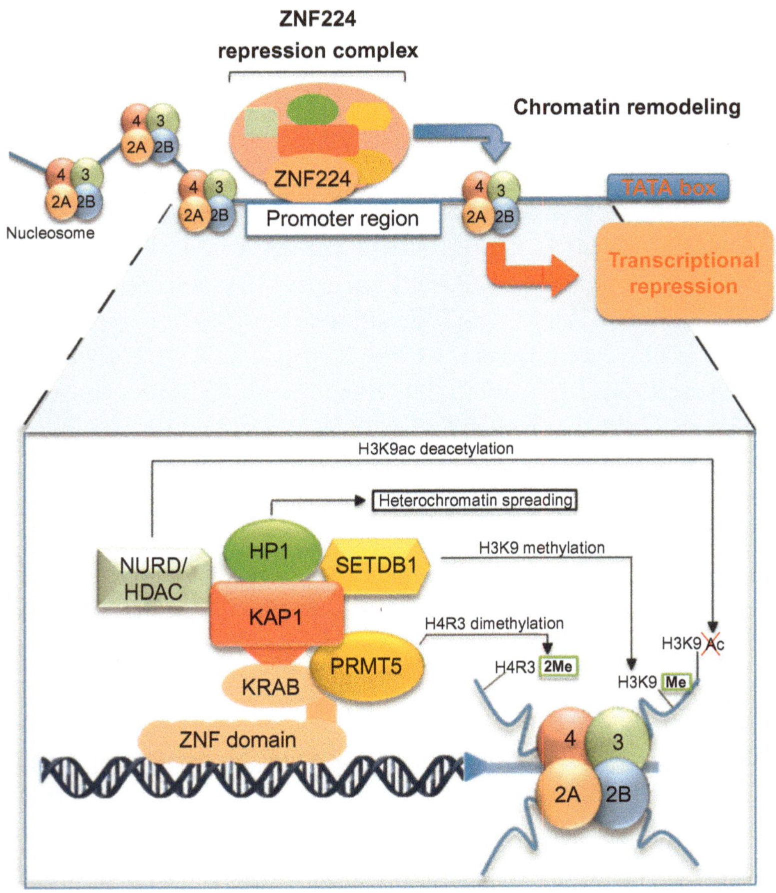

Fig. 3 Schematic representation of the ZNF224 repression complex, showing the role of ZNF224 interacting proteins in the chromatin-remodeling process that lead to transcriptional repression.

However, further investigations are required to clarify how different combinations of histone modifications lead to the recruitment of protein complexes that regulate KRAB-ZFPs-mediated transcriptional regulation.

Many studies reveal that KAP1 is subjected to multiple posttranslational modifications (Lee, Thomas, Yang, & Ann, 2007; White et al., 2006) that govern KAP1 function and are crucial to properly regulate the expression of KAP1 target genes in response to different extracellular stimuli (Kubota et al., 2013; Li et al., 2007).

Our more recent findings indicate that KAP1 interacts with PRMT5 and is a novel substrate for PRMT5 methylation (di Caprio, Ciano, Montano, Costanzo, & Cesaro, 2015). We suggest that the methylation of KAP1 arginine residues could prevent the recruitment of KAP1 on the ZNF224 transcriptional complex, thus promoting the interaction between ZNF224 and other cofactors and the switch of ZNF224 from a repression to an activation complex.

Indeed, although traditionally described as transcriptional repressors, some KRAB-ZFPs may display a dual function in gene repression and activation depending on specific cellular contests and interacting proteins (Chen et al., 2015; Hallen et al., 2011; Yi et al., 2004). Recent findings obtained in our laboratory and in other laboratories highlighted a role for ZNF224 as a transcriptional activator. In chronic lymphocytic leukemia (CLL) we demonstrated that ZNF224 binds cyclin D3 gene and positively modulates its expression, thus contributing to impaired growth of CLL cells (manuscript submitted for publication).

Cho and coauthors, instead, identified miR-663a as a target gene of ZNF224 transcriptional activation (Cho et al., 2016). They demonstrated that ZNF224 increases miR-663a transcription by binding to a ZNF224 consensus sequence in the miR-663a promoter. miR-663a in turn decreases p53 and p21 expression, thus inducing cell growth and apoptosis resistance in breast cancer (Cho et al., 2016).

However, the mechanisms underlying the transcriptional activation function of ZNF224 and the cofactors recruited by ZNF224 on the cyclin D3 and miR-663a genes are still largely unknown.

DEPDC1 (DEP domain-containing 1) protein is another molecular partner of ZNF224, identified in bladder cancer (Harada et al., 2010). In bladder cancer cells DEPDC1 acts as a transcriptional corepressor of ZNF224 (Harada et al., 2010). ZNF224 recruits DEPDC1 on the promoter of A20, a negative regulator of the NF-κB antiapoptotic pathway (Beyaert, Heyninck, & Van Huffel, 2000; Krikos, Laherty, & Dixit, 1992), thus repressing the A20 gene transcription.

All together, these findings indicate that ZNF224 is able to form different transcriptional complexes on its target genes by recruiting partners that act as coactivators or corepressors. The identification of the components taking part to ZNF224 multiprotein transcriptional complexes in different cell types will contribute to elucidate physiological and pathological functions of ZNF224.

Apart from its role as a DNA-binding transcription factor, ZNF224 can also act as a transcriptional cofactor of the Wilms' tumor 1 protein, WT1, through the interaction of ZNF224 with the WT1(-KTS) isoform (Florio et al., 2010). WT1 is a zinc finger protein that exerts an oncogenic role in a wide range of solid tumors and hematopoietic malignancies (Hohenstein & Hastie, 2006; Keilholz et al., 2005). The WT1(-KTS) transcriptional isoform can work as an activator, coactivator, or repressor depending on the cellular context and cofactor interactions (Larsson et al., 1995; Nakagama, Heinrich, Pelletier, & Housman, 1995; Roberts, 2006).

Our findings demonstrated that ZNF224, by acting as a transcriptional cofactor of WT1, cooperates with WT1 in the transcriptional regulation of WT1 apoptotic target genes in chronic myelogenous leukemia (CML) cells (Montano et al., 2016).

More in detail, by analyzing at the molecular level the function of ZNF224 as a cofactor of WT1, we identified at least three different mechanisms through which ZNF224 modulates WT1-dependent transcription: by acting as a coactivator of WT1 in the regulation of proapoptotic genes, by suppressing WT1-mediated transactivation of antiapoptotic genes (Montano et al., 2013), and by preventing WT1 repression of the tumor suppressor interferon regulatory factor 8 (IRF8) proapoptotic gene (Montano et al., 2016).

3. MULTIPLE ROLES OF KRAB-ZFPS IN CANCER

Alterations in transcriptional regulation and genomic and epigenomic aberrations are essential components of tumorigenesis (Baylin et al., 2001; Baylin & Ohm, 2006; Chakravarthi, Nepal, & Varambally, 2016; Esteller, 2007; Hanahan & Weinberg, 2011). In humans, KRAB-ZPFs constitute one of the largest family of transcriptional regulators (Urrutia, 2003) that play a role in epigenetic silencing. These proteins, through the recruitment of KAP1 and histone modifiers, lead to the spread of heterochromatin and long-range repression of promoters (Vogel et al., 2006). Functional modifications of KRAB-ZPFs may result in failure of heterochromatin formation and inappropriate gene expression. Therefore, it is not surprising that many KRAB-ZFPs have been found to be involved in neoplastic transformation (Lim & Park, 2015).

Several KRAB-ZFPs are frequently downregulated in cancer by promoter methylation and exhibit tumor-suppressive properties by controlling genes involved in cell proliferation and apoptosis. For instance, ZNF23, whose expression is reduced in human cancer, inhibits cell cycle progression via upregulation of p27-kip1 (Huang et al., 2007). In ovarian cancer cells, it has been demonstrated that ZNF23 also induces apoptosis to suppress tumor cell growth (Huang et al., 2008).

The KRAB-ZFP ZNF382 acts as a tumor suppressor in a variety of carcinomas, by repressing the expression of multiple oncogenes and by exerting a proapoptotic role through the inhibition of NF-κB and AP-1 signaling (Cheng et al., 2010).

ZNF545 is involved in the inhibition of tumorigenesis through at least two different mechanisms, deriving from its localization in both the nucleus and the nucleolus of some cells. Consequently, ZNF545 may act as a tumor suppressor partially regulating nucleolar functions, such as ribosome biogenesis, and partially through downregulation of target genes transcription, such as NF-κB and AP-1 (Cheng et al., 2012).

The KRAB-ZFP Apak (ATM and p53-associated KZNF protein) was identified as a negative regulator of p53-mediated apoptosis (Tian et al., 2009). Two different mechanisms are involved in this regulation. Apak binds to p53 through its KRAB domain and recruits KAP1–HDAC–ATM complex to attenuate p53 acetylation, thus repressing p53 activity. The p53 repression leads to the inhibition of proapoptotic genes expression (Tian et al., 2009).

In addition, Apak competes with p53 for the binding to the proapoptotic gene p53AIP1 to inhibit p53AIP1 expression (He & Zhang, 2012).

In addition to their role as transcriptional repressors, however, some KRAB-ZFPs act as activators of gene expression in cancer. The KRAB-ZFP ZNF300 is emerging as a novel oncogene in humans. Indeed, it has been demonstrated that ZNF300 promotes tumor progression by activating the NF-κB pathway. ZNF300 exerts a transcriptional activation on TRAF2, a member of the TRAF protein family, which in turn could further activate the NF-κB pathway. Moreover, ZNF300 physically interacts with IKKβ and is able to activate the kinase activity of IKKβ, thus further stimulating the NF-κB signaling pathway (Wang et al., 2012).

Interestingly, some KRAB-ZFPs have been shown to be involved in both activation and repression of gene transcription in cancer, depending on the cellular context and the interaction with cell-specific molecular partners. For example, the KRAB-zinc finger transcriptional repressor ZBRK1 (zinc finger and BRCA1-interacting protein with KRAB domain-1) modulates the transcription of gene involved in cell cycle and in DNA damage response, by interacting with two different corepressors, KAP1 and BRCA1 (Furuta et al., 2006; Lee et al., 2007). However, in addition to its role of transcriptional repressor, ZBRK1 may positively affect gene transcription. Indeed, it has been recently reported that numerous genes are transcriptionally activated by ZBRK1, although in many cases the molecular mechanisms and the role of potential ZBRK1 coactivators still remain unclear (Lin et al., 2010). These evidences suggest that ZBRK1 can function as repressor or activator according to the cancer type and, by controlling the transcription of different target genes, it can promote or impede cancer development (Furuta et al., 2006; Zheng et al., 2000).

Due to the high number of putative target genes and molecular partners of the approximately 400 KRAB-ZFPs in humans and the dual role that some KRAB-ZFPs may exert in transcriptional regulation (as repressor or activator), many efforts will be necessary to shed light on the cellular function of KRAB-ZFPs in cancer. These efforts are expected to provide support to the development of cancer therapeutics.

Recent findings have highlighted the emerging role of ZNF224 in cancer and have demonstrated a dual function for this KRAB-ZFP, which acts as oncogene in some types of cancer and as tumor suppressor in others. The different roles played by ZNF224 in cancer derive from its ability to behave as a multifunctional protein that exerts complex transcriptional regulatory activities depending on the cell context and the presence of specific interactors.

3.1 ZNF224 as Tumor Suppressor: ZNF224/WT1 Complex Is a Model of Transcriptional Control of Apoptosis in Leukemia

In chronic myelogenous leukemia (CML), ZNF224 plays a key proapoptotic and antiproliferative role. To exert these effects, ZNF224 works in a DNA-binding-independent mode of transcriptional regulation, acting as a WT1 transcriptional cofactor.

One of the main functions of WT1 in cancer seems to be related to the control of apoptosis, but the regulatory properties of WT1 are complex; indeed, the effects of WT1 on target gene expression appear to be isoform specific, strongly influenced by the cellular context and physical interactions with different cofactors (Davies et al., 1998; Richard, Schumacher, Royer, & Roberts, 2001; Roberts, 2006).

In leukemic cells WT1 exerts antiapoptotic effects (Ito et al., 2006; Jomgeow et al., 2006; Loeb, 2006) through the transcriptional regulation of several apoptotic genes, including the proapoptotic bcl-2 family members, Bax and Bak (Holloway, Kon, & De Lisi, 2008; Morrison, English, & Licht, 2005), bcl-2 itself (Mayo et al., 1999), and the antiapoptotic genes A1/Bfl1 (Simpson et al., 2006) and bag3 (Cesaro et al., 2010).

We had previously demonstrated that ZNF224 is a novel transcriptional cofactor of WT1 (Florio et al., 2010). Identification of a functional interaction between ZNF224 and WT1 that, in turn, is responsible of the increased transcription of the VDR gene was the first step toward the definition of a novel role for ZNF224 as a transcriptional cofactor. The modulation of VDR gene expression by the ZNF224/WT1 complex in the CML K562 cell line points out a very interesting function for this transcriptional complex in leukemia (Florio et al., 2010). In fact, leukemic cells in the majority of CMLs express high levels of WT1; on the other hand, vitamin D-induced pathway is really crucial to activate Hox-A10/MafB system and to control hematopoietic differentiation (Gemelli et al., 2008; Rosenfeld, Cheever, & Gaiger, 2003).

Starting from these data, we searched for other WT1 target genes potentially regulated by the ZNF224/WT1 complex. By chromatin immunoprecipitation experiments, we proved that ZNF224 is recruited by WT1(-KTS) on the promoter of apoptosis-regulating genes in CML cells.

A combination of overexpression and knockdown experiments in K562 cell line led us to highlight the dual effect of ZNF224 on the expression of WT1 target genes; ZNF224 acts as a coactivator of WT1 in the regulation of proapoptotic genes such as Bak, Bax, and VDR and suppresses WT1-mediated transactivation of antiapoptotic genes such as bag3 and A1/Bfl1 (Montano et al., 2013).

More recently, we have identified a third mechanism by which ZNF224 acts in fine-tuning of WT1-dependent control of gene expression. More in detail, we found that it positively modulates the expression of IRF8 gene (Montano et al., 2016). IRF8 is known as a tumor suppressor in some leukemias (Gurevich, Rosten, Schwieger, Stocking, & Humphries, 2006;

Hao & Ren, 2000; Konieczna et al., 2008), and its expression is repressed by WT1 (Vidovic et al., 2010). We demonstrated that ZNF224 is recruited by WT1 on the IRF8 promoter and takes part in the transcriptional regulation of IRF8, counteracting the repressive activity of WT1 (Montano et al., 2016; Fig. 4).

All together, our findings reveal that ZNF224 in CML, through the interaction with WT1, is able to shift the balance of antiapoptotic and proapoptotic signals in favor of the latter.

Moreover, consistent with a proapoptotic role of ZNF224, we also demonstrated that the DNA-damaging drug cytosine arabinoside (ara-C) induces the expression of ZNF224 in leukemic cells and this induction enhances cell apoptotic response to ara-C (Montano et al., 2013), while WT1 can confer resistance to this treatment (Vidovic et al., 2010). These findings disclose a novel role for ZNF224 as an important mediator of the ara-C-induced activation of proapoptotic genes expression, thus providing, for the first time, knowledge on the mechanisms of ara-C-mediated apoptosis which are still largely undetermined.

CML is a myeloproliferative disorder in which the rearranged fusion protein BCR–ABL possesses a constitutively activated ABL tyrosine kinase domain (Deininger, Goldman, & Melo, 2000) and supports malignant transformation, promoting uncontrolled proliferation and antiapoptotic processes (Bedi, Zehnbauer, Barber, Sharkis, & Jones, 1994; McGahon et al.,

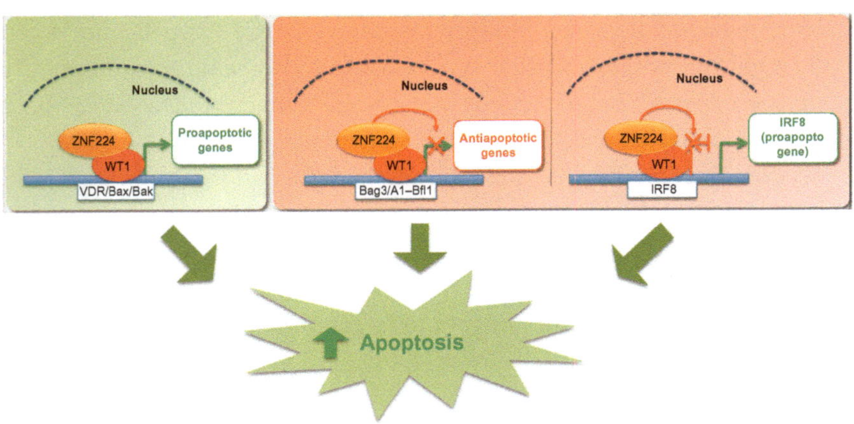

Fig. 4 Apoptosis induction by the ZNF224/WT1 complex in CML. ZNF224 acts as a coactivator on WT1 proapoptotic genes and prevents both WT1 activation on its antiapoptotic genes and WT1 repression on the proapoptotic gene IRF8.

1994). Nevertheless, the antiapoptotic and proliferative pathways triggered by BCR/ABL are still largely uncharacterized.

Our previous findings on the significant role exerted by ZNF224 in ara-C-mediated apoptosis prompted us to investigate the molecular mechanisms controlling ZNF224 expression in CML and to verify if ZNF224 is a downstream target of BCR–ABL kinase activity. Indeed, the identification of the pathways involved in ZNF224 downmodulation in CML provides the rationale for the development of new therapeutic approaches in CML.

We initially observed that ZNF224 gene expression is downmodulated both in BCR/ABL-positive cell lines and in primary CML samples, and that inhibition of BCR–ABL tyrosine kinase activity by imatinib and by second-generation tyrosine kinase inhibitors triggers upregulation of ZNF224 expression.

Interestingly, evaluation of ZNF224 mRNA expression in patients with BCR–ABL-positive chronic phase CML showed that ZNF224 is significantly downregulated in patients with a warning or failure response to imatinib treatment as compared to patients with an optimal response to the treatment, thus indicating that ZNF224 could represent a promising prognostic marker to predict possible response to imatinib therapy.

Successively, by investigating the molecular mechanisms by which BCR–ABL signaling represses ZNF224 expression, we identified WT1 as a transcriptional repressor of ZNF224 in CML cells, in agreement with the role of WT1 as BCR–ABL survival factor. Furthermore, starting from the observation that WT1 expression is induced by the oncogenic signaling triggered by BCR–ABL via the phosphatidylinositol-3 kinase (PI3K)–Akt pathway (Svensson et al., 2007), we showed that suppression of ZNF224 occurs via the PI3K–Akt signaling pathway (Montano et al., 2015; Fig. 5). Taken as a whole, these data, through the identification of ZNF224 as a downstream target of BCR–ABL kinase activity, add new insights into the molecular mechanisms underlying CML and strongly indicate that ZNF224 could represent a novel promising therapeutic target in CML.

The identification of WT1 as a transcriptional repressor of ZNF224 gene in CML could also contribute to explain the ZNF224 induction by ara-C and the consequent apoptosis; indeed, we previously observed that ara-C induces a downmodulation of WT1 in K562 cells (Montano et al., 2013). This finding is consistent with the ZNF224 upregulation that we found following treatment with either ara-C or imatinib, thus leading us to

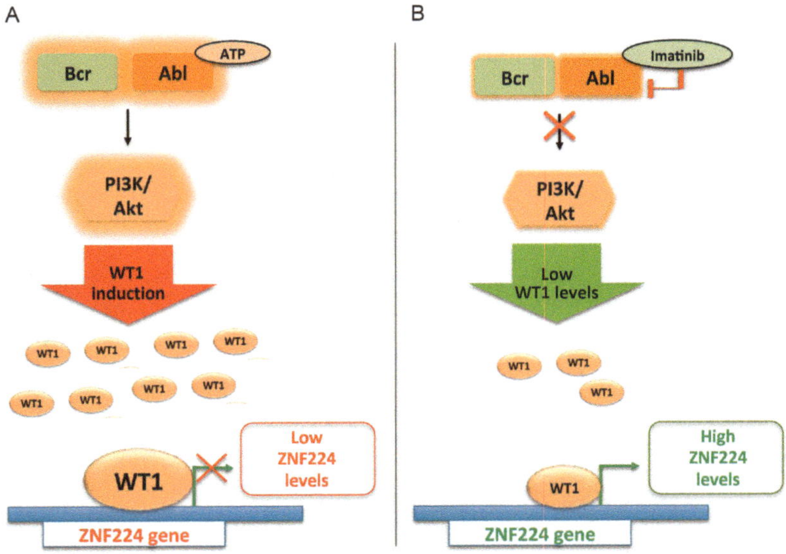

Fig. 5 Representation of the Bcr–Abl downstream pathway involved in ZNF224 transcriptional repression. (A) Bcr–Abl/PI3K pathway increases WT1 expression, which is involved in the transcriptionally repression of ZNF224 gene expression. (B) Inhibition of Bcr/Abl by imatinib increases ZNF224 expression via downmodulation of WT1 expression.

hypothesize that the distinct signaling cascades induced by ara-C and imatinib could converge on ZNF224 induction through the downmodulation of WT1.

Our molecular results also suggest a regulatory circuit between WT1 and ZNF224 in CML. Indeed, the transcriptional repression that WT1 exerts on ZNF224 gene could prevent the proapoptotic effect of the ZNF224–WT1 complex itself, thus contributing to the WT1 prosurvival role in leukemia.

An example of a similar regulatory pathway is provided by the role played by WT1 in kidney development, in which WT1 protein inhibits the expression of Pax-2 (Ryan, Steele-Perkins, Morris, Rauscher, & Dressler, 1995), a transcription factor that, in turn, activates the WT1 gene (Dehbi, Ghahremani, Lechner, Dressler, & Pelletier, 1996).

The data reported so far highlight a proapoptotic role for ZNF224 in CML, due to its ability to interact with WT1. Nevertheless, we cannot rule out that ZNF224 may perform this role also by operating as a DNA-binding transcriptional repressor. Indeed, we have recently identified some target genes involved in oncogenic transformation that are direct targets of ZNF224 transcriptional repression activity (manuscript in preparation).

3.2 ZNF224 as Oncogene in Antiapoptotic and Proliferative Pathways

KRAB-ZNFs are involved in important cellular functions, such as cell differentiation, cell proliferation, and apoptosis, and often result overexpressed in most tumors, playing an important role in cancer outgrowth (Addison et al., 2015). These transcription factors directly or indirectly regulate the expression of different genes, such as CCND1, VEGF, integrin-β4, and NF-κB, all of which are known to be involved in tumorigenesis, tumor progression, and/or metastasis (Jen & Wang, 2016; Lupo et al., 2013). Accordingly, several findings suggest that KRAB-ZNFs may be a potential therapeutic target for cancer treatment (Kawahara et al., 2016).

The role of ZNF224 as potential oncogene has been described for the first time in bladder carcinogenesis, where the DEPDC1 (DEP domain-containing 1) protein behaves as a molecular partner of ZNF224. DEPDC1 is a highly conserved protein, overexpressed in a variety of tumor types including bladder cancer, breast cancer, multiple myeloma, and hepatocellular carcinomas (Kanehira et al., 2007; Kassambara et al., 2013; Kretschmer et al., 2011; Yuan, Liao, Yang, Huang, & Huang, 2014). Interestingly, this protein is not expressed in any normal human organs with the exception of testis, thus leading to consider DEPDC1 as a novel cancer/testis antigen. However, the mechanism of action of DEPDC1 in cancer cells is still poorly understood. In multiple myeloma cell lines it has been observed that DEPDC1 depletion promotes growth delay, associated with a block in G2 phase of the cell cycle; furthermore, there is an increase of p53 phosphorylation and p21Cip1 induction that are involved in the maintenance of G2 arrest (Kassambara et al., 2013). Also, the DEPDC1 knockdown in the human cervical cancer HeLa cells causes mitotic defects with abnormal multiple nuclei and multipolar spindle structures, thus leading to suppose that DEPDC1 plays a key role in cell cycle progression (Mi et al., 2015). These findings suggest that DEPDC1 could correlate with adverse prognosis of cancer patients, by inhibiting malignant cells differentiation and promoting cell cycle progression.

The involvement of DEPDC1 in bladder carcinogenesis was demonstrated by the fact that downregulation of DEPDC1 expression resulted in suppression of bladder cancer cell growth (Kanehira et al., 2007). DEPDC1 is essential for the survival of bladder cancer cells but it is not able to sustain the transforming activity by itself, thus leading to suppose that interaction with other proteins may be required for proper oncogenic function of DEPDC1.

Interestingly, investigations on the molecular mechanisms of DEPDC1 signaling in bladder cancer cells have demonstrated nuclear interactions between DEPDC1 and ZNF224 and the role of DEPDC1 as a transcriptional corepressor of ZNF224 (Harada et al., 2010).

The A20 gene, a negative regulator of the NF-κB antiapoptotic pathway, was identified among the target genes of the ZNF224–DEPDC1 complex (Beyaert et al., 2000; Krikos et al., 1992). The repression of A20 gene transcription by DEPDC1/ZNF224 complex leads to the nuclear translocation of NF-κB protein and the subsequent activation of NF-κB-mediated antiapoptotic pathway in bladder cancer cells. The identification of the DEPDC1 domain required for the interaction with ZNF224 allowed the design of a cell-permeable dominant-negative peptide, inhibiting the DEPDC1–ZNF224 complex formation. This resulted in the activation of A20 gene transcription, consequent inactivation of the NF-κB pathway, and apoptosis induction (Fig. 6).

The NF-κB family of transcription factors has an essential role in inflammation and innate immunity. The constitutive activation of NF-κB occurs in a significant percentage of human cancers and is associated with chronic inflammation and tumor development. In the latest years, many data support the idea that the inflammation contributes to provide conditions leading to cancer (Fernandes et al., 2015). In fact, overexpression of proinflammatory cytokines is sufficient to induce hyperplasia due to the creation of a favorable microenvironment for tumor progression and metastasis (Hoesel & Schmid, 2013). It is now accepted that NF-κB plays a crucial role in many steps of cancer initiation and progression. During these processes, NF-κB physically interacts with different partners. The functional cross talk between NF-κB pathway and other signaling molecules and pathways, such as STAT3 and p53 or the ETS-related gene ERG, can influence the transcriptional and DNA-binding activities of NF-κB (Hoesel & Schmid, 2013). The aberrant NF-κB activation promotes cellular transformation, sustains cancer survival, and contributes to cancer progression by controlling epithelial-to-mesenchymal transition and metastasis (Huber et al., 2004).

The transcriptional repression of A20 gene by ZNF224/DEPDC1 complex further contributes to highlight the role played by ZNF224 in pathological conditions; in particular, a new function for ZNF224 is emerging to connect inflammation and cancer development through modulation of the NF-κB pathway.

Moreover, a role for ZNF224 as oncogene cells has been recently reported in breast cancer cells, in which ZNF224 increases the miR-663a

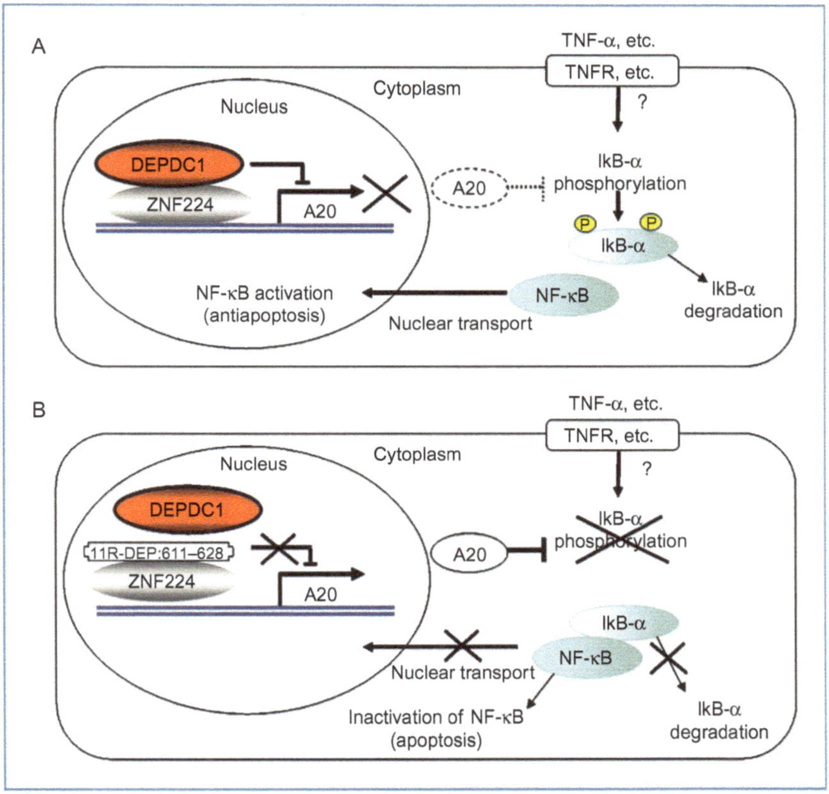

Fig. 6 Role of the DEPDC1/ZNF224 complex in bladder carcinogenesis. (A) Schematic representation of the DEPDC1/ZNF224 complex signaling pathway in bladder cancer cells, resulting in NF-κB activation. (B) Inhibition of the DEPDC1/ZNF224 complex formation by a dominant-negative peptide results in inactivation of NF-κB. *By permission of American Association for Cancer Research.*

transcription. miR-663a is a noncoding small RNA, belonging to the primate-specific miRNAs family that govern many important biological processes including development, inflammatory response, apoptosis, and carcinogenesis (Zang et al., 2015). The function of miR-663a in tumor progression is controversial. Indeed, many data provide strong evidence that miR-663a may acts both as oncogene, playing crucial roles in the initiation and progression of the malignancy of lung cancer, nasopharyngeal carcinoma, and breast cancer, and as tumor suppressor in gastric cancer, colorectal carcinoma, prostate cancer, and acute lymphoblastic leukemia (Zhang et al., 2016). Recently, it has been reported that miR-663a directly targets the 3′

UTR of p21 and p53 mRNAs and downregulates their expression levels (Cho et al., 2016; Yi et al., 2012).

In breast cancer cells, ZNF224 binds and activates miR-663a gene transcription; the resulting miR-663a upregulation has been found associated to reduced p53 and p21 mRNA levels. So, ZNF224 decreases p53 and p21 expression via transcriptional activation of miR-663a. Therefore, in breast cancer ZNF224 overexpression induces apoptosis resistance following the treatment with DNA-damaging agents, and affects cell proliferation and cell cycle progression by reducing p21 and p53 expression via miR-663a (Cho et al., 2016). However, it cannot be excluded that ZNF224 could be a negative regulator of p21 and p53 gene expression by direct binding to their promoter sequences.

Conversely, recent findings show low miR-663 expression levels in CML patients and indicate that overexpression of miR-663 suppresses proliferation of the K562 cell line, in part by enhancing cell apoptosis (Yang et al., 2013). Nevertheless, low levels of ZNF224 expression in bone marrow of CML patients have been observed (Montano et al., 2015) and the proapoptotic function of ZNF224 in CML has been clearly demonstrated (Montano et al., 2013, 2016, 2015). Hence, the expression levels and the functional roles of ZNF224 and miR-663a in different tumors seem to overlap, allowing us to assume that miR-663a may provide a functional link to mediate ZNF224 activity on different target genes. These data thus allow us to speculate that one of the mechanisms through which ZNF224 may operate as oncogene or tumor suppressor is the transcriptional regulation of miR-663a.

Furthermore, we have recently highlighted an oncogenic role for ZNF224 in hematological malignancies. Indeed, high ZNF224 expression levels in CLL patients have been found to correlate with high cyclin D3 expression levels and absolute lymphocyte count, both factors being related with poor prognosis. The aberrant ZNF224 expression affects cell cycle progression in CLL, potentially through the transcriptional control of cyclin D3, a cell cycle regulator factor with protumorigenic function. Furthermore, we demonstrated that ZNF224 is involved in apoptosis resistance in CLL cells (manuscript submitted for publication) (Fig. 7).

According to these findings, ZNF224 may be placed among the factors involved in the onset and tumor progression, via cell cycle deregulation and impaired apoptosis. In particular, ZNF224 might influence carcinogenesis by affecting the activity of cell cycle regulatory components.

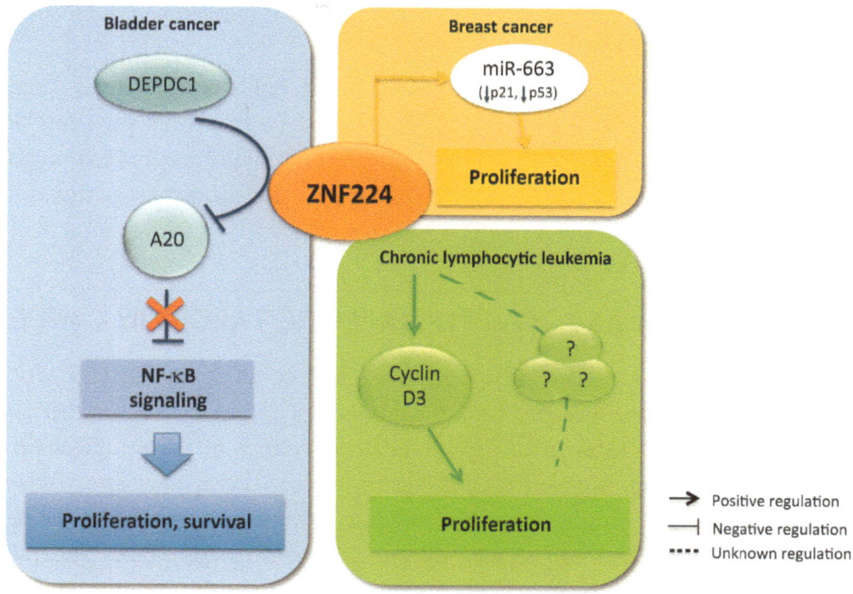

Fig. 7 Schematic representation of the three tumoral contexts so far identified in which ZNF224 induces proliferation and survival processes, by activating different oncogenic pathways.

ZNF224 may perform its function of oncogene through direct control of genes that are known to be involved in tumor progression, cell cycle, and apoptosis, and/or through the direct or indirect control of sentinel molecules, such as p53. Overexpression of ZNF224 may alter the balance between the various molecular processes required to maintain cellular homeostasis, thus leading to the escape from the control mechanisms involved in normal cellular physiology.

Moreover, ZNF224 could play its oncogenic role via physical interactions with protein cofactor linked to tumor-promoting activities, such as PRMT5. Newly accumulating evidences implicate that PRMT5 and its methyltransferase activity are involved in carcinogenesis (Stopa, Krebs, & Shechter, 2015) and propose PRMT5 as a marker of poor clinical outcome.

In some human cancers the PRMT5/ZNF224 interaction could modulate the ZNF224 activity through alteration of its transcriptional complex; in particular, PRMT5 could facilitate ZNF224 switch from repressor to activator role, by reducing the binding affinity for its corepressor KAP1.

Therefore, in pathological conditions PRMT5 could promote the ZNF224 function as activator of specific target genes.

The molecular mechanisms by which ZNF224 switches from tumor suppressive to oncogenic role are unclear, and further investigations are required to understand the mechanisms regulating the differential expression and function of ZNF224 in various cancers and the molecular pathway in which ZNF224 is involved.

4. ZNF224 AS A NEW THERAPEUTIC TARGET IN CANCER

Over the last years a large body of evidence has accumulated which indicates that transcriptional factors, regulatory RNAs, protein/protein, and protein/DNA interactions involved in a variety of pathological processes could provide novel attractive and promising therapeutic approaches. Despite the initial enthusiasm, most of these factors have proven to be difficult to target and, consequently, for a long time they have been generally considered as undruggable, largely owing to the lack of specificity and the potential redundancy of many targets that could generate undesired adverse reactions.

More recently, improved high-throughput screening procedures and new technologies have facilitated the discovery of agents that provide more effective strategies to manipulate transcription by disrupting specific DNA/protein and protein/protein interactions, including dimerization of transcriptional factors and/or interactions with transcriptional coregulators. It is clear that these targeting strategies require precharacterization of the mechanisms responsible for aberrant expression of disease-related genes and mainly rely on the identification of small molecules able to modulate specific target functions.

ZNF224 provides an interesting example of this approach. Indeed, some recent reports have highlighted the oncogenic and/or oncosuppressive properties of protein complexes involving ZNF224 interactions, thus suggesting a promising use of this factor as therapeutic target in a variety of disorders, including hematological malignancies and solid tumors (Harada et al., 2010; Montano et al., 2015).

Among leukemia disorders, an excellent example of the potential role of ZNF224 as a therapeutic target has been exploited in CML, where ZNF224 exerts a specific proapoptotic role through the modulation of WT1-dependent transcription of apoptotic genes. Interestingly, in K562 cells expression levels of ZNF224 have been found increased following treatment with the proapoptotic drug cytosine arabinoside (ara-C), thus suggesting that ZNF224 can be a mediator of DNA damage-induced apoptosis in leukemia cells (Montano et al., 2013, 2015).

More recently, Montano et al. (2015) disclosed one of the pathways activated by BCR–ABL that leads to ZNF224 transcriptional repression. Indeed, they showed that the PI3K/Akt/mTOR pathway, a BCR–ABL downstream pathway which is deregulated in a large fraction of human cancers and is implicated in the survival of leukemic cells (Martelli, Evangelisti, Chiarini, & McCubrey, 2010), is involved in downmodulation of ZNF224 expression and consequently in the inhibition of apoptosis in CML. Therefore, the identification of molecules able to prevent the transcriptional repression of ZNF224 gene expression, targeting signaling pathways downstream of BCR/ABL, could lead to the development of new therapeutic approaches in CML and could provide promising means to overcome imatinib resistance in CML. Indeed, the development of imatinib resistance in CML patients is a dramatic issue in clinical practice and a key topic currently addressed in CML research.

In the above-mentioned case, ZNF224 acts as a tumor suppressor factor; therefore, a therapeutic strategy should be based on the induction of its expression. A different approach, on the other hand, is required when ZNF224 functions as an oncoprotein, as recently reported by Harada et al. (2010). These authors described the key role played by the DEPDC1/ZNF224 complex in bladder carcinogenesis that functions as a repressor complex of A20, a negative regulator of the NF-κB signaling pathway that, in turn, induces the expression of antiapoptotic proteins (Lee et al., 2000; Pekalski et al., 2013). As also reported earlier in this chapter, a successful strategy was designed based on the disruption of the DEPDC1/ZNF224 interaction with cell-permeable dominant-negative peptides (see Fig. 6). Inhibition of DEPDC1/ZNF224 complex formation results in growth suppression of bladder cancer cells. Furthermore, knockdown of ZNF224 gene expression by small interfering RNA results in growth inhibition and/or cell death of bladder cancer cells. This study thus offers a promising strategy for the treatment of bladder cancer.

Cho et al. found that ZNF224 overexpression in breast cancer induces apoptosis resistance by downmodulating the expression of p21 and p53 via of miR-663a. In these cells, indeed, ZNF224 binds to the promoter of miR-663 and increases its transcription (Cho et al., 2016; see Fig. 6).

Very recently, ZNF224 oncogenic activity has been also described in CLL, where ZNF224 is able to promote growth and proliferation of malignant B cells by, at least in part, positively modulating cyclin D3 gene expression. Moreover, high ZNF224 expression levels prevent the drug-induced apoptosis contributing to drug resistance in CLL cells (manuscript submitted for publication) (see Fig. 6).

In light of these recent findings, it is conceivable that characterization of the transcriptional activation complex recruited by ZNF224 on miR-663a and cyclin D promoters will lead to the identification of molecules able to inhibit the DNA-binding ability of ZNF224 or its interaction with transcriptional cofactors that could potentially be used for cancer therapy. Although more studies are required to better define the tumorigenic role of ZNF224 in these systems, it is expected that strategies based on the suppression of ZNF224 activity should contribute to the identification of novel and potential therapeutic tools.

Anyway, it is evident that, given the opposite role displayed by ZNF224 as tumor suppressor or oncogene in cell- and/or cancer type-dependent manner, there is not a unique therapeutic strategy to successfully target this factor. Therefore, taken as a whole, the promising results so far achieved further strengthen the need of a more comprehensive knowledge of the tissue- and tumor-specific transcriptional complexes involving ZNF224. This would contribute to better clarify their role in pathogenetic mechanisms to define novel and more effective therapeutic approaches.

5. CONCLUSIONS

In recent years, a growing body of research offers compelling evidence regarding the prominent role of KRAB-ZFPs in cancer, according to their ability to modulate the expression of genes involved in differentiation, cell cycle, apoptosis, and metabolism.

ZNF224 is an excellent example of a KRAB-ZFP originally identified as a classical transcriptional repressor of genes involved in energy metabolism. More recently, it has been indicated as a key protein in some human malignancies. Interestingly, current understanding of how ZNF224 is associated with cancer suggests for this protein a dual role of oncogene or tumor suppressor in different cancers. Indeed, molecular studies show that ZNF224 is a multifunctional protein that functions not only as a transcriptional repressor but also as a cofactor of WT1 in chronic myeloid leukemia, thus exerting a proapoptotic and antiproliferative function. On the contrary, it has been proven that in bladder cancer ZNF224 behaves as an oncogene, through its nuclear interaction with DEPDC1 and the subsequent negative modulation of A20 expression. Finally, recent findings have revealed the ability of ZNF224 to function as an oncogene, by positively modulating the expression of its target genes in breast cancer and CLL.

All together, the findings presented here show the flexibility of ZNF224 both in terms of DNA-binding and in protein–protein interaction. The assembly of different multiprotein complexes in different cell types provides the mechanistic explanation for ZNF224 differential control of gene expression. Elucidating how alterations of these complexes could affect gene expression in tumor cells will shed further light into the complex role of ZNF224 in cancer.

REFERENCES

Addison, J. B., Koontz, C., Fugett, J. H., Creighton, C. J., Chen, D., Farrugia, M. K., et al. (2015). KAP1 promotes proliferation and metastatic progression of breast cancer cells. *Cancer Research, 75*(2), 344–355.

Baylin, S. B., Esteller, M., Rountree, M. R., Bachman, K. E., Schuebel, K., & Herman, J. G. (2001). Aberrant patterns of DNA methylation, chromatin formation and gene expression in cancer. *Human Molecular Genetics, 10*(7), 687–692.

Baylin, S. B., & Ohm, J. E. (2006). Epigenetic gene silencing in cancer—A mechanism for early oncogenic pathway addiction? *Nature Reviews. Cancer, 6*(2), 107–116.

Bedi, A., Zehnbauer, B. A., Barber, J. P., Sharkis, S. J., & Jones, R. J. (1994). Inhibition of apoptosis by BCR-ABL in chronic myeloid leukemia. *Blood, 83*(8), 2038–2044.

Beyaert, R., Heyninck, K., & Van Huffel, S. (2000). A20 and A20-binding proteins as cellular inhibitors of nuclear factor-κB-dependent gene expression and apoptosis. *Biochemical Pharmacology, 60*(8), 1143–1151.

Cesaro, E., De Cegli, R., Medugno, L., Florio, F., Grosso, M., Lupo, A., et al. (2009). The Kruppel-like zinc finger protein ZNF224 recruits the arginine methyltransferase PRMT5 on the transcriptional repressor complex of the aldolase A gene. *The Journal of Biological Chemistry, 284*(47), 32321–32330.

Cesaro, E., Montano, G., Rosati, A., Crescitelli, R., Izzo, P., Turco, M. C., et al. (2010). WT1 protein is a transcriptional activator of the antiapoptotic bag3 gene. *Leukemia, 24*(6), 1204–1206.

Chakravarthi, B. V., Nepal, S., & Varambally, S. (2016). Genomic and epigenomic alterations in cancer. *The American Journal of Pathology, 186*(7), 1724–1735.

Chen, K., Yu, G., Gumireddy, K., Li, A., Yao, W., Gao, L., et al. (2015). ZBRK1, a novel tumor suppressor, activates VHL gene transcription through formation of a complex with VHL and p300 in renal cancer. *Oncotarget, 6*(9), 6959–6976.

Cheng, Y., Geng, H., Cheng, S. H., Liang, P., Bai, Y., Li, J., et al. (2010). KRAB zinc finger protein ZNF382 is a proapoptotic tumor suppressor that represses multiple oncogenes and is commonly silenced in multiple carcinomas. *Cancer Research, 70*(16), 6516–6526.

Cheng, Y., Liang, P., Geng, H., Wang, Z., Li, L., Cheng, S. H., et al. (2012). A novel 19q13 nucleolar zinc finger protein suppresses tumor cell growth through inhibiting ribosome biogenesis and inducing apoptosis but is frequently silenced in multiple carcinomas. *Molecular Cancer Research, 10*(7), 925–936.

Cho, J. G., Park, S., Lim, C. H., Kim, H. S., Song, S. Y., Roh, T. Y., et al. (2016). ZNF224, Krüppel like zinc finger protein, induces cell growth and apoptosis-resistance by down-regulation of p21 and p53 via miR-663a. *Oncotarget*, *7*(21), 31177–31190.

Cote, F., Boisvert, F., Grondin, B., Bazinet, M., Goodyer, C., Bazett-Jones, D., et al. (2001). Alternative promoter usage and splicing of ZNF74 multifinger gene produce protein isoforms with a different repressor activity and nuclear partitioning. *DNA and Cell Biology*, *20*, 159–173.

Davies, R. C., Calvio, C., Bratt, E., Larsson, S. H., Lamond, A. I., & Hastie, N. D. (1998). WT1 interacts with the splicing factor U2AF65 in an isoform-dependent manner and can be incorporated into spliceosomes. *Genes and Development*, *12*(20), 3217–3225.

Dehbi, M., Ghahremani, M., Lechner, M., Dressler, G., & Pelletier, J. (1996). The paired-box transcription factor, PAX2, positively modulates expression of the Wilms' tumor suppressor gene (WT1). *Oncogene*, *13*(3), 447–453.

Deininger, M. W., Goldman, J. M., & Melo, J. V. (2000). The molecular biology of chronic myeloid leukemia. *Blood*, *96*(10), 3343–3356.

di Caprio, R., Ciano, M., Montano, G., Costanzo, P., & Cesaro, E. (2015). KAP1 is a novel substrate for the arginine methyltransferase PRMT5. *Biology*, *4*(1), 41–49.

Emerson, R. O., & Thomas, J. H. (2010). Adaptive evolution in zinc finger transcription factors. *PLoS Genetics*, *5*, e1000325.

Esteller, M. (2007). Epigenetic gene silencing in cancer: The DNA hypermethylome. *Human Molecular Genetics*, *16*(Spec No. 1), R50–R59.

Fernandes, J. V., Cobucci, R. N., Jatobá, C. A., Fernandes, T. A., de Azevedo, J. W., & de Araújo, J. M. (2015). The role of the mediators of inflammation in cancer development. *Pathology Oncology Research*, *21*(3), 527–534.

Florio, F., Cesaro, E., Montano, G., Izzo, P., Miles, C., & Costanzo, P. (2010). Biochemical and functional interaction between ZNF224 and ZNF255, two members of the Kruppel-like zinc-finger protein family and WT1 protein isoforms. *Human Molecular Genetics*, *19*(18), 3544–3556.

Friedman, J. R., Fredericks, W. J., Jensen, D. E., Speicher, D. W., Huang, X. P., Neilson, E. G., et al. (1996). KAP-1, a novel corepressor for the highly conserved KRAB repression domain. *Genes & Development*, *10*(16), 2067–2078.

Furuta, S., Wang, J. M., Wei, S., Jeng, Y. M., Jiang, X., Gu, B., et al. (2006). Removal of BRCA1/CtIP/ZBRK1 repressor complex on ANG1 promoter leads to accelerated mammary tumor growth contributed. *Cancer Cell*, *10*(1), 13–24.

Gaston, K., & Jayaraman, P. S. (2003). Transcriptional repression in eukaryotes: Repressors and repression mechanisms. *Cellular and Molecular Life Sciences*, *60*(4), 721–741.

Gemelli, C., Orlandi, C., Marani, T. Z., Martello, A., Vignudelli, T., Ferrari, F., et al. (2008). The vitamin D3/Hox-A10 pathway supports MafB function during the monocyte differentiation of human CD34+ hemopoietic progenitors. *Journal of Immunology*, *181*(8), 5660–5672.

Gurevich, R. M., Rosten, P. M., Schwieger, M., Stocking, C., & Humphries, R. K. (2006). Retroviral integration site analysis identifies ICSBP as a collaborating tumor suppressor gene in NUP98-TOP1-induced leukemia. *Experimental Hematology*, *34*(9), 1192–1201.

Hallen, L., Klein, H., Stoschek, C., Wehrmeyer, S., Nonhoff, U., Ralser, M., et al. (2011). The KRAB containing zinc-finger transcriptional regulator ZBRK1 activates SCA2 gene transcription through direct interaction with its gene product, ataxin-2. *Human Molecular Genetics*, *20*(1), 104–114.

Hanahan, D., & Weinberg, R. A. (2011). Hallmarks of cancer: The next generation. *Cell*, *144*(5), 646–674.

Hao, S. X., & Ren, R. (2000). Expression of interferon consensus sequence binding protein (ICSBP) is downregulated in Bcr-Abl-induced murine chronic myelogenous leukemia-like disease, and forced coexpression of ICSBP inhibits Bcr-Abl-induced myeloproliferative disorder. *Molecular and Cellular Biology, 20*(4), 1149–1161.

Harada, Y., Kanehira, M., Fujisawa, Y., Takata, R., Shuin, T., Miki, T., et al. (2010). Cell-permeable peptide DEPDC1-ZNF224 interferes with transcriptional repression and oncogenicity in bladder cancer cells. *Cancer Research, 70*(14), 5829–5839.

He, F., & Zhang, L. (2012). Apak competes with p53 for direct binding to intron 1 of p53AIP1 to regulate apoptosis. *EMBO Reports, 13*(4), 363–370.

Hoesel, B., & Schmid, J. A. (2013). The complexity of NF-κB signaling in inflammation and cancer. *Molecular Cancer, 12*, 86. http://dx.doi.org/10.1186/1476-4598-12-86.

Hohenstein, P., & Hastie, N. D. (2006). The many facets of the Wilms' tumour gene, WT1. *Human Molecular Genetics, 15*, R196–R201.

Holloway, D. T., Kon, M., & De Lisi, C. (2008). In silico regulatory analysis for exploring human disease progression. *Biology Direct, 18*, 3–24.

Hu, G., Kim, J., Xu, Q., Leng, Y., Orkin, S. H., & Elledge, S. J. (2009). A genome-wide RNAi screen identifies a new transcriptional module required for self-renewal. *Genes & Development, 23*, 837–848.

Huang, C., Jia, Y., Yang, S., Chen, B., Sun, H., Shen, F., et al. (2007). Characterization of ZNF23, a KRAB-containing protein that is downregulated in human cancers and inhibits cell cycle progression. *Experimental Cell Research, 313*(2), 254–263.

Huang, C., Yang, S., Ge, R., Sun, H., Shen, F., & Wang, Y. (2008). ZNF23 induces apoptosis in human ovarian cancer cells. *Cancer Letters, 266*(2), 135–143.

Huber, M. A., Azoitei, N., Baumann, B., Grünert, S., Sommer, A., Pehamberger, H., et al. (2004). NF-κB is essential for epithelial-mesenchymal transition and metastasis in a model of breast cancer progression. *The Journal of Clinical Investigation, 114*(4), 569–581.

Huntley, S., Baggott, D. M., Hamilton, A. T., Tran-Gyamfi, M., Yang, S., Kim, J., et al. (2006). A comprehensive catalog of human KRAB-associated zinc finger genes: Insights into the evolutionary history of a large family of transcriptional repressors. *Genome Research, 16*(5), 669–677.

Iacobazzi, V., Infantino, V., Convertini, P., Vozza, A., Agrimi, G., & Palmieri, F. (2009). Transcription of the mitochondrial citrate carrier gene: Identification of a silencer and its binding protein ZNF224. *Biochemical and Biophysical Research Communications, 386*(1), 186–191.

Ito, K., Oji, Y., Tatsumi, N., Shimizu, S., Kanai, Y., Nakazawa, T., et al. (2006). Anti-apoptotic function of 17AA(+)WT1 (Wilms' tumor gene) isoforms on the intrinsic apoptosis pathway. *Oncogene, 25*, 4217–4229.

Jakobsson, J., Cordero, M. I., Bisaz, R., Groner, A. C., Busskamp, V., Bensadoun, J. C., et al. (2008). KAP1-mediated epigenetic repression in the forebrain modulates behavioral vulnerability to stress. *Neuron, 60*(5), 818–831.

Jen, J., & Wang, Y. (2016). Zinc finger proteins in cancer progression. *Journal of Biomedical Science, 23*(1), 53.

Jomgeow, T., Oji, Y., Tsuji, N., Ikeda, Y., Ito, K., Tsuda, A., et al. (2006). Wilms' tumor gene WT1 17AA(-)/KTS(-) isoform induces morphological changes and promotes cell migration and invasion in vitro. *Cancer Science, 97*(4), 259–270.

Kanehira, M., Harada, Y., Takata, R., Shuin, T., Miki, T., Fujioka, T., et al. (2007). Involvement of upregulation of DEPDC1 (DEP domain containing 1) in bladder carcinogenesis. *Oncogene, 26*(44), 6448–6455.

Kassambara, A., Schoenhals, M., Moreaux, J., Veyrune, J. L., Rème, T., Goldschmidt, H., et al. (2013). Inhibition of DEPDC1A, a bad prognostic marker in multiple myeloma, delays growth and induces mature plasma cell markers in malignant plasma cells. *PLoS One, 8*(4). e62752.

Kassiotis, G. (2014). Endogenous retroviruses and the development of cancer. *Journal of Immunology, 192*(4), 1343–1349.
Katoh, I., & Kurata, S. (2013). Association of endogenous retroviruses and long terminal repeats with human disorders. *Frontiers in Oncology, 3*, 234.
Kawahara, T., Inoue, S., Ide, H., Kashiwagi, E., Ohtake, S., Mizushima, T., et al. (2016). ZKSCAN3 promotes bladder cancer cell proliferation, migration, and invasion. *Oncotarget*, 7(33), 53599–53610. http://dx.doi.org/10.18632/oncotarget.10679.
Keilholz, U., Menssen, H. D., Gaiger, A., Menke, A., Oji, Y., Oka, Y., et al. (2005). Wilms' tumour 1 (WT1) in human neoplasia. *Leukemia, 19*(8), 1318–1323.
Kim, S. S., Chen, Y. M., O'Leary, E., Witzgall, R., Vidal, M., & Bonventre, J. V. (1996). A novel member of the RING finger family, KRIP-1, associates with the KRAB-A transcriptional repressor domain of zinc finger proteins. *Proceedings of the National Academy of Sciences of the United States of America, 93*(26), 15299–15304.
Konieczna, I., Horvath, E., Wang, H., Lindsey, S., Saberwal, G., Bei, L., et al. (2008). Constitutive activation of SHP2 in mice cooperates with ICSBP deficiency to accelerate progression to acute myeloid leukemia. *The Journal of Clinical Investigation, 118*(3), 853–867.
Kretschmer, C., Sterner-Kock, A., Siedentopf, F., Schoenegg, W., Schlag, P. M., & Kemmner, W. (2011). Identification of early molecular markers for breast cancer. *Molecular Cancer, 10*(1), 15.
Krikos, A., Laherty, C. D., & Dixit, V. M. (1992). Transcriptional activation of the tumor necrosis factor alpha-inducible zinc finger protein, A20, is mediated by kappa-B elements. *The Journal of Biological Chemistry, 267*(25), 17971–17976.
Kubota, S., Fukumoto, Y., Aoyama, K., Ishibashi, K., Yuki, R., Morinaga, T., et al. (2013). Phosphorylation of KRAB-associated protein 1 (KAP1) at Tyr-449, Tyr-458, and Tyr-517 by nuclear tyrosine kinases inhibits the association of KAP1 and heterochromatin protein 1 (HP1) with heterochromatin. *The Journal of Biological Chemistry, 288*(24), 17871–17883.
Lamprecht, B., Walter, K., Kreher, S., Kumar, R., Hummel, M., Lenze, D., et al. (2010). Derepression of an endogenous long terminal repeat activates the CSF1R protooncogene in human lymphoma. *Nature Medicine, 16*(5), 571–579.
Larsson, S. H., Charlieu, J. P., Miyagawa, K., Engelkamp, D., Rassoulzadegan, M., Ross, A., et al. (1995). Subnuclear localization of WT1 in splicing or transcription factor domains is regulated by alternative splicing. *Cell, 81*(3), 391–401.
Lechner, M. S., Begg, G. E., Speicher, D. W., & Rauscher, F. J., 3rd (2000). Molecular determinants for targeting heterochromatin protein 1-mediated gene, silencing: Direct chromoshadow domain-KAP-1 corepressor interaction is essential. *Molecular and Cellular Biology, 20*(17), 6449–6465.
Lee, E. G., Boone, D. L., Chai, S., Libby, S. L., Chien, M., et al. (2000). Failure to regulate TNF-induced NF-κB and cell death responses in A20-deficient mice. *Science, 289*(5488), 2350–2354.
Lee, Y. K., Thomas, S. N., Yang, A. J., & Ann, D. K. (2007). Doxorubicin down-regulates Kruppel-associated box domain-associated protein 1 sumoylation that relieves its transcription repression on p21WAF1/CIP1 in breast cancer MCF-7 cells. *The Journal of Biological Chemistry, 282*(2), 1595–1606.
Li, X., Lee, Y. K., Jeng, J. C., Yen, Y., Schultz, D. C., Shih, H. M., et al. (2007). Role for KAP1 serine 824 phosphorylation and sumoylation/desumoylation switch in regulating KAP1-mediated transcriptional repression. *The Journal of Biological Chemistry, 282*(50), 36177–36189.
Lim, K. H., & Park, S. G. (2015). Transcriptional regulation of KRAB-ZFPs in cancer. *Molecular & Cellular Toxicology, 11*(4), 389–394.

Lin, L. F., Chuang, C. H., Li, C. F., Liao, C. C., Cheng, C. P., Cheng, T. L., et al. (2010). ZBRK1 acts as a metastatic suppressor by directly regulating MMP9 in cervical cancer. *Cancer Research*, *70*(1), 192–201.

Loeb, D. M. (2006). WT1 influences apoptosis through transcriptional regulation of Bcl-2 family members. *Cell Cycle*, *5*(12), 1249–1253.

Lupo, A., Cesaro, E., Montano, G., Zurlo, D., Izzo, P., & Costanzo, P. (2013). KRAB-zinc finger proteins: A repressor family displaying multiple biological functions. *Current Genomics*, *14*(4), 268–278.

Lupo, A., Izzo, P., & Costanzo, P. (2012). ZNF224: Structural and functional insights into a KRAB zinc finger protein. In R. Ciofani & L. Makrlík (Eds.), *Zinc fingers: Structure, properties and applications* (pp. 101–120). New York: Nova Publishers.

Martelli, A. M., Evangelisti, C., Chiarini, F., & McCubrey, J. A. (2010). The phosphatidylinositol 3-kinase/Akt/mTOR signaling network as a therapeutic target in acute myelogenous leukemia patients. *Oncotarget*, *1*(2), 89–103.

Mayo, M. W., Wang, C. Y., Drouin, S. S., Madrid, L. V., Marshall, A. F., Reed, J. C., et al. (1999). WT1 modulates apoptosis by transcriptionally upregulating the bcl-2 proto-oncogene. *EMBO Journal*, *18*(4), 3990–4003.

McGahon, A., Bissonnette, R., Schmitt, M., Cotter, K. M., Green, D. R., & Cotter, D. J. (1994). Bcr-Abl maintains resistance of chronic myelogenous leukemia cells to apoptotic cell death. *Blood*, *83*(5), 1179–1187.

Medugno, L., Costanzo, P., Lupo, A., Monti, M., Florio, F., Pucci, P., et al. (2003). A novel zinc finger transcriptional repressor, ZNF224, interacts with the negative regulatory element (AldA-NRE) and inhibits gene expression. *FEBS Letters*, *534*(1–3), 93–100.

Medugno, L., Florio, F., Cesaro, E., Grosso, M., Lupo, A., Izzo, P., et al. (2007). Differential expression and cellular localization of ZNF224 and ZNF255, two isoforms of the Krüppel-like zinc-finger protein family. *Gene*, *403*(1–2), 125–131.

Medugno, L., Florio, F., Costanzo, P., Lupo, A., De Cegli, R., Grosso, M., et al. (2005). The Krüppel-like zinc-finger protein ZNF224 represses aldolase A gene transcription by interacting with the KAP-1 corepressor protein. *Gene*, *10*(359), 35–43.

Messerschmidt, D. M., de Vries, W., Ito, M., Solter, D., Ferguson-Smith, A., & Knowles, B. B. (2012). Trim28 is required for epigenetic stability during mouse oocyte to embryo transition. *Science*, *335*(6075), 1499–1502.

Mi, Y., Zhang, C., Bu, B., Zhang, Y., He, L., Li, H., et al. (2015). DEPDC1 is a novel cell cycle related gene that regulates mitotic progression. *BMB Reports*, *48*(7), 413–418.

Montano, G., Cesaro, E., Fattore, L., Vidovic, K., Palladino, C., Crescitelli, R., et al. (2013). Role of WT1-ZNF224 interaction in the expression of apoptosis-regulating genes. *Human Molecular Genetics*, *22*(9), 1771–1782.

Montano, G., Ullmark, T., Jernmark-Nilsson, H., Sodaro, G., Drott, K., Costanzo, P., et al. (2016). The hematopoietic tumor suppressor interferon regulatory factor 8 (IRF8) is upregulated by the antimetabolite cytarabine in leukemic cells involving the zinc finger protein ZNF224, acting as a cofactor of the Wilms' tumor gene 1 (WT1) protein. *Leukemia Research*, *40*, 60–67.

Montano, G., Vidovic, K., Palladino, C., Cesaro, E., Sodaro, G., Quintarelli, C., et al. (2015). WT1-mediated repression of the proapoptotic transcription factor ZNF224 is triggered by the BCR-ABL oncogene. *Oncotarget*, *6*(29), 28223–28237.

Morrison, D. J., English, M. A., & Licht, J. D. (2005). WT1 induces apoptosis through transcriptional regulation of the proapoptotic Bcl-2 family member Bak. *Cancer Research*, *65*(18), 8174–8182.

Nakagama, H., Heinrich, G., Pelletier, J., & Housman, D. E. (1995). Sequence and structural requirements for high-affinity DNA binding by the WT1 gene product. *Molecular and Cellular Biology*, *15*(3), 1489–1498.

Nowick, K., Hamilton, A. T., Zhang, H., & Stubbs, L. (2010). Rapid sequence and expression divergence suggest selection for novel function in primate-specific KRAB-ZNF genes. *Molecular Biology and Evolution, 27*(11), 2606–2617.

Payankaulam, S., Li, L. M., & Arnosti, D. N. (2010). Transcriptional repression: Conserved review and evolved features. *Current Biology: CB, 20*(17), R764–R771.

Pekalski, J., Zuk, P. J., Kochanczyk, M., Junkin, M., Kellogg, R., Tay, S., et al. (2013). Spontaneous NF-κB activation by autocrine TNF-α signaling: A computational analysis. *PLoS One, 8*(11), e78887.

Peng, H., Begg, G. E., Harper, S. L., Friedman, J. R., Speicher, D. W., & Rauscher, F. J. (2000). Biochemical analysis of the Kruppel-associated box (KRAB) transcriptional repression domain. *The Journal of Biological Chemistry, 275*(24), 18000–18010.

Quenneville, S., Verde, G., Corsinotti, A., Kapopoulou, A., Jakobsson, J., Offner, S., et al. (2011). In embryonic stem cells, ZFP57/KAP1 recognize a methylated hexanucleotide to affect chromatin and DNA methylation of imprinting control regions. *Molecular Cell, 44*(3), 361–372.

Richard, D. J., Schumacher, V., Royer, P. B., & Roberts, S. G. (2001). Par4 is a coactivator for a splice isoform-specific transcriptional activation domain in WT1. *Genes and Development, 15*(3), 328–339.

Roberts, S. G. E. (2006). The modulation of WTI transcription function by cofactors. *Biochemical Society Symposium, 73*, 191–201.

Rosenfeld, C., Cheever, M. A., & Gaiger, A. (2003). WT1 in acute leukemia, chronic myelogenous leukemia and myelodysplastic syndrome: Therapeutic potential of WT1 targeted therapies. *Leukemia, 17*(7), 1301–1312.

Ryan, G., Steele-Perkins, V., Morris, J. F., Rauscher, F. J., 3rd, & Dressler, G. R. (1995). Repression of Pax-2 by WT1 during normal kidney development. *Development, 121*(3), 867–875.

Santoni de Sio, F. (2014). Kruppel-associated box (KRAB) proteins in the adaptive immune system. *Nucleus, 5*(2), 138–148.

Santoni de Sio, F. R., Massacand, J., Barde, I., Offner, S., Corsinotti, A., Kapopoulou, A., et al. (2012). KAP1 regulates gene networks controlling mouse B-lymphoid cell differentiation and function. *Blood, 119*(20), 4675–4685.

Schultz, D. C., Ayyanathan, K., Negorev, D., Maul, G. G., & Rauscher, F. J., III (2002). SETDB1: A novel KAP1-associated histone H3, lysine 9-specific methyltransferase that contributes to HP1-mediated silencing of euchromatic genes by KRAB zinc finger proteins. *Genes & Development, 16*(8), 919–932.

Schultz, D. C., Friedman, J. R., & Rauscher, F. J., III (2001). Targeting histone deacetylase complexes via KRAB-zinc finger proteins: The PHD and bromodomains of KAP1 form a cooperative unit that recruits a novel isoform of the Mi-2alpha subunit of NuRD. *Genes & Development, 15*(4), 428–443.

Seki, Y., Kurisaki, A., Watanabe-Susaki, K., Nakajima, Y., Nakanishi, M., Arai, Y., et al. (2010). TIF1beta regulates the pluripotency of embryonic stem cells in a phosphorylation-dependent manner. *Proceedings of the National Academy of Sciences of the United States of America, 107*(24), 10926–10931.

Shannon, M., Hamilton, A. T., Gordon, L., Branscomb, E., & Stubbs, L. (2003). Differential expansion of zinc finger transcription factor loci in homologous human and mouse gene clusters. *Cytogenetic and Genome Research, 13*(6A), 1097–1110.

Shibata, M., Blauvelt, K. E., Liem, K. F., Jr., & García-García, M. J. (2011). TRIM28 is required by the mouse KRAB domain protein ZFP568 to control convergent extension and morphogenesis of extra-embryonic tissues. *Development, 138*(24), 5333–5343.

Shin, J. H., Ko, H. S., Kang, H., Lee, Y., Lee, Y. I., Pletinkova, O., et al. (2011). PARIS (ZNF746) repression of PGC-1α contributes to neurodegeneration in Parkinson's disease. *Cell, 144*(5), 689–702.

Simpson, L. A., Burwell, E. A., Thompson, K. A., Shahnaz, S., Chen, A. R., & Loeb, D. M. (2006). The antiapoptotic gene A1/BFL1 is a WT1 target gene that mediates granulocytic differentiation and resistance to chemotherapy. *Blood*, *107*(12), 4695–4702.

Sripathy, S. P., Stevens, J., & Schultz, D. C. (2006). The KAP1 corepressor functions to coordinate the assembly of de novo HP1 demarcated microenvironments of heterochromatin required for KRAB zinc finger protein mediated transcriptional repression. *Molecular and Cellular Biology*, *26*(22), 8623–8638.

Stopa, N., Krebs, J. E., & Shechter, D. (2015). The PRMT5 arginine methyltransferase: Many roles in development, cancer and beyond. *Cellular and Molecular Life Sciences*, *72*(11), 2041–2059.

Sun, L., Gu, S., Li, N., Zheng, D., Sun, Y., Li, D., et al. (2005). A novel zinc finger gene ZNF468 with two co-expressional splice variants, ZNF468.1 and ZNF468.2. *Biochemical Genetics*, *43*, 271–286.

Svensson, E., Vidovic, K., Lassen, C., Richter, J., Olofsson, T., Fioretos, T., et al. (2007). Deregulation of the Wilms' tumour gene 1 protein (WT1) by BCR/ABL1 mediates resistance to imatinib in human leukaemia cells. *Leukemia*, *21*(12), 2485–2494.

Tian, C., Xing, G., Xie, P., Lu, K., Nie, J., Wang, J., et al. (2009). KRAB-type zinc-finger protein Apak specifically regulates p53-dependent apoptosis. *Nature Cell Biology*, *11*(5), 580–591.

Trono, D. (2016). Transposable elements, polydactyl proteins, and the genesis of human-specific transcription networks. *Cold Spring Harbor Symposia on Quantitative Biology*, *80*, 281–288.

Urrutia, R. (2003). KRAB-containing zinc-finger repressor proteins. *Genome Biology*, *4*(10), 231–238.

Vidovic, K., Svensson, E., Nilsson, B., Thuresson, B., Olofsson, T., Lennartsson, A., et al. (2010). Wilms' tumor gene 1 protein represses the expression of the tumor suppressor interferon regulatory factor 8 in human hematopoietic progenitors and in leukemic cells. *Leukemia*, *24*(5), 992–1000.

Vogel, M. J., Guelen, L., de Wit, E., Peric-Hupkes, D., Lodén, M., Talhout, W., et al. (2006). Human heterochromatin proteins form large domains containing KRAB-ZNF genes. *Genome Research*, *16*(12), 1493–1504.

Wang, T., Wang, X. G., Xu, J. H., Wu, X. P., Qiu, H. L., Yi, H., et al. (2012). Overexpression of the human ZNF300 gene enhances growth and metastasis of cancer cells through activating NF-κB pathway. *Journal of Cellular and Molecular Medicine*, *16*(5), 1134–1145.

White, D. E., Negorev, D., Peng, H., Ivanov, A. V., Maul, G. G., & Rauscher, F. J., 3rd. (2006). KAP1, a novel substrate for PIKK family members, colocalizes with numerous damage response factors at DNA lesions. *Cancer Research*, *66*(24), 11594–11599.

Wolf, G., Greenberg, D., & MacFarland, T. S. (2015). Spotting the enemy: Targeted silencing of foreign DNA in mammalian genomes by the Kruppel-associated box protein family. *Mobile DNA*, *6*, 17.

Yang, Y., Wang, L. L., Wang, H. X., Guo, Z. K., Gao, X. F., Cen, J., et al. (2013). The epigenetically-regulated miR-663 targets H-ras in K-562 cells. *FEBS Journal*, *280*, 5109–5117.

Yi, Z., Li, Y., Ma, W., Li, D., Zhu, C., Luo, J., et al. (2004). A novel KRAB zinc-finger protein, ZNF480, expresses in human heart and activates transcriptional activities of AP-1 and SRE. *Biochemical and Biophysical Research Communications*, *320*(2), 409–415.

Yi, C., Wang, Q., Wang, L., Huang, Y., Li, L., Liu, L., et al. (2012). MiR-663, a microRNA targeting p21(WAF1/CIP1), promotes the proliferation and tumorigenesis of nasopharyngeal carcinoma. *Oncogene*, *31*(41), 4421–4433.

Yuan, S. G., Liao, W. J., Yang, J. J., Huang, G. J., & Huang, Z. Q. (2014). DEP domain containing 1 is a novel diagnostic marker and prognostic predictor for hepatocellular carcinoma. *Asian Pacific Journal of Cancer Prevention*, *15*(24), 10917–10922.

Zang, W., Wang, Y., Wang, T., Du, Y., Chen, X., Li, M., et al. (2015). miR-663 attenuates tumor growth and invasiveness by targeting eEF1A2 in pancreatic cancer. *Molecular Cancer, 14*, 37.

Zhang, Y., Xu, X., Zhang, M., Wang, X., Bai, X., Li, H., et al. (2016). MicroRNA-663a is downregulated in non-small cell lung cancer and inhibits proliferation and invasion by targeting JunD. *BMC Cancer, 16*, 315.

Zheng, L., Pan, H., Li, S., Flesken-Nikitin, A., Chen, P. L., Boyer, T. G., et al. (2000). Sequence-specific transcriptional corepressor function for BRCA1 through a novel zinc finger protein, ZBRK1. *Molecular Cell, 6*(4), 757–768.

CHAPTER SEVEN

Transcription Factors in Breast Cancer—Lessons From Recent Genomic Analyses and Therapeutic Implications

E. Zacksenhaus*[,†,1], J.C. Liu*, Z. Jiang*, Y. Yao[‡], L. Xia[‡], M. Shrestha*[,†], Y. Ben-David[‡,1]

*Toronto General Research Institute, University Health Network, Toronto, ON, Canada
[†]University of Toronto, Toronto, ON, Canada
[‡]The Key Laboratory of Chemistry for Natural Products of Guizhou Province and Chinese Academy of Sciences, Guizhou Medical University, Guiyang, Guizhou, PR China
[1]Corresponding authors: e-mail address: eldad.zacksenhaus@utoronto.ca; yaacovbendavid@hotmail.com

Contents

Advances in Protein Chemistry and Structural Biology, Volume 107
ISSN 1876-1623
http://dx.doi.org/10.1016/bs.apcsb.2016.10.003

Abstract

Multiplatform genomic analyses have identified 93 frequently altered genes in breast cancer. Of these, as many as 49 genes are directly or indirectly involved in transcription. These include constitutive and inducible DNA-binding transcription factors (DB-TFs, 13 genes), corepressors/coactivators (14 genes), epigenetic (10), and mediator/splicing/rRNA (3) factors. At least nine additional genes are immediate upstream regulators of transcriptional cofactors. G:profiler analysis reveals that these alterations affect cell cycle, development/differentiation, steroid hormone, and chromatin modification pathways. A notable observation is that DB-TFs that mediate major oncogenic signaling (e.g., WNT, receptor tyrosine kinase (RTK), NOTCH, and HIPPO), which switch from default repression (signal OFF) to transcriptional activation (signal ON), are not altered in breast cancer. Instead, corepressors (e.g., pRb for E2F1 downstream of various proliferation signals) or upstream factors (e.g., APC and AXIN for TCF, downstream of canonical WNT signaling) are lost, or coactivators (e.g., NOTCH1/2 for CSL/RBPJk) are induced. In contrast, constitutive (MYC, TBX3) and signal-induced (TP53, FOXA1) DB-TFs that do not mediate default repression are directly altered in breast cancer. Some of these TFs have been implicated in the establishment of super-enhancers and positive transcriptional elongation. In addition, oncogenic transcription is induced by mutations affecting regulatory elements or chromatin conformation that create new TF-binding sites in promoters and enhancers of oncogenic genes to promote tumorigenesis. Here we review these diverse oncogenic alterations in TFs in BC and discuss implications for therapy.

1. INTRODUCTION

1.1 Breast Cancer Heterogeneity and Oncogenic Transcription Factors

Breast cancer (BC) is a major cause of malignancy-induced mortality in women worldwide. It includes multiple subtypes that can be divided by histopathology into three major groups: estrogen receptor alpha (ERα) positive, HER2 positive, and triple negative (TNBC). The latter tumors do not express high levels of ERα, progesterone receptor, or HER2. Molecular classification subdivides $ER\alpha^{+}$ tumors into luminal A and luminal B, and TNBCs into basal-like, claudin-low/mesenchymal-like, and other subtypes (Herschkowitz et al., 2007; Perou et al., 2000; Prat, Parker, Fan, & Perou, 2012; Prat et al., 2010; Prat & Perou, 2011). Pathway signature analyses

(Gatza et al., 2010; Lehmann et al., 2011) and copy number alterations (Curtis et al., 2012) identify multiple additional subtypes, highlighting the histology and genomic heterogeneity of this disease.

BC heterogeneity stems from three major variables: (A) cell of origin, (B) epigenetic, and (C) genetic alterations. (A) The cell of origin of different BC subtypes is a contentious issue. It was suggested that claudin-low/mesenchymal-like BC originates from mammary stem cells (MaSC); basal-like BC, including BRCA1-deficient tumors, from luminal progenitors; and luminal A, luminal B, and HER2 subtypes from mature luminal cells (Prat & Perou, 2009; Visvader, 2011; Visvader & Stingl, 2014). The oncogenic networks that drive tumorigenesis can also affect the tumor phenotype irrespective of the cell of origin (Jones et al., 2016; Liu et al., 2014; Melchor et al., 2014; Van Keymeulen et al., 2015; Wang, Liu, Kim, Datti, & Zacksenhaus, 2016). In reality, both cell of origin and oncogenic signaling likely dictate tumor subtype.

(B) Accumulative evidence suggests that tumors are organized in hierarchy with cancer stem cells (CSCs), capable of self-renewal and tumor initiation (following engraftment into immune-deficient mice) at its apex (Dick, 2003; Magee, Piskounova, & Morrison, 2012). In addition to self-renewal, CSCs can also give rise to non-CSCs, which form the tumor bulk, through aberrant differentiation and loss of self-renewal/tumorigenic capacity. This was shown in human BC (Al-Hajj, Wicha, Benito-Hernandez, Morrison, & Clarke, 2003) and subsequently in mouse models of BC (Jones et al., 2016; Liu, Deng, Lehal, Kim, & Zacksenhaus, 2007; Liu et al., 2014; Vaillant et al., 2008), including high-resolution analysis with single-cell transplantation assays (Liu et al., 2012). It was postulated that the relationship between a CSC and non-CSC is driven by epigenetic changes, and, therefore, non-CSCs can revert, albeit at low frequency, back to tumorigenic CSCs (Kreso & Dick, 2014). Thus, although a CSC can arise from transformation of any cell within the stem cell hierarchy through acquisition of self-renewal capacity, its relationship to non-CSC progenies is similar to that observed between normal stem cells and their descendants, and both are based on epigenetic, not genetic alterations.

(C) Actual genetic alterations in oncogenes and tumor suppressors fuel clonal evolution, leading to genetically distinct tumor populations, each with its unique CSC and propensity to grow locally and/or migrate and disseminate to distal sites (Campbell & Polyak, 2007; Shackleton, Quintana, Fearon, & Morrison, 2009). The complex heterogeneity in BC entails that each tumor subtype, indeed each individual lesion, carries a unique set of oncogenic alterations. While these oncogenic alterations underlie heterogeneity, proliferation, invasion, and metastatic potential, as well as drug resistance and relapse, they also offer targets for therapeutic intervention.

With the advent of parallel DNA sequencing and powerful bioinformatic algorithms to manage large sequencing data sets, a number of groups have identified the landscape of genetic alterations in BC (Curtis et al., 2012; Ellis et al., 2012; Koboldt et al., 2012; Lawrence et al., 2014; Shah et al., 2012; Stephens et al., 2012). These alterations were identified through exome sequencing, copy number variation, mRNA, and proteomic analyses of thousands of BC samples representing all major subtypes. A recent paper compiling whole genome sequencing data of 560 BC patients with the previously identified genetic alterations has reached a consensus set of cancer-driven alterations in 93 genes (Nik-Zainal et al., 2016) (Table 1).

In this review, we describe the transcription factors (TFs), cofactors, and immediate regulators affected by genetic alterations within the 93 gene sets in BC. We will not discuss alterations in microRNA expression in BC, some of which target transcription regulators (Dvinge et al., 2013; Wang & Wang, 2012). However, the consensus 93 altered genes in BC likely reflect the effects of aberrant microRNA levels on gene and protein expression. Finally, we note that the rate-limiting step in protein expression is not transcription but protein translation (Truitt & Ruggero, 2016; Vogel & Marcotte, 2012). However, it is likely that alterations in TFs in BC are "translated" into effects on protein levels. This is because, as described later, the PI3K pathway, which promotes protein translation via mTOR, is a major target of many genetic alterations in BC, and because certain TFs induced in BC, such as MYC, directly stimulate the protein translation machinery (Chu, Cargnello, Topisirovic, & Pelletier, 2016; Lin, Malina, & Pelletier, 2009; Pelletier, Graff, Ruggero, & Sonenberg, 2015).

2. FEATURES OF ONCOGENIC ALTERATIONS IN TFs

2.1 TFs Are Frequently Altered in BC to Control the Cell Cycle

The human genome was initially curated manually to have 1391 sequence-specific DNA-binding transcription factors (DB-TFs) and then expanded to include over 2000 TFs and cofactors (Vaquerizas, Kummerfeld, Teichmann, & Luscombe, 2009). Currently, the animal TF database (http://www.bioguo.org/AnimalTFDB/species.php?spe=Homo_sapiens) contains 1544 DB-TFs in 71 families, 124 chromatin-remodeling factors,

Table 1 93 Altered Genes in Breast Cancer

Gene	Full Name
TP53	Tumor protein p53
PIK3CA	Phosphoinositide-3-kinase catalytic alpha
MYC	v-myc myelocytomatosis viral oncogene homolog
CCND1	Cyclin D1
PTEN	Phosphatase and tensin homolog
ERBB2/HER2	Erb-B2 receptor tyrosine kinase 2/human epidermal growth factor receptor 2
ZNF703/ FGFR1	Zinc finger protein 703/fibroblast growth factor receptor 1
GATA3	GATA-binding protein 3
RB1	Retinoblastoma 1
MAP3K1	Mitogen-activated protein kinase kinase kinase 1, E3 ubiquitin
MAP2K4	Mitogen-activated protein kinase kinase 4
ZNF217	Zinc finger protein 217
CDH1	Cadherin 1, type 1/E-cadherin
MLL3/KMT2C	Myeloid/lymphoid or mixed-lineage leukemia protein 3/lysine (K)-specific methyltransferase 2C
ARID1B	AT-rich interactive domain 1B (SWI1-like)
CDKN2A	Cyclin-dependent kinase inhibitor 2A (melanoma, p16-INK4A, inhibits CDK4)
MLLT4	Myeloid/lymphoid or mixed-lineage leukemia; translocated to, 4
AKT1/PKBa	V-Akt murine thymoma viral oncogene homolog 1/protein kinase B alpha
FBXW7	F-box and WD repeat domain-containing 7, E3 ubiquitin protein
ARID1A	AT-rich interactive domain 1A (SWI-like)
CCND3	Cyclin D3
CBFB	Core-binding factor, beta subunit
MDM2	Mdm2 p53-binding protein homolog

Continued

Table 1 93 Altered Genes in Breast Cancer—cont'd

Gene	Full Name
CCNE1	Cyclin E1
CDKN2B	CDKN2B/p15 INK4b protein
NCOR1	Nuclear receptor corepressor 1
SF3B1	Splicing factor 3b, subunit 1
SPEN	Spen homolog, transcriptional regulator
TBX3	T-box 3
IGF1R	Insulin-like growth factor 1 receptor
BRCA2	Breast cancer 2, early onset
NF1	Neurofibromin 1
PIK3R1	Phosphoinositide-3-kinase, regulatory subunit 1
EGFR	Epidermal growth factor receptor
KRAS	v-Ki-ras2 Kirsten rat sarcoma viral oncogene homolog
ESR1	Estrogen receptor 1
FOXA1	Forkhead box A1
MED23	Mediator complex subunit 23
CDK6	Cyclin-dependent kinase
NOTCH2	Notch homolog 2
AKT2/PKBb	V-Akt murine thymoma viral oncogene homolog/protein kinase B beta
BRCA1	Breast cancer 1, early onset
CTCF	CCCTC-binding factor (zinc finger protein)
KDM6A	Lysine (K)-specific demethylase 6A/histone demethylase UTX
SETD2	SET domain-containing 2
CREBBP	CREB-binding protein
DNMT3A	DNA (cytosine-5-)-methyltransferase 3 alpha
FOXP1	Forkhead box P1
MLL2	Myeloid/lymphoid or mixed-lineage leukemia 2
RUNX1	Runt-related transcription factor 1

Table 1 93 Altered Genes in Breast Cancer—cont'd

Gene	Full Name
USP9X	Ubiquitin-specific peptidase 9, X-linked
XBP1	X-box-binding protein 1a
PDGFRA	Platelet-derived growth factor receptor, alpha
ATR	Ataxia telangiectasia and Rad3-related protein
ERBB3	Erb-B2 receptor tyrosine kinase 3
FGFR2	Fibroblast growth factor receptor 2
PALB2	Partner and localizer of BRCA2—recruit BRCA2 and RAD51 to DNA breaks
RHOA	Ras homolog family member A
SMAD4	SMAD family member 4
ATM	Ataxia telangiectasia mutated
ATRX	Alpha thalassemia/mental retardation syndrome X-linked
AXIN1	Axin 1
BCOR	BCL6 corepressor
CDKN1B	Cyclin-dependent kinase inhibitor 1B (p27, Kip1)
CUX1	Cut-like homeobox 1
GNAS	GNAS complex locus
MLH1	MutL homolog 1
NOTCH1	Notch homolog 1, translocation-associated
PHF6	PHD finger protein 6
SMARCA4	SWI/SNF-related, matrix-associated, actin-dependent regulator of chromatin, subfamily a, member 4
STAG2	Stromal antigen 2
ZFP36L1	ZFP36 ring finger protein-like 1
APC	Adenomatous polyposis coli
CASP8	Caspase 8, apoptosis-related cysteine peptidase
CBLB	Cbl proto-oncogene B, E3 ubiquitin protein ligase
CNOT3	CCR4-NOT transcription complex, subunit 3 (complex mRNA regulation, etc.)

Continued

Table 1 93 Altered Genes in Breast Cancer—cont'd

Gene	**Full Name**
ECT2L	Epithelial cell-transforming 2 like—may act as a guanine nucleotide exchange factor (GEF)
MEN1	Multiple endocrine neoplasia I
MSH2	MutS homolog 2
NRAS	Neuroblastoma RAS viral (v-ras) oncogene homolog
PBRM1	Polybromo 1
PMS2	PMS1 homolog 2, mismatch repair system component
STK11	Serine/threonine kinase 11
TET2	Tet methylcytosine dioxygenase 2
ASXL1	Additional sex combs-like transcriptional regulator 1
BRAF	v-raf murine sarcoma viral oncogene homolog B1
BUB1B	BUB1 mitotic checkpoint serine/threonine kinase B
CIC	Capicua transcriptional repressor
ERCC4	Excision repair cross-complementation group 4
HRAS	v-Ha-ras Harvey rat sarcoma viral oncogene homolog
NF2	Neurofibromin 2 (Merlin)
PRDM1	PR domain-containing 1
PREX2	Phosphatidylinositol-3,4,5-trisphosphate-dependent Rac exchange factor 2

and 302 transcription cofactors. We will classify TFs as described by Brivanlou and Darnell (2002).

As noted, compilation of multiple genomic analyses has identified genetic alterations in 93 BC genes (Nik-Zainal et al., 2016) (Table 1). Nine of these (*TP53*, *PIK3CA*, *MYC*, *CCND1*, *PTEN*, *ERBB2/HER2*, *ZNF703/FGFR1*, *GATA3*, *RB1*), highlighted in bold in Table 1, are frequently altered, while others marked in bold underline, and unbold underlined are affected at progressively lower frequencies. Remarkably, of these 93 genes, 48 are either directly (DB-TFs, epigenetic, mediator/splicing/rRNA factors) or indirectly (cofactors, immediately upstream regulators) involved in transcription. Specifically, 7 TFs are constitutive (e.g., MYC; Table 2A); 6 are induced/signal dependent (e.g., TP53; Table 2B); 14 are transcription

Table 2A Constitutive TFs

Gene	Full Name
MYC	v-myc myelocytomatosis viral oncogene homolog
GATA3	GATA-binding protein 3
ZNF217	Zinc finger protein 217
TBX3	T-box 3
RUNX1	Runt-related transcription factor 1
CIC	Capicua transcriptional repressor
PRDM1	PR domain-containing 1

Table 2B Inducible TFs

Gene	Full Name
TP53	Tumor protein p53
ESR1	Estrogen receptor 1
FOXA1	Forkhead box A1
FOXP1	Forkhead box P1
XBP1	X-box-binding protein 1a
SMAD4	SMAD family member 4

cofactors (e.g., pRb; Table 3); 9 epigenetic modifiers (e.g., MLL3/KMT2C; Table 4); and 3 are mediator/splicing/rRNA factors (Table 5). Eleven additional genes are immediate upstream (e.g., cyclins D3 and E1; Table 6). This classification is not rigid because certain genes can be classified into more than one category. For example, CYCLIN D1 and CDK6 are immediate upstream regulators of pRb phosphorylation and also act in noncanonical pathways as transcription cofactors in other contexts (Bienvenu et al., 2010; Hydbring, Malumbres, & Sicinski, 2016; Kollmann et al., 2013). Thus, these genes are shown in both Tables 3 and 6. The remaining 44 genes are not directly involved in transcription. However, many, such as ERBB2/RAS/MAPK, stimulate downstream transcriptional programs. Thus, overall, alterations in transcriptional output represent a major oncogenic event in the course of breast tumorigenesis.

Cancer cells exhibit alterations in several Hallmark pathways that disrupt normal homeostasis and promote unrestrained cellular growth (Hanahan &

Table 3 Cofactors

Gene	Full Name
CCND1[a]	Cyclin D1
RB1	Retinoblastoma 1
CBFB	Core-binding factor, beta subunit
NCOR1	Nuclear receptor corepressor 1
SPEN	Spen homolog, transcriptional regulator
BRCA1	Breast cancer 1, early onset
CDK6[a]	Cyclin-dependent kinase
NOTCH2	Notch homolog 2
CTCF	CCCTC-binding factor (zinc finger protein)
CREBBP	CREB-binding protein
BCOR	BCL6 corepressor
CUX1	Cut-like homeobox 1
NOTCH1	Notch homolog 1, translocation-associated
ASXL1	Additional sex combs-like transcriptional regulator 1

[a]Also function as "Immediate Upstream Regulators" (Table 6).

Table 4 Chromatin Modifiers

Gene	Full Name
MLL3/ KMT2C	Myeloid/lymphoid or mixed-lineage leukemia protein 3/lysine (K)-specific methyltransferase 2C
ARID1B	AT-rich interactive domain 1B (SWI1-like)
KDM6A	Lysine (K)-specific demethylase 6A/histone demethylase UTX
DNMT3A	DNA (cytosine-5-)-methyltransferase 3 alpha
MLL2	Myeloid/lymphoid or mixed-lineage leukemia 2
MEN1	Multiple endocrine neoplasia I
PBRM1	Polybromo 1
SMARCA4	SWI/SNF-related, matrix-associated, actin-dependent regulator of chromatin, subfamily a, member 4
TET2	Tet methylcytosine dioxygenase 2

Table 5 Basal/RNA

Gene	Full Name
SF3B1	Splicing factor 3b, subunit 1
MED23	Mediator complex subunit 23
PHF6	PHD finger protein 6

Table 6 Immediate Upstream Regulators

Gene	Full Name
CCND1[a]	Cyclin D1
FBXW7	F-box and WD repeat domain-containing 7, E3 ubiquitin protein
CCND3	Cyclin D3
MDM2	Mdm2 p53-binding protein homolog
CCNE1	Cyclin E1
CDK6[a]	Cyclin-dependent kinase
APC	Adenomatous polyposis coli
USP9X	Ubiquitin-specific peptidase 9, X-linked
AXIN1	Axin 1
CDKN1B	Cyclin-dependent kinase inhibitor 1B (p27, Kip1)
NF2	Neurofibromin 2 (merlin)

[a]Also function as "cofactors" (Table 3).

Weinberg, 2011; Iorio et al., 2016). Of the major 13 oncogenic pathways that drive cancer, 7 are commonly disrupted in BC. These include cell cycle, DNA repair, PI3K/AKT/mTOR signaling, RAS/RAF/MEK/ERK/JNK, RTK (receptor tyrosine kinase) signaling, TP53 signaling, and TF regulation (Iorio et al., 2016). Histone modifiers, telomerase maintenance, and other growth/proliferative signaling pathways are also altered, albeit at lower frequency.

To gauge the signaling pathways driven by the 93 altered genes in BC, specifically those involved in transcription; we performed G:profiler pathway analysis as described (Liu et al., 2014). Interestingly, when all 93 genes were considered, the two major signaling pathways were "PI3K/AKT" and "cell cycle" (Fig. 1). Additional pathways included "mammary epithelium," "nuclear localization," and "hemopoiesis" as well as "UV response," "pol II,"

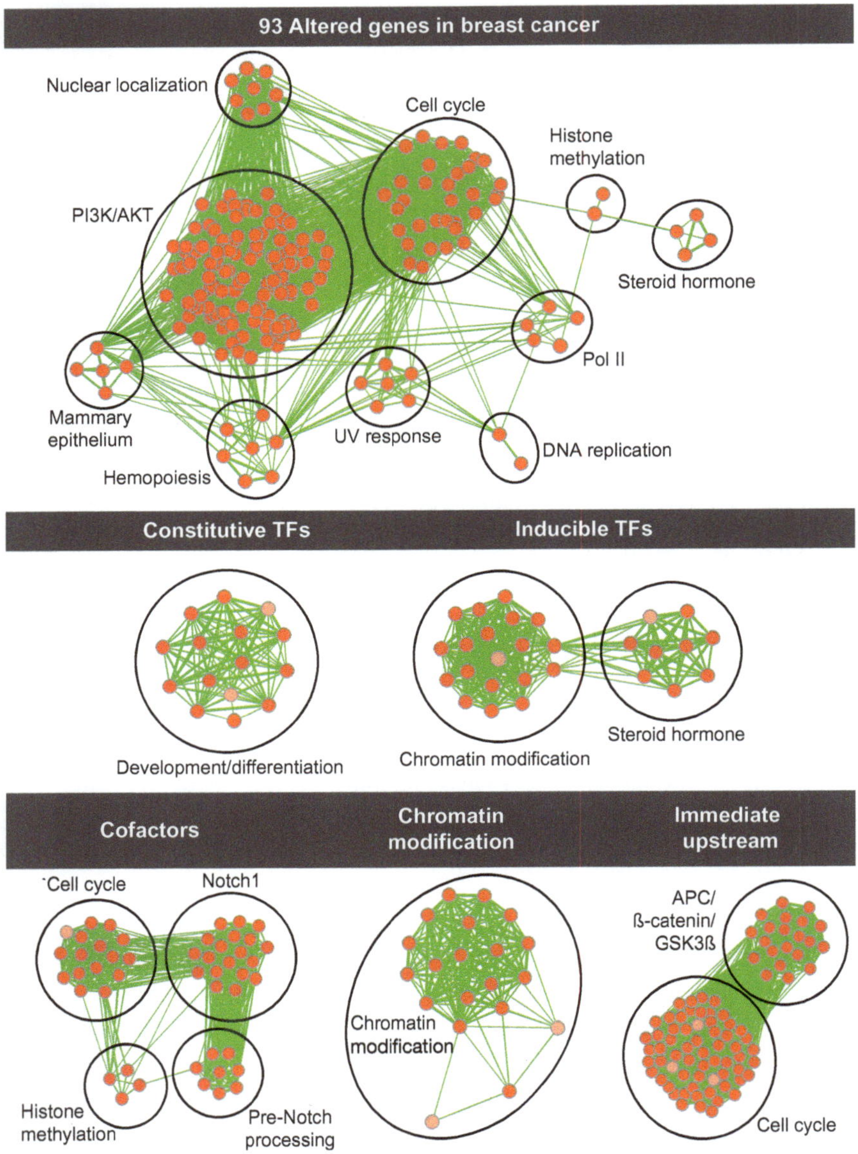

Fig. 1 G:profile analysis of altered genes in BC, classified as per Tables 1–6. For all 93 genes, we used high stringent Bonferroni correction of $P < 1 \times 10^{-4}$. For all other gene sets, Bonferroni correction was $P < 0.05$.

and "steroid hormones." When only constitutive TFs were examined, the major pathway was "development/differentiation"; for induced TFs the major pathways were "chromatin modification" and "steroid hormone"; for transcription cofactors, "cell cycle," "Notch1," "Pre-Notch processing," and "histone methylation"; for immediate upstream factors, "cell cycle" and "APC/β-catenin/GSK3β"; and, as expected, for epigenetic genes, "chromatin modification." Thus, "cell cycle," development/differentiation, chromatin modification, and others but not PI3K/AKT signaling are the major pathways induced by oncogenic alterations in TFs and their regulators in BC.

2.2 No Alterations Occur in TFs That Mediate Default Repression Downstream of Oncogenic Signaling

Signaling pathways that control embryogenesis such as WNT, TGFβ, Hedgehog (HH), RTK, NOTCH, and HIPPO have long been implicated in cancer. These signaling pathways share common features: default repression, activator insufficiency, and cooperative activation (Barolo & Posakony, 2002) (Fig. 2). In the absence of a signal, TFs mediating each of these signaling pathways interact with their cognate binding sites at the promoters and recruit corepressors, thereby establishing default repression (OFF position). In BC, the TF downstream of WNT signaling, TCF, recruits the corepressor Groucho; for RTK and other pathways that promote cell proliferation, E2F1, E2F2, and E2F3a recruit the tumor suppressor pRb, which forms a complex with histone deacetylases (HDACs) or methyltransferases to silence gene expression; for Notch signaling, CSL recruits N-COR to suppress basal transcription; and finally, for the HIPPO pathway, TEAD (TEA domain transcription factor) recruits a repressor to shut down gene expression (Fig. 2). Upon signaling, the corepressors are replaced by transcriptional coactivators (β-catenin for WNT; histone acetyltransferase (HAT) for activating E2Fs; a Notch-mastermind complex for Notch signaling; and YAP/TAZ for the HIPPO pathway), thereby potentiating gene expression (ON position). Signaling activation is often insufficient for robust transcription (activator insufficiency), and cooperation with additional TFs is required (cooperative activation), thus enabling input from other signaling pathways (Barolo & Posakony, 2002).

Importantly, none of the genetic alterations found in BC affect TFs that mediate default repression. This is likely because amplification/overexpression of such TFs would further repress transcription in the absence of signaling, resulting in growth suppression. Only when additional mutations that induce a TF involved in default repression occur, its amplification/overexpression would further stimulate transcription. While the latter scenario is rare in BC, in bladder cancer, 6p22.3 amplification encompassing E2F3 occurs exclusively together with deletion of the tumor suppressor pRb or of its upstream regulator $p16^{ink4a}$ (Hurst, Tomlinson, Williams,

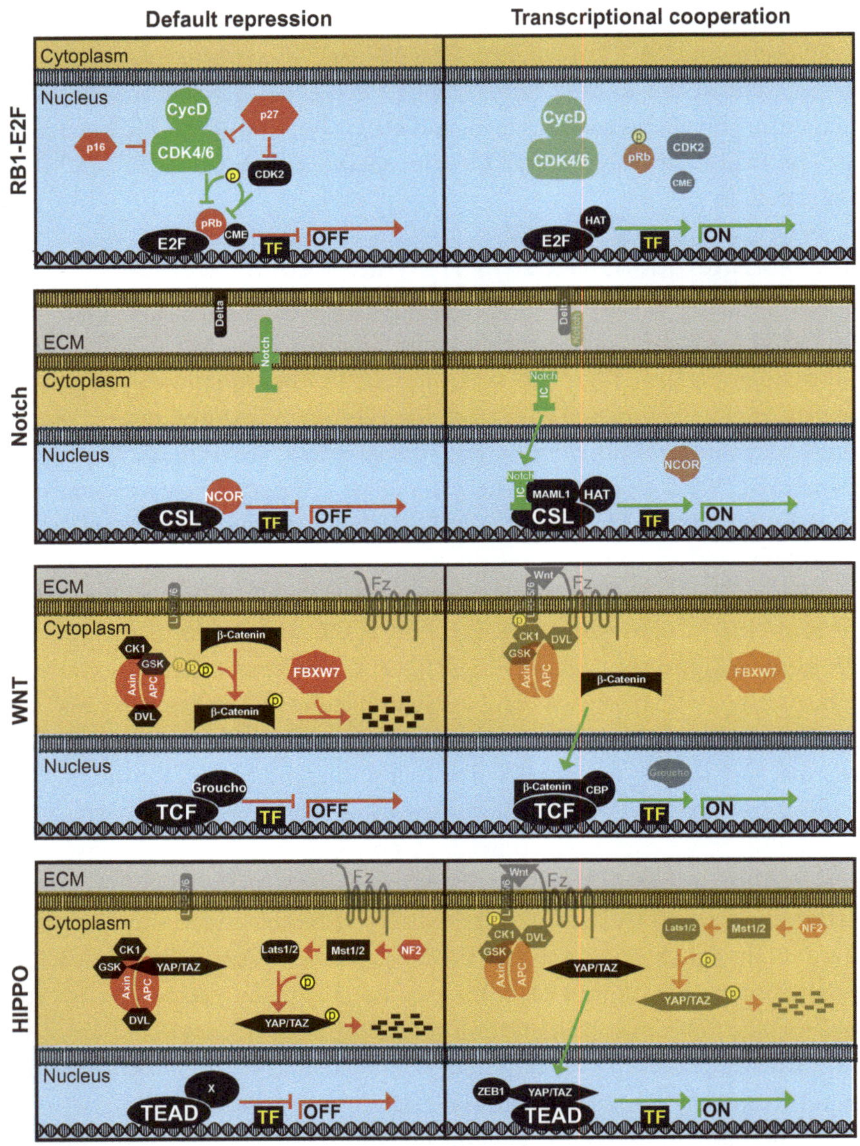

Fig. 2 Oncogenic signaling mediated by default repression in BC: E2F1, NOTCH, WNT, and HIPPO. Genes in *black* are not frequently altered in BC; genes in green and red are oncogenes and tumor suppressors, respectively. See text for details.

Platt, & Knowles, 2008). Thus, in this subset of bladder tumors, inactivation of p16^{ink4a} or RB1 cooperates with E2F3 overexpression to induce cell proliferation, but E2F3 amplification does not occur without loss of these negative regulators. For signaling pathways that are controlled by default repression in BC (RB-E2F, Notch, Wnt, Hippo), the TF is not activated directly, e.g., by amplification, but through alterations in cofactors or upstream regulators as depicted in Fig. 2 (Section 2.4).

2.3 Alterations in TFs That Do Not Mediate Default Repression

Direct alterations in bona fide TFs that constitutively induce or repress transcription include amplification of the *MYC* oncogene (Sections 3 and 8). *GATA3* forms a positive regulatory loop with ERα to specify luminal epithelial cell fate and luminal tumors (Eeckhoute et al., 2007), though no statistically significant differences in recurrence or survival rates were found between GATA3-positive and GATA3-negative $ER\alpha^{+}$ BC (Ciocca, Daskalakis, Ciocca, Ruiz-Orrico, & Palazzo, 2009; Voduc, Cheang, & Nielsen, 2008). Other TFs in this category include *ZNF217* (zinc finger protein 217); *TBX3* (T-box 3); *RUNX1* (runt-related transcription factor 1); *CIC* (capicua homolog), an HMG-box superfamily of transcriptional repressors; and *PRDM1* (PR domain-containing 1), a repressor of β-interferon gene expression (Table 2A).

Signal-dependent inducible TFs include the *TP53* tumor suppressor and *ESR1* (estrogen receptor 1) (see Section 8). The *FOXA1* (forkhead box A1) and *FOXP1* (forkhead box P1) behave as oncogenes or tumor suppressors in different contexts (Bernardo et al., 2010; Li et al., 2013); *XBP1* (X-box-binding protein 1a) regulates the unfolded protein response upon endoplasmic reticulum (ER) stress (Yoshida, Matsui, Yamamoto, Okada, & Mori, 2001) as well as the expression of HIF1α targets via the recruitment of RNA pol II in TNBC (Chen et al., 2014). Finally, *SMAD4* (SMAD family member 4) is a TF downstream of TGFβ/BMP signaling (Table 2B).

2.4 Alterations in Transcription Cofactors or Immediate Upstream Regulators Affecting TFs Involved in Default Repression

Table 3 depicts transcription cofactors altered in BC, some of which are involved in default repression. These include *RB1*, the corepressor of E2F1/2/3, and *CYCLIN D1* and *CDK6*, which phosphorylate and inactivate pRb (Fig. 2). As noted, in addition to controlling pRb (i.e., being immediately upstream regulators), CYCLIN D1 and CDK6 can act in non-canonical pathways as co-TFs independently of pRb-E2F (Bienvenu et al., 2010; Hydbring et al., 2016; Kollmann et al., 2013). Induction of *NOTCH1* and *NOTCH2* coactivators or loss of *NCOR1* (nuclear receptor corepressor 1), a corepressor of CSL, deregulates the NOTCH pathway (Fig. 2). Additional cofactors involved in various transcriptional regulations are *CBFB* (core-binding factor, beta subunit); the coactivator/repressor *SPEN* (spen homolog, transcriptional regulator); CTCF (CCCTC-binding factor, zinc finger protein), which acts as a transcriptional activator when bound to HAT-containing complexes or as a repressor when bound to HDAC-containing complexes; *CREBBP* (CREB-binding protein), a transcriptional coactivator of many different TFs; *BCOR* (BCL6 corepressor); and *ASXL1* (additional sex combs-like transcriptional regulator 1), which directly interacts with BAP1, KDM1A (LSD1), NCOA1, and nuclear hormone receptors (NHRs), such as retinoic acid, estrogen, and androgen receptors. Finally, the tumor suppressor *BRCA1* (breast cancer 1, early onset) maintains genomic stability through multiple mechanisms including cotranscriptional regulation (Section 8) (Huen, Sy, & Chen, 2010; Wang et al., 2013; Wu, Lu, & Yu, 2010).

Alterations in immediate upstream regulators (Table 6) include *CYCLIN D1*, *CDK6* (both also acting as transcription cofactors in other contexts), *CYCLIN D3* and *CYCLIN E1* effecting the RB-E2F proliferation pathway, and *CDKN1B* (cyclin-dependent kinase inhibitor 1B, or $p27^{Kip1}$), which suppresses several CDKs. *APC* (adenomatous polyposis coli) and *AXIN1* are part of the destruction complex that degrades β-catenin to block WNT signaling (Fig. 2). *FBXW7*, an E3 ligase for β-catenin (as well as for MYC (see below), NOTCH1, and CYCLIN E), is frequently lost in BC leading to stimulation of these pathways (Fujii et al., 2006; Jiang et al., 2016; Yumimoto et al., 2015). The APC and AXIN1 tumor suppressors also regulate YAP/TAZ by sequestering it in the cytoplasm (Azzolin et al., 2014) and together with the tumor suppressor *NF2* (Neurofibromin 2/Merlin) regulate YAP/TAZ stability and thereby Hippo signaling via the LATS1/2 kinase (Halder, Dupont, & Piccolo, 2012; Johnson & Halder, 2014) (Fig. 2). *USP9X* (ubiquitin-specific peptidase 9, X-linked) is a deubiquitinase that regulates turnover of multiple proteins including SMAD4 on the TGFβ/BMP pathway. Finally, *MDM2* amplification induces ubiquitination and degradation of TP53.

2.5 Alterations in Epigenetic Regulators

These include three distinct subgroups (Table 4).

Histone modifiers: *MLL3/KMT2C* (myeloid/lymphoid or mixed-lineage leukemia protein 3), a lysine (K)-specific methyltransferase 2C; *MLL2/KMT2D*, a histone-lysine *N*-methyltransferase 2D; *KDM6A* (lysine-specific demethylase 6A), which demethylates Lys-27 of histone H3

(H3K27 demethylase); *SETD2* (SET domain-containing 2), a histone methyltransferase (HMT) that specifically trimethylates Lys-36 of histone H3 (H3K36me3) using dimethylated Lys-36 (H3K36me2) as substrate; and *MEN1* (multiple endocrine neoplasia I), a component of an MLL/SET1 HMT complex that specifically methylates Lys-4 of histone H3 (H3K4).

Chromatin remodeling: *ARID1B* (AT-rich interactive domain 1B, SWI1-like); *PBRM1* (polybromo 1), an SWI/SNF chromatin-remodeling complex gene that negatively regulates cell proliferation through ligand-dependent transcriptional activation by NHRs; and *SMARCA4* (SWI/SNF-related, matrix-associated, actin-dependent regulator of chromatin, subfamily a, member 4).

DNA modifiers: *DNMT3A* (DNA (cytosine-5-)-methyltransferase 3 alpha); and *TET2* (Tet methylcytosine dioxygenase 2), which converts 5-methylcytosine (5mC) into 5-hydroxymethylcytosine (5hmC), thereby promoting DNA demethylation and gene expression.

Chromatin-modifying enzymes are druggable, thus offering a great opportunity for therapeutic intervention (Arrowsmith, Bountra, Fish, Lee, & Schapira, 2012). Although some epigenetic inhibitors exhibit pleiotropic, hence nonspecific effects, many are being assessed for specific applications/therapies and are extensively reviewed elsewhere (Ahuja, Easwaran, & Baylin, 2014; Dawson & Kouzarides, 2012; De Carvalho, You, & Jones, 2010; Jones & Baylin, 2002).

2.6 Alterations in Mediator, Splicing mRNA, and Ribosomal RNA Factors

Three additional factors implicated in RNA regulation include SF3B1 (splicing factor 3b, subunit 1), a component of the splicing machinery; *PHF6* (PHD finger protein 6), which suppresses ribosomal RNA (rRNA) transcription; and *MED23* (mediator complex subunit 23), a constituent of the mediator complex, which connects TFs to RNA pol II (Section 3; Table 5).

3. ALTERED TFs INVOLVED IN SUPER-ENHANCERS AND POSITIVE TRANSCRIPTIONAL ELONGATION

Some of the TFs altered in BC are major drivers (MYC, CSL/RBPJk) or implicated (MED23) in establishing super-enhancers. Enhancers are bidirectional, *cis*-acting elements located as far as a million base pairs upstream or downstream from target transcriptional start sites. They induce transcription by attracting activator proteins, which bind the mediator complex and, through looping of the chromatin, contact the promoter region to recruit components of positive transcription elongation factor b (P-TEFb), and thereby RNA pol II. Super-enhancers are defined as clusters of enhancers at close proximity with high level of mediator 1 (MED1) binding and open chromatin marks such as H2K4me1 and H3K27ac (Pott & Lieb, 2015; Sur & Taipale, 2016). There are ~10,000–150,000 putative enhancers and <1000 super-enhancers per cell.

In normal cells, MYC binds the core promoter of target genes. By contrast, in tumor cells with MYC amplification, this oncogene seems to accumulate in both promoter and enhancer regions of genes that are already actively expressed to induce transcriptional amplification. Thus, the effect of MYC is tumor specific—and when MYC is induced in preneoplastic tumor cells, it promotes tumorigenesis by amplifying existing oncogenic pathways, not by inducing a new oncogenic program. Interestingly, a recent paper shows that cancer cells, particularly ER-negative, are more transcriptionally active than intermixed stromal cells (Shlien et al., 2016). Thus, not only ER-negative tumors exhibit transcriptional amplification, but also mutations, rearrangements, and other alterations affecting mRNA are actively transcribed.

Transcription is characterized by a rate-liming initiation step, which, in the absence of promoter escape, leads to abortive initiation and the release of short <10 nt transcripts. Promoter escape and elongation (pause release) require phosphorylation on Ser2 of RNA polymerase II (pol II) carboxyl-terminal domain (CTD) by CDK9, a component of P-TEFb complex (Jonkers & Lis, 2015; Kwak & Lis, 2013; Sims, Belotserkovskaya, & Reinberg, 2004). RNA pol II is also phosphorylated and activated by CDK7, a subunit of TFIIH. These CDKs play pivotal roles in MYC and CSL/RBPJk-induced transcriptional activation and can be targeted for therapy (see below).

4. INDUCTION OF TFs THROUGH MUTATIONS IN DNA-BINDING SITES AND ALTERED CHROMATIN CONFORMATION

Genome-wide association studies (GWAS) identified over 400 single-nucleotide polymorphisms (SNPs) that predispose to cancer. Most of these SNPs are outside coding regions and presumed to create new cognate binding sites for TFs. Bioinformatic analysis of BC-specific SNPs revealed that they modulate the affinity of chromatin for FOXA1 at distal regulatory elements, thereby leading to aberrant gene expression of various genes such as *TOX3* (TOX high mobility group box family member 3) (Cowper-Sal lari et al., 2012). In accordance with the GWAS, there is evidence for single-nucleotide alterations that create binding sites for TFs in individual cancer genomes leading to unscheduled transcription and oncogenic transformation. An example of this was provided in a study of T-ALL in which heterozygote somatic mutations upstream of the *TAL1* oncogene were found to generate binding sites for the MYB TF. This leads to recruitment of CBP, GATA3, RUNX1, and TAL1 itself and the creation of a super-enhancer that induces TAL1 expression. MYB/CBP/GATA3/RUNX1/TAL1 complexes occupy other super-enhancers in the T-ALL genome. Thus, the somatic mutations upstream of TAL1 create sites for an oncogenic transcriptional complex that is already actively engaged in transcriptional activation of other genes. A similar scenario drives expression of the estrogen receptor α (*ESR1*) gene in 7% of ERα$^+$ BC (Bailey et al., 2016). Specifically, somatic mutations are enriched within a set of regulatory elements including the enhancer region of ESR1, leading to enhanced *ESR1* gene expression.

It is important to recognize that not all consensus DNA-binding sites identified by sequencing are actually occupied by TFs even when activated. Indeed, genome-wide *ch*romatin *immuno-p*recipitation (ChIP)-sequencing analyses reveal that TFs bind many thousands of genomic regions, many of which lack bona fide consensus binding sites for the specific TF, and conversely, not all binding sites actually recruit the TF (Spitz & Furlong, 2012). Effectiveness of binding motifs is governed by nucleotide sequence and torsional flexibility (Jordan et al., 2012), as well as by nucleosome occupancy and chromatin conformation that in turn depend on the effect of other TFs and chromatin modifiers (Barozzi et al., 2014; Cohanim & Haran, 2009; Slattery et al., 2014; Tong et al., 2016). However, it is likely that as with T-ALL described earlier, somatic mutations that create new binding sites in BC are selected for TFs like MYC, FOXA1, and ESR1 that are already altered in a particular tumor.

5. ADDITIONAL TFs IMPLICATED IN BC BUT NOT IDENTIFIED BY RECENT GENOMIC ANALYSES

While the recent genomic analyses described earlier point to a number of TFs and transcriptional pathways that are altered in BC, certain factors previously implicated in this disease are clearly missing. This can be attributed at least in part to the fact that ~65% BC samples are EPα$^+$, 20% are HER2$^+$, and only 10–15% are TNBC. Thus, altered TFs in HER2$^+$, TNBC, and other rare subtypes are likely underrepresented or completely overlooked. Indeed, a recent sequencing analysis of 64 HER2$^+$ BC samples identified additional genes in this subtype, some of which affect transcription (Ferrari et al., 2016). These include *ZNF703* (zinc finger elbow-related proline domain protein 1), *HMGA2* (high-mobility group AT-hook 2), *CDK12*, which phosphorylates RNA pol II CTD, thereby regulating transcription elongation, and possibly super-enhancers, as well as the epigenetic factors *KMT2D* (lysine methyltransferase 2D) and *KAT7* (lysine acetyltransferase 7). Other genes such as *STAT5* and *STAT3*, which are elevated/activated in BC, are not altered directly but are induced by alterations in upstream signaling pathways.

BCl11A (B-cell CLL/lymphoma 11A) is a C2H2-type zinc finger TF not listed in the 93 consensus altered genes but was found to be overexpressed in ~38% of TNBC samples (Khaled et al., 2015). Knockdown of BCL11A in TNBC cell lines suppressed tumorigenic potential in xenograft models pointing to this TF as a potential therapeutic target for TNBC.

NFIB (nuclear factor I B) is another TF not listed in the 93 consensus BC genes yet is elevated in ER-negative BC samples (Moon et al., 2011). Genetic knockdown of NFIB via RNA interference in a HER2$^+$ BC cell line induced apoptosis, suggesting tumor addiction. However, whether high NFIB expression is due to a primary genetic alteration as opposed to the consequence of upstream oncogenic signals, and whether it is a bona fide driver of ER-negative BC have yet to be established.

NFkB (nuclear factor kappa B). The NF-kappa-B family of TFs is activated by multiple oncogenic pathways to induce the expression of genes involved in inflammation, cell survival, and other biological processes (Hayden & Ghosh, 2012; Shen & Hahn, 2011). NFkB subunits have pleiotropic effects and can promote or inhibit cancer in different contexts (Perkins, 2012). A recent paper demonstrates that induction of noncanonical p52/NFkB in BRCA1 mutant cells promotes cell proliferation and thereby increased genomic instability at the replication fork, ultimately leading to neoplastic transformation (Sau et al., 2016). NFkB is elevated in many BC cells, not by direct genetic alterations but through as yet not fully defined upstream regulators. For example, the noncanonical NFkB kinase, IKBKE (IKKε) gene locus, 1q32, is frequently amplified and may drive NFkB signaling in these tumors (Boehm et al., 2007). However, 1q gains are most common in BC and *IKBKE* may be only one of several genes targeted by these amplifications. Nonetheless, combined inhibition of IKBKE and its related factor TBK1 suppresses the growth of many human BC lines. TBK1 was identified in a synthetic lethal screen for shRNAs that can suppress the growth of mutant RAS-expressing lung tumors vs parental cells (Barbie et al., 2009). It was also independently identified following a kinome shRNA screen as the most critical kinase for growth of mouse Her2/neu tumorspheres (Deng et al., 2014). TBK1/IKBKE inhibitors were then shown to potently suppress the growth of multiple HER2$^+$ BC cells (Deng et al., 2014). TBK1/IKBKE inhibition induces cellular senescence accompanied by inactivation of p65-NFkB and cooperates with the HER1(EGFR1)/HER2 kinase antagonist, lapatinib, to accelerate apoptosis. Inhibition of TBK1/IKBKE was also implicated in growth suppression of an immune-activated subset of TNBC through inhibition of the pro-tumorigenic cytokines CCL5 and IL-6 (Barbie et al., 2014). Thus, inhibition of NFkB via its upstream kinases is an attractive approach to suppress the growth of certain BC subtypes such as HER2$^+$, RTK-induced, BRCA1-mutant, and immunogenic TNBC.

GLI-1 (GLI family zinc finger 1). Hedgehog signaling has been implicated in BC through multiple studies in human and mice (Kasper, Jaks, Fiaschi, & Toftgard, 2009; Lewis et al., 1999). A recent paper identified overexpression of sonic hedgehog (SHH) in 25–50% TNBC (Noman et al., 2016). Gli-1, the TF downstream of SHH signaling, is highly expressed in inflammatory BC cells, and SHH pathway inhibition induces apoptosis and suppresses migration (Thomas et al., 2011). It is important to emphasize the intricate interconnection that exists between different signaling pathways (McNeill & Woodgett, 2010). Thus, while the above observations suggest that the HH pathway may be activated in certain BCs, this could be due to induction of other pathways that cross talk with and induce HH signaling in BC rather than bona fide genetic alterations in components of this important pathway.

6. THERAPEUTIC IMPLICATIONS

Current treatments for BC are based on cytotoxic antineoplastic agents for TNBC, chemotherapy plus endocrine therapies for ER$^+$, and chemotherapy plus anti-HER2 inhibitors for HER2$^+$ patients. These standard treatments overlook the enormous complexity of genetic alterations in BC and are likely to be replaced with personalized (precision) medicine, in which therapy is tailored to each individual patient based on the specific combinations of oncogenic alterations in each specific tumor. Thus, once a successful therapy for each of these genetic alterations is established, precision therapy may be prescribed targeting these alterations with a large therapeutic window. Personalized therapy has its critiques (Biankin, Piantadosi, & Hollingsworth, 2015; Swanton et al., 2016; Tannock & Hickman, 2016). In addition to incomplete drug response, emergence of drug-resistant clones, etc., we note here three major caveats: (i) parallel evolution, (ii) low-frequency mutations, and (iii) possible alternative approaches such as targeting oncogenic pathways rather than individual genes. (i) Parallel evolution (parallel progression) refers to accumulative evidence that tumor dissemination often occurs early, and, therefore, metastatic clones acquire many oncogenic alterations that are different from those found in the primary site as they promote the growth at different environments (Husemann et al., 2008; Klein, 2009). Thus, with the exception of some common founding mutations that drive the initial neoplastic transformation, metastatic clones have different landscapes of genetic alterations rendering them irresponsive to therapies designed to treat the primary lesion. Identifying and targeting oncogenic alterations driving metastatic growth is a formidable task but is the focus of extensive studies around the globe and may be resolved in the future. (ii) Oncogenic alterations that occur at low frequency are unlikely to be pursued, even if they promote the growth of particular tumors, because of the enormous financial burden of drug development and clinical trials. Thus, targeted therapies will realistically be designed toward TFs and oncogenic pathways that are frequently altered in BC. (iii) The alternative to precision therapy is the identification of oncogenic pathways (as opposed to genes) that drive a particular cancer and the design of general therapies for these pathways rather than specific treatments for each altered gene. In reality, therapies may include both approaches, i.e., targeted therapy against specific genetic alterations (e.g., endocrine therapies, anti-HER2, anti-PIK3CA) in combination with therapies against major oncogenic vulnerabilities critical for tumor growth (cell proliferation, survival, protein translation, autophagy, super-enhancers, glycolysis/OXPHOS, angiogenesis, etc.). Below, we discuss approaches to target-specific genetic alterations in TFs and then focus on those that appear frequently in BC.

7. APPROACHES TO TARGET-SPECIFIC GENETIC ALTERATIONS IN TFs

Targeting oncogenic alterations in cancer relies conceptually on the idea of oncogenic addiction, where tumor growth and survival depend on continuous expression of the oncogenic alterations that drive them (Luo, Solimini, & Elledge, 2009; Weinstein & Joe, 2008). This may not always be the case as certain mutator genes, such as *MSH2* involved in different forms of DNA repair and the cytidine deaminase *APOBEC3B* (Alexandrov et al., 2013; Cescon, Haibe-Kains, & Mak, 2015; Fishel et al., 1993; Roberts et al., 2013), may act via a hit-and-run mechanism to allow the accumulation of other oncogenic mutations, but their continuous presence may not be necessary for survival of a tumor once formed. However, for most oncogenic drivers and TFs, oncogenic addiction is probably the rule.

TFs are notoriously difficult to target for therapy. The simplest approach to inhibit their activity is to target upstream positive regulators (e.g., CDK4/6 inhibitors for the RB1/E2F pathway) or the interface between TFs and positive cofactors such as between β-catenin and CBP (see below). Here we discuss biological approaches to inhibit TFs: (a) targeting downstream pathways, (b) over-the-cliff strategy, and (c) synthetic lethality. Other approaches including screening natural products, novel drug designs, chemical targeting of TFs for proteolysis, RNAi/nanoparticle delivery, and computational approaches have been reviewed elsewhere (Bauer, Wurst, & Tan, 2010; Bhagwat & Vakoc, 2015; Gayvert et al., 2016; Libermann & Zerbini, 2006). In this regard, the recent development of a drug for the transcription repressor BCL6 using in silico drug design is encouraging (Cardenas et al., 2016):

a. *Downstream pathways.* If a TF is not actionable, the genetic program it induces may still be targetable. We have recently used this rationale to identify mitochondrial protein translation (MPT) as a targetable pathway downstream of RB loss (see below). A similar approach may be adapted to identify vulnerabilities that are druggable downstream of other TFs. While such approaches may not be entirely TF specific, they may target important nodes of regulation and have dramatic therapeutic outcome.

b. *Over-the-cliff approaches: pressing the accelerator instead of the brake pedal.* To survive and maintain oncogenic transformation, tumor cells rewire various pathways to reestablish homeostasis. Indeed, aberrant cell proliferation often induces proapoptotic signals, ATP deprivation, hypoxia, autophagy, reactive oxygen species (ROS), genotoxic stress, immune response, a shift to anabolic metabolism, and high protein translation rate that must be counteracted by induction of survival pathways and evasion of immune response. Ultimately, tumors cells reach a balance that enables net growth, but they are "on the edge" and can readily be "pushed" over the cliff toward cell demise. This "push" can be induced by various cytotoxic therapies, increasing ROS, blocking autophagy, or survival signals using, for example, inhibitors for MDM2 (see below), Bcl-2 (Merino, Lok, Visvader, & Lindeman, 2016), or XIAP (Schimmer et al., 2004). An alternative approach is to further accelerate proliferation or consequences thereof by pushing the cells over the cliff. Thus, inhibition of eEF2K, which phosphorylates and blocks the activity of the protein translation elongation factor eEF2, leads to increased protein translation and cell death under nutrient deprivation or in AKT pathway-elevated BC cells (Leprivier et al., 2013; Liu et al., 2014).

For TFs, this strategy has been illustrated for retinoic acid, which induces differentiation of cultured neuroblasts and leukemic cells via its nuclear receptors (RARα family) (Liu et al., 2000; Pahlman, Ruusala, Abrahamsson, Mattsson, & Esscher, 1984; Zhang, Gu, Wang, Chen, & Chen, 2000).

Interestingly, RAR and the ETS TF Fli-1 (friend leukemia virus integration 1) (Li, Luo, Liu, Zacksenhaus, & Ben-David, 2015) reciprocally suppress each other (Darby, Meissner, Ruhlmann, Mueller, & Scheibe, 1997). Inhibitors of Fli-1 suppress leukemia through the induction of erythroid differentiation (Li et al., 2012). Remarkably, we have recently found that agonists of Fli-1 transcriptional activation, which stimulate its upstream regulator PKC, also suppress leukemia by inducing megakaryocytic differentiation (Liu T.J.J. et al., submitted for publication). Thus, while overexpression of Fli-1 induces leukemia, inhibition of this TF promotes erythroid differentiation and further transcriptional activation (i.e., over the cliff) induces megakaryocyte differentiation. Fli-1 is elevated in certain TNBC cells (Scheiber et al., 2014; Song et al., 2015). Whether Fli-1 agonists and antagonists promote differentiation of BC cells with elevated Fli-1 expression awaits further investigation.

c. *Synthetic lethality*. Targeting the escape routes/alternative pathways that compensate for an oncogenic driver (Downward, 2015; Fece de la Cruz, Gapp, & Nijman, 2015). Perhaps the best example for this approach is the use of poly(ADP) ribose polymerase (PARP) inhibitors for BRCA1-deficient tumors (Ashworth, 2008; Lord, Tutt, & Ashworth, 2015). The advantage of this approach is that it bypasses the need to target TFs, and it is readily amenable to high-throughput (HTP) screens. Moreover, inhibitors that emerge from such screens are selected to have only marginal effects when administered alone on tumor growth and, presumably, on normal cells that do not carry these oncogenic alterations. The screen setup requires the generation of isogenic lines with overexpression or shRNA knockdown/CRISPR knockout of a TF that can drive tumorigenesis. Then, two screens are performed in parallel on the parental and modified isogenic lines. Drugs that have minimal effect on the parental cells but block the growth (or other quantifiable parameters such as migration) of the modified cells are candidate synthetic lethal (or other parameter) inhibitors. Some of these inhibitors may directly suppress/reverse the effect of the oncogenic TF, while others may target compensating pathways that permit survival in its presence.

8. TARGETING FREQUENTLY ALTERED TFs IN BC

The identification of TFs and cofactors as major drivers of BC as reviewed herein should intensify efforts to identify therapeutic drugs for these factors. Here we describe efforts and avenues for the most frequently altered TFs in BC.

8.1 pRb/E2F1–3

The tumor suppressor pRb exerts its effect on cell growth through interaction with other factors such as activating E2F (E2F1, E2F2, and E2F3) TFs (Burkhart & Sage, 2008; Dyson, 2016). pRb binds E2F1–3 during most G1 phase of the cell cycle and silences gene expression by recruiting HDACs or HMTs. Upon phosphorylation by CDKs, or in response to RB1 loss or sequestration by viral oncoproteins, the pRb–E2F complex is disrupted and free E2F recruits HATs to stimulate gene expression (Bremner et al., 1995; Weintraub, Prater, & Dean, 1992; Zacksenhaus, Jiang, Phillips, & Gallie, 1996). Thus, E2F complexes switch from repressors to activators through a default repression mechanism (Fig. 2). pRb–E2F complexes coordinate expression of genes involved in cell cycle and apoptosis (Chen, Tsai, & Leone, 2009; Dyson, 2016; Jones et al., 2016). The induction of proapoptotic genes serves as a safeguard to prevent cells from dividing in the absence of sufficient survival signals.

In ER^+ tumors in which RB1 is often intact, the pRb protein is inactivated via hyperphosphorylation. In BC, pRb hyperphosphorylation is induced by overexpression of cyclin-dependent kinases (CDK4 and CDK6); amplification/overexpression of cyclinD1, cyclinD3, and cyclinE1; and loss of the CDK4/6 inhibitors $p16^{INK4a}$ (CDKN2A) and $p15^{INK4b}$ (CDKN2B) or the CDK4/6–CDK2 inhibitor $p27^{Kip1}$ (CDKN1B)

(Tables 1 and 3). Tumors in which the drivers for pRb phosphorylation are loss of p16^{INK4a} or p15^{INK4b}, or overexpression of CDK4 and CDK6 or cyclin D1/D3, are highly sensitive to CDK4/6 inhibitors (Asghar, Witkiewicz, Turner, & Knudsen, 2015; McCain, 2015). Thus, PD 0332991, an anti-CDK4/6 inhibitor (Palbociclib, Pfizer, New York City, USA), exhibits strong cytostatic effects on luminal (but not TNBC) in vitro (Finn et al., 2009). Strong efficacy and safety in clinical trials of luminal BC patients have been reported for PD 0332991 and Abemaciclib (https://clinicaltrials.gov/ct2/results?term=PD+0332991&Search=Search) (O'Leary, Finn, & Turner, 2016; Patnaik et al., 2016). Briciclib (prodrug of ON-013100), which blocks cyclin D mRNA translation and protein levels, is also in Phase I trial for solid tumors. In contrast, tumors in which the drivers for pRb phosphorylation is driven by cyclin E–CDK2 or loss of p27^{Kip1} are likely to be more sensitive to CDK2-specific inhibitors or to combination of CDK4/6 plus CDK2 inhibitors, respectively.

While these approaches for RB-proficient tumors are promising, RB1 deficiency, mostly occurring in TNBC (Koboldt et al., 2012; Nik-Zainal et al., 2016), is not amenable to therapeutic targeting. We recently tackled this problem by probing for pathways downstream of Rb loss in TNBC, which can be exploited therapeutically. Briefly, we compared very similar mammary tumors induced by inactivation of Rb/p53 or p53 alone and identified MPT as a pathway that is specifically induced in the double mutant lesions (Jones et al., 2016). We then showed that pRb–E2F1/E2F3 complexes bind and transcriptionally regulate MPT gene expression, thereby controlling MPT. A nonbiased repurposing drug screen performed in parallel identified tigecycline, an FDA-approved antibiotic and MPT inhibitor, as a potent drug against Rb/p53-deficient mammary tumors. We further demonstrated that human RB1^{-}/TP53^{-} TNBC cells are more sensitive to tigecycline than RB1^{+}/TP53^{-} tumors, both in vitro and in xenografts (Jones et al., 2016). Thus, although deregulated E2F1/E2F3 are not readily actionable, a downstream metabolic pathway they regulate is amenable to therapeutic intervention.

In a different avenue, a synthetic lethal screen was conducted in the fruit fly to identify genes that are essential for survival of Rb-deficient cells. This screen showed that loss of the tumor suppressor TSC2, a negative regulator of mTOR, induced synergistic cell death with RB1 loss in both drosophila developing tissues and human cancer cells (Danos, Liao, Li, & Du, 2013; Li, Gordon, Du, Xu, & Du, 2010). It was further shown that synergy was due to accumulation of genotoxic stress, suggesting that inactivation of TSC2 together with chemotherapies that induce such stress may be combined to treat RB-deficient BC. Synthetic lethal screens of RB1-mutant cells are urgently needed especially for RB1-deficient TNBC.

8.2 TP53

The tumor suppressor TP53 is kept at low levels in unstressed cells by proteasomal degradation induced by the MDM2/HDM2 (and MDM4/MDMX) E3 ligase. Upon various types of genotoxic stress, such as DNA damage, TP53 is phosphorylated by stress-sensing kinases such as ATM/ATR or CHK1/2 leading to inhibition of its interaction with MDM2 (Levine & Oren, 2009). Stabilized TP53 then binds as a tetramer to promoters, such as the CDK inhibitor $p21^{cip1}$ involved in cell cycle arrest, and induces gene expression, enabling cells to pause and counteract the stress through DNA damage repair (Aylon & Oren, 2007). Under high or continuous stress signal, TP53 transcriptionally stimulates proapoptotic BCL2 family members such as BAX (Miyashita & Reed, 1995), resulting in cytochrome *c* release, caspase activation, and cell demise. Differential induction of cell cycle arrest genes vs apoptotic genes has been attributed to high- and low-affinity interactions of TP53 with the corresponding promoters (Aylon & Oren, 2007). Thus, TP53 loss enables accumulation of oncogenic mutations and thereby neoplastic transformation. It is important to note that mutations that significantly but not completely compromise transcriptional activation by TP53 do not degrade its ability to exert tumor suppression in diverse lineages in vivo (Jiang, Brady, et al., 2011). Only additional mutations that fully disrupt p53 transcriptional regulation abrogate tumor suppression. Thus, other functions of p53, for example, in the cytosol (e.g., inhibition of autophagy; induction of apoptosis via mitochondrial outer membrane permeabilization), may have important role in tumor suppression independently of its transcriptional function (Green & Kroemer, 2009).

TP53 is functionally inactivated in cancer by amplification of MDM2, or genetically by truncations/deletions or point mutations (Koboldt et al., 2012). To overcome MDM2-induced TP53 degradation, compounds such as Nutlins, which bind MDM2 and inhibit its interaction with p53, or RITA, which bind TP53 and prevent its interaction with MDM2, have been developed (Issaeva et al., 2004; Vassilev et al., 2004). TP53 loss has been exploited to generate oncolytic viruses like Onyx, an adenovirus engineered to lack the E1B 55 kDa protein, which binds and inactivates TP53 (Heise et al., 1997; Khuri et al., 2000; Kirn,

Hermiston, & McCormick, 1998). Onyx can therefore only replicate in and lyse TP53-deficient tumors, sparing normal somatic cells.

While the consequences of functional/genetic loss of TP53 are indicative of a tumor suppressor function, p53 point mutations in its DB domain are also oncogenic as they confer gain-of-function and/or dominant-negative effects through sequestration of the remaining wild-type TP53 allele or its relatives p63 and p73 as well as through other mechanisms (Adorno et al., 2009; Godar et al., 2008; Muller et al., 2009; Powell, Piwnica-Worms, & Piwnica-Worms, 2014; Senoo, Pinto, Crum, & McKeon, 2007; Wang et al., 2009). Thus, efforts are being invested to induce p63 and p73, which share many functions with TP53 including induction of proapoptotic genes such as Puma and Noxa (Di Como, Gaiddon, & Prives, 1999; Murray-Zmijewski, Lane, & Bourdon, 2006; Rocco, Leong, Kuperwasser, DeYoung, & Ellisen, 2006). Additionally, there is evidence that tumors driven by TP53 mutations are addicted to continuous expression of these oncogenic forms of TP53. This in turn can be exploited using approaches to destabilize mutant p53. For example, heat-shock protein 90 (HSP90) is required for appropriate folding of mutant p53. Ganetespib, a potent HSP90 inhibitor, preferentially targets p53-mutant vs p53-deletion tumor cells both in vitro and in xenografts (Alexandrova et al., 2015). However, in addition to mutant TP53, HSP90 inhibition destabilizes several other factors, e.g., HER2 (Friedland et al., 2014; Grieve, Gao, Hall, Hu, & Greer, 2016), and p53-deficient tumors are also sensitive to this drug, albeit to a lesser extent. Although this is therapeutically advantageous, ganetespib is not p53-mutant-specific inhibitor per se and, as such, its effect is not entirely predictable. Finally, a synthetic lethal screen showed that TP53 deficiency increases sensitivity to loss of the TF ETV1 and the DNA damage kinase ATR (Xie et al., 2012). These findings point to a potentially new therapeutic avenue for p53-null tumors and should encourage additional screens to identify synthetic lethality with p53 deficiency.

8.3 MYC

The *MYC* oncogene is a basic helix-loop-helix (bHLH) leucine zipper (LZ) TF that heterodimerizes with Max (and Miz1) and binds E-box containing promoters to induce genes involved in cell proliferation, survival, and metabolism (Adhikary & Eilers, 2005; Tu et al., 2015). MYC–MAX heterodimers compete with MAD–MAX repressor complexes; so that when MYC accumulates MYC–MAX complexes prevail. MYC also interacts with the SWI/SNF chromatin-remodeling complex, INI1

(SMARCB1/hSNF5/BAF47), a tumor suppressor that antagonizes MYC (Stojanova et al., 2016).

MYC itself is overexpressed in cancer through copy number gains as well as various oncogenic signaling pathways that induce its expression or stability. Proteasomal degradation of MYC is induced by FBXW7-mediated polyubiquitination of Thr58-phosphorylated MYC when Ser62 is not phosphorylated. Formation of the MYC degradation complex is coordinated by the scaffold protein AXIN (Farrell & Sears, 2014). RAS signaling protects MYC from proteasomal degradation by inhibiting GSK3-mediated phosphorylation of Thr58 through PI3K/AKT, and by promoting Ser62 phosphorylation by MAPK/ERK (Darnell, 2002). MYC stability is also regulated by the E3 ligase Mule/Huwe1/Arf-BP1, but relevance to BC is not clear as Mule is elevated, not lost, in this malignancy (Inoue et al., 2013). In sum, MYC is stabilized in BC by alterations that induce RAS signaling, MAPK, PI3K, and AKT, and those that disrupt AXIN and FBXW7.

MYC is induced in cancer primarily by amplification of 8q24.21, which also includes the long noncoding RNA gene PVT1 gene. Indeed, in nearly all human cancers, high MYC expression coincides with high PVT1 levels. Remarkably, PVT1 forms a ribonucleoprotein complex with MYC and protects it from Thr58 phosphorylation and degradation. Moreover, knockdown of PVT1 diminishes MYC expression (Tseng et al., 2014). Thus, PVT1 offers a novel target for MYC-driven tumors. Finally, 8q24 is a risk locus that predisposes individuals to breast as well as prostate and colon malignancy. The locus exhibits marks of open chromatin characteristics of enhancers and induces the expression of the nearby MYC proto-oncogene (Ahmadiyeh et al., 2010).

An obvious, yet challenging approach to target MYC relies on direct inhibition of MYC–MAX interaction. Although multiple such compounds were identified in vitro (e.g., 10059-F4 (Zirath et al., 2013)), none has so far exhibited sufficient pharmacokinetics and pharmacodynamics properties to warrant clinical trials. Another approach is based on the finding that MYC promotes anabolic metabolism not only through stimulating expression of glycolytic genes but also by increasing glutamine usage through transcriptional activation of glutaminase (GLS), an enzyme that converts glutamine to glutamate (Hensley, Wasti, & DeBerardinis, 2013; Hsieh, Walton, Altman, Stine, & Dang, 2015; Wise & Thompson, 2010). The latter is then converted to α-ketoglutarate, which enters the tricarboxylic acid cycle to generate intermediates for lipid, amino acid, and nucleotide synthesis, as well as NADH for OXPHOS (Pavlova & Thompson, 2016). In addition, glutamine is used by antiporters for exchange of other amino acids (Wise & Thompson, 2010). For example, the LAT1 antiporter imports the essential amino acid leucine in exchange for efflux of glutamine. Interestingly, while luminal BC cells are addicted to glucose, TNBC cells exhibit greater dependency on glutamine (Timmerman et al., 2013). Thus, these tumors, particularly those with MYC overexpression, may be more sensitive to therapies that block glutamine usage through, for example, GLS inhibitors or antagonists of the glutamine transporter SLC1A5/ASCT2 (Altman, Stine, & Dang, 2016; van Geldermalsen et al., 2016). Notably, glutamate is exchanged with cystine by the glutamate/cystine antiporter xCT, allowing further conversion to cysteine. The later plus glutamate are used to generate the antioxidant glutathione. Inhibition of xCT with the FDA-approved antiinflammatory prodrug sulfasalazine reduces intracellular glutathione, leading to increased ROS, cell cycle arrest, and attenuation of xenograft growth (Jones et al., 2016; Timmerman et al., 2013).

Various synthetic lethal screens for MYC have been attempted (Cermelli, Jang, Bernard, & Grandori, 2014). A recent screen identified the core spliceosome as a therapeutic vulnerability in MYC-driven tumors (Hsu et al., 2015). In particular, BUD31, a component of the core spliceosome, as well as splicing factor SF3B1, which is altered in human BC (Table 5), and U2AF1 are required for survival of MYC-induced tumors and, therefore, provide new therapeutic targets for MYC-induced malignancies.

Perhaps the most promising therapeutic approach for MYC came from studies on the role of this oncogene in the assembly and activation of super-enhancers, leading to transcriptional amplification (Lin et al., 2012). P-TEFb enables RNA pol II to transition from promoter-initiation/pausing to productive elongation. CDK7, which is required to phosphorylate and induce pol II CTD, plays a critical role in MYC-induced transcriptional elongation (Nilson et al., 2015). Pharmacological inhibition of CDK7 by THZ1 suppresses MYC-induced transcription and blocks tumor growth (Chipumuro et al., 2014; Loven et al., 2013). THZ1 was found to inhibit the growth of nearly half of ~1000 different human cancer cell lines, suggesting that CDK7 is a critical target in both MYC-dependent and -independent tumors, through, for example, RUNX1 in certain leukemia (Kwiatkowski et al., 2014). The BET bromodomain factor BRD4 co-occupies thousands of enhancers associated with active genes in MYC-induced tumors, and in accordance, the BRD4 inhibitor, JQ1, exerts potent growth inhibition on multiple tumors (Delmore et al., 2011; Filippakopoulos et al., 2010; Sahai, Redig, Collier, Eckerdt, & Munshi,

2016). Recent reports also showed that JQ1 preferentially and efficiently suppresses TNBC, leading to growth arrest and differentiation (Shu et al., 2016). Interestingly, tumor cells can resist JQ1 inhibition through MYC amplification, further underlying the role of JQ1 in blocking super-enhancers. Another study showed synergy between inhibition of CDK9, another CDK that recruits P-TEFb, and BRD4 in effectively killing cancer cells (Lu et al., 2015). Thus, CDK7/CDK9 and BRD4 provide new avenues to target MYC-induced BC, and perhaps a large percentage of TNBC.

Although we have limited this review to discussing altered TFs in BC in isolation, there is a considerable cross talk between these factors. Here we briefly consider the above three TFs (MYC, E2F, TP53). E2F2 and E2F3 are required for MYC-induced S phase entry, whereas E2F1 is required for MYC-induced apoptosis, which involves the stabilization of p53 via ARF (Leone et al., 2001; Polager & Ginsberg, 2009). Apoptosis in some (CNS, lens) but not all (e.g., muscle) RB-deficient embryos is suppressed by loss of p53 (Ho, Li, Hakem, Mak, & Zacksenhaus, 2004; Jiang et al., 2000; Macleod, Hu, & Jacks, 1996; Morgenbesser, Williams, Jacks, & DePinho, 1994). Conversely, cell cycle arrest induced by stabilization of TP53 requires intact RB1 (Polager & Ginsberg, 2009). Thus, RB1 and TP53 are frequently lost together in cancer in general and in TNBC in particular (Jiang, Jones, et al., 2011). While proliferation of E2F1–3-null fibroblasts is severely impaired (Wu et al., 2001), E2F1/2/3 triple KO retinal progenitor cells can divide through MYCN (Chen, Pacal, et al., 2009). Moreover, a rare subtype of $RB^{+/+}$ retinoblastoma is driven by MYCN amplification, indicating that RB1 loss and MYCN amplification may be interchangeable in certain contexts. In a murine intestinal model, E2F1–3 and MYC control an S–G2, not G1/S, transcriptional program. Following Rb loss, both Myc and E2F3 are redeployed to induce G1/S gene expression in an MYC-dependent manner, leading to widespread ectopic cell proliferation (Liu et al., 2015). The cooperation and interchangeability between MYC and activating E2Fs, as is the case for other oncogenic pathways, may enable tumor cells to rewire and escape therapies directed at one of these genes/pathways, leading to relapse.

8.4 NOTCH/CSL

Notch family comprises Notch 1 and 2, which are frequently induced in BC, as well as Notch 3 and 4 (Callahan & Egan, 2004). These factors have different effects on development and tumorigenesis. While Notch signaling induces T-cell leukemia and certain epithelial tumors including BC by promoting progenitor cell proliferation, Notch 1 acts as a tumor suppressor in other contexts such as skin cancer where it is required for differentiation and its loss is oncogenic (Radtke & Raj, 2003). Notch signaling can be suppressed by γ-secretase inhibitors (GSIs) (Curry et al., 2005; Wong et al., 2004), which block its cleavage into the oncogenic Notch-IC fragment at the cell membrane upon signaling, and by antagonists of Notch trafficking and processing in the ER (Kramer et al., 2013). Pan Notch inhibitors such as GSI induce on-target intestinal toxicity, attributed to Notch 1 and 2 inhibition. Using monoclonal antibodies specific to individual Notch factors, it was shown that Notch 1 inhibition maintained leukemic tumor suppression function, yet had only mild intestinal toxicity (Wu, Cain-Hom, et al., 2010). Clinical trials of Notch pathway inhibitors, mostly GSIs, in patients with solid tumors including BC are ongoing/have been reported (Andersson & Lendahl, 2014; Takebe et al., 2015). Other directions may include the generation of Notch paralogue-specific antagonists that disrupt activated intracellular NOTCHs by interfering with their interaction with CSL or coactivators.

Notwithstanding the discussion earlier, a recent paper showed that knocked-down of CSL/RBPJk has a much more potent effect on suppression of brain CSC proliferation than treatment with the GSI DAPT (Xie et al., 2016). CSL/RBPJk induces transcription via CDK9, the RNA pol II CTD kinase, and, accordingly, shRNA-mediated knockdown of CDK9 curtailed tumor growth in xenograft assays. These findings, in conjunction with the potent effects of CDK7 inhibitors on MYC-driven tumors, point to the transcriptional elongation apparatus as an effective therapeutic target downstream of multiple activated TFs in BC and other malignancies (Takebe et al., 2015).

8.5 ESR1

About two-thirds of BC patients express ERα. The importance of estrogens in BC has been demonstrated as early as 1896 when George Beatson showed that bilateral oophorectomy which depletes E2 from premenopausal women significantly improves overall survival (Clemons & Goss, 2001). Current endocrine therapy is based on selective estrogen receptor modulators (SERMs), which are ER ligands that in some tissues (i.e., bone and cardiovascular system) act like estrogens, while in others including BC function act as estrogen antagonists (Lewis & Jordan, 2005; Maximov, Lee,

& Jordan, 2013). The first approved SERM for BC was tamoxifen followed by toremifene for postmenopausal patients (Hayes et al., 1995). In addition to the ovary, estrogen is synthesized locally in adipose tissue by aromatase, which catalyzes the aromatization of androgen into estrogen. Aromatase inhibitors such as letrozole are effective for postmenopausal ER^+ BC patients (Goss et al., 2003, 2016).

Mutations in *ESR1* were identified upon relapse in women treated with tamoxifen (Robinson et al., 2013). These mutations block ligand binding and render the mutant receptor independent of E2 and constitutively active. Much effort is therefore invested on alternative approaches to target ER signaling. One direction is to target other components of the ERα transcriptional complex. The histone demethylase JMJD2B (KDM4B) is recruited to ERα-target sites through interaction with the receptor. Once recruited, it demethylates H3K9me3 and stimulates transcription of ER-responsive genes including *MYB*, *MYC*, and *CCND1*. Importantly, knockdown of JMJD2B blocked estrogen-induced cell proliferation in vitro and tumor formation of engrafted BC cells in vivo, pointing to this chromatin modifier as potentially new therapeutic target for ER^+ BC (Kawazu et al., 2011).

These and other approaches attempt to target ESR1-mediated transcription directly. In addition, combinations of endocrine therapies plus other cell cycle inhibitory drugs are used to block alternative proliferation pathways in ER^+ BC cells. Indeed, a recent phase II trial revealed that the combination of letrozole plus palbociclib (PD0332991) significantly improved progression-free survival in women with advanced ER^+; HER2-negative BC (Finn et al., 2015). A phase III trial is currently underway (https://clinicaltrials.gov/ct2/results?term=PD+0332991&Search=Search). Patients are also actively recruited to trials of letrozole plus the PI3K inhibitor BYL719 (https://clinicaltrials.gov/ct2/results?term=PI3K&Search=Search), and for tamoxifen plus LY2228820, a p38 MAPK inhibitor (https://clinicaltrials.gov/ct2/show/NCT02322853?term=tamoxifen&rank=5).

We note that forkhead TFs, FOXA1 and FOXP1, appear to function as oncogenes or tumor suppressors in different contexts (Bernardo et al., 2010; Choi et al., 2016; Ijichi et al., 2012; Li et al., 2013; Pereira et al., 2016; Shigekawa et al., 2011). Thus, while FOXP1 promotes CSC-like features in ovarian cancer cells, homozygous deletion analysis in BC identified this gene as a tumor suppressor driver (Pereira et al., 2016). Likewise, while FOXA1 is a key regulator of ERα expression and luminal cell fate, its loss promotes basal-like tumors, which are even more aggressive, therefore questioning this TF as a viable therapeutic target (Bernardo et al., 2013).

8.6 WNT/β-Catenin/TCF

The canonical WNT pathway is activated in BC through the loss of AXIN, APC, or FBXW7 (Fig. 2). This pathway is a major regulator of CSCs, at least in part by inducing expression of epithelial–mesenchymal transition (EMT) TFs such as SNAIL, TWIST, and ZEB (Pattabiraman & Weinberg, 2014). The pathway can be inhibited at multiple levels including sequestration of WNT itself, its receptors, or blocking β-catenin–CBP interaction. To identify inhibitors for β-catenin-induced transcription, Michael Kahn's group performed a screen for inhibitors of a β-catenin-responsive reporter and identified ICG-001, an antagonist of cyclic AMP response element-binding protein (CBP, also known as CREBBP) (Emami et al., 2004). ICG-001 selectively induces apoptosis in transformed but not normal cells. This drug specifically inhibits the interaction of β-catenin with CBP, which is required for cell proliferation, but not with p300, which promotes differentiation (Teo, Ma, Nguyen, Lam, & Kahn, 2005). A functionally similar molecule, PRI-724 (Prism Pharma, Yokohama, Japan; Eisai Pharmaceuticals, Tokyo, Japan), was tested in phase I on patients with solid tumors, and tolerable doses were tested in combination with chemotherapy (2013 ASCO Annual Meeting). Other compounds inhibiting the interaction of β-catenin with TCF, such as PKF115-584, CGP049090, and FH53, are also being developed (Lepourcelet et al., 2004). FH53 inhibits recruitment of β-catenin but not the corepressor NCoR to TCF (Vilchez, Turcios, Marti, & Gedaly, 2016). However, it also antagonizes PPARgamma and PPARdelta ligand-dependent activation (Handeli & Simon, 2008).

FOXP1, which is altered in BC (Table 2B), is a positive transcriptional cofactor of β-catenin. In diffuse large B-cell lymphoma (DLBCL), FOXP1 promotes the acetylation of β-catenin by CBP, and this potentiates β-catenin-dependent transcription (Walker et al., 2015). Both knockdown of FOXP1 and inhibition of β-catenin transcriptional activity suppressed xenograft tumor growth. Whether FOXP1 inhibition can suppress WNT signaling in BC and block tumor growth remains to be seen.

8.7 BRCA1

Although this tumor suppressor is not frequently altered in sporadic BC, it plays an important role in hereditary BC (Narod, 2010; Nielsen, van Overeem Hansen, & Sorensen, 2016; Turner & Reis-Filho, 2006). Inactivation of Brca1 is lethal in normal cells (Gorrini et al., 2013; Hakem et al., 1996), but it can cooperate with other oncogenic events such as p53 loss to promote basal-like BC both in mouse and in human (Foulkes et al., 2003; Liu, Holstege, et al., 2007). BRCA1 tumors show defects in repair of double-strand DNA breaks by homologous recombination and are therefore highly dependent on repair of DNA single-strand breaks via the enzyme PARP. This can be exploited in a synthetic lethal treatment of BRCA1 (and other lesions such as BRCA2, leading to similar defects referred to as BRCAness) using PARP inhibitors (Ashworth, 2008; Lord et al., 2015). Recent advances identified additional approaches based on other dependencies of BRCA1 tumors. Gorrini et al. have uncovered a mechanism that can explain the tissue specificity of BRCA1 loss in breast and ovary epithelium vs other tissues. They showed that BRCA1 interacts with and stabilizes NRF2/NFE2L2 (nuclear factor, erythroid 2 like-2), thereby diminishing intracellular ROS levels and promoting survival (Gorrini et al., 2013). NRF2 is a TF that binds to antioxidant response elements in promoter regions of target genes such as GSTA2 (glutathione-S-transferase alpha 2) and NQO1 (NAD(P)H quinone dehydrogenase 1) (Nguyen, Nioi, & Pickett, 2009). Most tissues lacking BRCA1 exhibit low NRF2 levels and increased cell senescence or cell death. In BRCA1-deficient ER^+ epithelial cells, NRF2 is induced transcriptionally by ESR1 and transnationally via an ESR1–PI3K–mTOR signaling, leading to cell survival and ultimately cancer (Gorrini et al., 2014). Thus, inhibition of NRF2-dependent transcription may prove effective in preventing or treating BRCA1 patients. Finally, NFkB expression in BRCA1-mutant cells promotes accumulation of DNA damage and increased genomic instability at the replication fork and ultimately neoplastic transformation, thus pointing to NFkB-mediated transcription as a potential prophylactic target for BRCA1 mutant carriers (Sau et al., 2016).

9. CONCLUDING REMARKS

In this review, we described oncogenic alterations affecting TFs in BC. We made the following key observations:

(i) Alterations affecting TFs and transcriptional regulation are frequent (49/93) in BC.

(ii) No alterations are found in TFs that mediate default repression downstream of major oncogenic signaling.

(iii) Frequent alterations occur in transcriptional cofactors or upstream regulators of oncogenic signaling, in inducible or constitutive TFs that do not mediate default repression, and in epigenetic genes/chromatin modifiers.

(iv) Certain TFs such as MYC are the target of diverse oncogenic alterations including copy number change, mutations in upstream cofactors or their regulators, or mutations that create DNA-binding sites in enhancers, stimulating their expression.

Many of these TFs and cofactors are amenable to therapeutic intervention, and some are already assessed in clinical trials. New approaches to directly or indirectly target TFs in combination with other therapies may improve personalized treatment and remission rates in BC patients.

ACKNOWLEDGMENTS

The Zacksenhaus lab is supported by research grants from the Canadian Breast Cancer Foundation (Ontario Chapter), Canadian Institute of Health Research, the Israel-Cancer Research Foundation, US Army Department of Defense Breast Cancer Research Program, and the Terry-Fox Foundation. The Ben-David lab is supported by grants from the Science and Technology Department of Guizhou Province Innovation and Project Grant (2013-6012), Thousand Talent Program of China (WQ20135200171), and The Natural Science Foundation of China (81472609).

REFERENCES

Adhikary, S., & Eilers, M. (2005). Transcriptional regulation and transformation by Myc proteins. *Nature Reviews Molecular Cell Biology, 6*(8), 635–645. http://dx.doi.org/10.1038/nrm1703.

Adorno, M., Cordenonsi, M., Montagner, M., Dupont, S., Wong, C., Hann, B., … Piccolo, S. (2009). A mutant-p53/Smad complex opposes p63 to empower TGFbeta-induced metastasis. *Cell, 137*(1), 87–98.

Ahmadiyeh, N., Pomerantz, M. M., Grisanzio, C., Herman, P., Jia, L., Almendro, V., … Freedman, M. L. (2010). 8q24 prostate, breast, and colon cancer risk loci show tissue-specific long-range interaction with MYC. *Proceedings of the National Academy of Sciences of the United States of America, 107*(21), 9742–9746. http://dx.doi.org/10.1073/pnas.0910668107.

Ahuja, N., Easwaran, H., & Baylin, S. B. (2014). Harnessing the potential of epigenetic therapy to target solid tumors. *Journal of Clinical Investigation, 124*(1), 56–63. http://dx.doi.org/10.1172/JCI69736.

Alexandrov, L. B., Nik-Zainal, S., Wedge, D. C., Aparicio, S. A., Behjati, S., Biankin, A. V., … Stratton, M. R. (2013). Signatures of mutational processes in human cancer. *Nature, 500*(7463), 415–421. http://dx.doi.org/10.1038/nature12477.

Alexandrova, E. M., Yallowitz, A. R., Li, D., Xu, S., Schulz, R., Proia, D. A., … Moll, U. M. (2015). Improving survival by exploiting tumour dependence on stabilized mutant p53 for treatment. *Nature, 523*(7560), 352–356. http://dx.doi.org/10.1038/nature14430.

Al-Hajj, M., Wicha, M. S., Benito-Hernandez, A., Morrison, S. J., & Clarke, M. F. (2003). Prospective identification of tumorigenic breast cancer cells. *Proceedings of the National Academy of Sciences of the United States of America, 100*(7), 3983–3988.

Altman, B. J., Stine, Z. E., & Dang, C. V. (2016). From Krebs to clinic: Glutamine metabolism to cancer therapy. *Nature Reviews Cancer, 16*, 619–634. http://dx.doi.org/10.1038/nrc.2016.71.

Andersson, E. R., & Lendahl, U. (2014). Therapeutic modulation of Notch signalling—Are we there yet? *Nature Reviews Drug Discovery, 13*(5), 357–378. http://dx.doi.org/10.1038/nrd4252.

Arrowsmith, C. H., Bountra, C., Fish, P. V., Lee, K., & Schapira, M. (2012). Epigenetic protein families: A new frontier for drug discovery. *Nature Reviews Drug Discovery, 11*(5), 384–400. http://dx.doi.org/10.1038/nrd3674.

Asghar, U., Witkiewicz, A. K., Turner, N. C., & Knudsen, E. S. (2015). The history and future of targeting cyclin-dependent kinases in cancer therapy. *Nature Reviews Drug Discovery, 14*(2), 130–146. http://dx.doi.org/10.1038/nrd4504.

Ashworth, A. (2008). A synthetic lethal therapeutic approach: Poly(ADP) ribose polymerase inhibitors for the treatment of cancers deficient in DNA double-strand break repair. *Journal of Clinical Oncology, 26*(22), 3785–3790. http://dx.doi.org/10.1200/JCO.2008.16.0812.

Aylon, Y., & Oren, M. (2007). Living with p53, dying of p53. *Cell, 130*(4), 597–600.

Azzolin, L., Panciera, T., Soligo, S., Enzo, E., Bicciato, S., Dupont, S., … Piccolo, S. (2014). YAP/TAZ incorporation in the beta-catenin destruction complex orchestrates the Wnt response. *Cell, 158*(1), 157–170. http://dx.doi.org/10.1016/j.cell.2014.06.013.

Bailey, S. D., Desai, K., Kron, K. J., Mazrooei, P., Sinnott-Armstrong, N. A., Treloar, A. E., … Lupien, M. (2016). Noncoding somatic and inherited single-nucleotide variants converge to promote ESR1 expression in breast cancer. *Nature Genetics, 48*, 1260–1266. http://dx.doi.org/10.1038/ng.3650.

Barbie, T. U., Alexe, G., Aref, A. R., Li, S., Zhu, Z., Zhang, X., … Gillanders, W. E. (2014). Targeting an IKBKE cytokine network impairs triple-negative breast cancer growth. *Journal of Clinical Investigation, 124*(12), 5411–5423. http://dx.doi.org/10.1172/JCI75661.

Barbie, D. A., Tamayo, P., Boehm, J. S., Kim, S. Y., Moody, S. E., Dunn, I. F., … Hahn, W. C. (2009). Systematic RNA interference reveals that oncogenic KRAS-driven cancers require TBK1. *Nature, 462*(7269), 108–112. http://dx.doi.org/10.1038/nature08460.

Barolo, S., & Posakony, J. W. (2002). Three habits of highly effective signaling pathways: Principles of transcriptional control by developmental cell signaling. *Genes & Development, 16*(10), 1167–1181. http://dx.doi.org/10.1101/gad.976502.

Barozzi, I., Simonatto, M., Bonifacio, S., Yang, L., Rohs, R., Ghisletti, S., & Natoli, G. (2014). Coregulation of transcription factor binding and nucleosome occupancy through DNA features of mammalian enhancers. *Molecular Cell, 54*(5), 844–857. http://dx.doi.org/10.1016/j.molcel.2014.04.006.

Bauer, R. A., Wurst, J. M., & Tan, D. S. (2010). Expanding the range of 'druggable' targets with natural product-based libraries: An academic perspective. *Current Opinion in Chemical Biology, 14*(3), 308–314. http://dx.doi.org/10.1016/j.cbpa.2010.02.001.

Bernardo, G. M., Bebek, G., Ginther, C. L., Sizemore, S. T., Lozada, K. L., Miedler, J. D., … Keri, R. A. (2013). FOXA1 represses the molecular phenotype of basal breast cancer cells. *Oncogene, 32*(5), 554–563. http://dx.doi.org/10.1038/onc.2012.62.

Bernardo, G. M., Lozada, K. L., Miedler, J. D., Harburg, G., Hewitt, S. C., Mosley, J. D., … Keri, R. A. (2010). FOXA1 is an essential determinant of ERalpha expression and mammary ductal morphogenesis. *Development, 137*(12), 2045–2054. http://dx.doi.org/10.1242/dev.043299.

Bhagwat, A. S., & Vakoc, C. R. (2015). Targeting transcription factors in cancer. *Trends in Cancer, 1*(1), 53–65. http://dx.doi.org/10.1016/j.trecan.2015.07.001.

Biankin, A. V., Piantadosi, S., & Hollingsworth, S. J. (2015). Patient-centric trials for therapeutic development in precision oncology. *Nature, 526*(7573), 361–370. http://dx.doi.org/10.1038/nature15819.

Bienvenu, F., Jirawatnotai, S., Elias, J. E., Meyer, C. A., Mizeracka, K., Marson, A., … Sicinski, P. (2010). Transcriptional role of cyclin D1 in development revealed by a genetic-proteomic screen. *Nature, 463*(7279), 374–378.

Boehm, J. S., Zhao, J. J., Yao, J., Kim, S. Y., Firestein, R., Dunn, I. F., … Hahn, W. C. (2007). Integrative genomic approaches identify IKBKE as a breast cancer oncogene. *Cell, 129*(6), 1065–1079.

Bremner, R., Cohen, B. L., Sopta, M., Hamel, P. A., Ingles, C. J., Gallie, B. L., & Phillips, R. A. (1995). Direct transcriptional repression by pRb and its reversal by specific cyclins. *Molecular and Cellular Biology, 15*, 3256–3265.

Brivanlou, A. H., & Darnell, J. E., Jr. (2002). Signal transduction and the control of gene expression. *Science, 295*(5556), 813–818. http://dx.doi.org/10.1126/science.1066355.

Burkhart, D. L., & Sage, J. (2008). Cellular mechanisms of tumour suppression by the retinoblastoma gene. *Nature Reviews Cancer, 8*(9), 671–682.

Callahan, R., & Egan, S. E. (2004). Notch signaling in mammary development and oncogenesis. *Journal of Mammary Gland Biology and Neoplasia, 9*(2), 145–163.

Campbell, L. L., & Polyak, K. (2007). Breast tumor heterogeneity: Cancer stem cells or clonal evolution? *Cell Cycle, 6*(19), 2332–2338.

Cardenas, M. G., Yu, W., Beguelin, W., Teater, M. R., Geng, H., Goldstein, R. L., … Melnick, A. M. (2016). Rationally designed BCL6 inhibitors target activated B cell diffuse large B cell lymphoma. *Journal of Clinical Investigation, 126*(9), 3351–3362. http://dx.doi.org/10.1172/JCI85795.

Cermelli, S., Jang, I. S., Bernard, B., & Grandori, C. (2014). Synthetic lethal screens as a means to understand and treat MYC-driven cancers. *Cold Spring Harbor Perspectives in Medicine, 4*(3), 18 pp. http://dx.doi.org/10.1101/cshperspect.a014209.

Cescon, D. W., Haibe-Kains, B., & Mak, T. W. (2015). APOBEC3B expression in breast cancer reflects cellular proliferation, while a deletion polymorphism is associated with immune activation. *Proceedings of the National Academy of Sciences of the United States of America, 112*(9), 2841–2846. http://dx.doi.org/10.1073/pnas.1424869112.

Chen, X., Iliopoulos, D., Zhang, Q., Tang, Q., Greenblatt, M. B., Hatziapostolou, M., … Glimcher, L. H. (2014). XBP1 promotes triple-negative breast cancer by controlling the HIF1alpha pathway. *Nature, 508*(7494), 103–107. http://dx.doi.org/10.1038/nature13119.

Chen, D., Pacal, M., Wenzel, P., Knoepfler, P. S., Leone, G., & Bremner, R. (2009). Division and apoptosis of E2f-deficient retinal progenitors. *Nature, 462*(7275), 925–929. http://dx.doi.org/10.1038/nature08544.

Chen, H. Z., Tsai, S. Y., & Leone, G. (2009). Emerging roles of E2Fs in cancer: An exit from cell cycle control. *Nature Reviews Cancer, 9*(11), 785–797.

Chipumuro, E., Marco, E., Christensen, C. L., Kwiatkowski, N., Zhang, T., Hatheway, C. M., … George, R. E. (2014). CDK7 inhibition suppresses super-enhancer-linked oncogenic transcription in MYCN-driven cancer. *Cell, 159*(5), 1126–1139. http://dx.doi.org/10.1016/j.cell.2014.10.024.

Choi, E. J., Seo, E. J., Kim, D. K., Lee, S. I., Kwon, Y. W., Jang, I. H., … Kim, J. H. (2016). FOXP1 functions as an oncogene in promoting cancer stem cell-like characteristics in ovarian cancer cells. *Oncotarget*, 7(3), 3496–3509.

Chu, J., Cargnello, M., Topisirovic, I., & Pelletier, J. (2016). Translation initiation factors: Reprogramming protein synthesis in cancer. *Trends in Cell Biology, 26*, 918–933, http://dx.doi.org/10.1016/j.tcb.2016.06.005.

Ciocca, V., Daskalakis, C., Ciocca, R. M., Ruiz-Orrico, A., & Palazzo, J. P. (2009). The significance of GATA3 expression in breast cancer: A 10-year follow-up study. *Human Pathology, 40*(4), 489–495. http://dx.doi.org/10.1016/j.humpath.2008.09.010.

Clemons, M., & Goss, P. (2001). Estrogen and the risk of breast cancer. *The New England Journal of Medicine, 344*(4), 276–285.

Cohanim, A. B., & Haran, T. E. (2009). The coexistence of the nucleosome positioning code with the genetic code on eukaryotic genomes. *Nucleic Acids Research*, *37*(19), 6466–6476. http://dx.doi.org/10.1093/nar/gkp689.

Cowper-Sal lari, R., Zhang, X., Wright, J. B., Bailey, S. D., Cole, M. D., Eeckhoute, J., ... Lupien, M. (2012). Breast cancer risk-associated SNPs modulate the affinity of chromatin for FOXA1 and alter gene expression. *Nature Genetics*, *44*(11), 1191–1198. http://dx.doi.org/10.1038/ng.2416.

Curry, C. L., Reed, L. L., Golde, T. E., Miele, L., Nickoloff, B. J., & Foreman, K. E. (2005). Gamma secretase inhibitor blocks Notch activation and induces apoptosis in Kaposi's sarcoma tumor cells. *Oncogene*, *24*(42), 6333–6344. http://dx.doi.org/10.1038/sj.onc.1208783.

Curtis, C., Shah, S. P., Chin, S. F., Turashvili, G., Rueda, O. M., Dunning, M. J., ... Aparicio, S. (2012). The genomic and transcriptomic architecture of 2,000 breast tumours reveals novel subgroups. *Nature*, *486*(7403), 346–352. http://dx.doi.org/10.1038/nature10983.

Danos, A. M., Liao, Y., Li, X., & Du, W. (2013). Functional inactivation of Rb sensitizes cancer cells to TSC2 inactivation induced cell death. *Cancer Letters*, *328*(1), 36–43. http://dx.doi.org/10.1016/j.canlet.2012.09.016.

Darby, T. G., Meissner, J. D., Ruhlmann, A., Mueller, W. H., & Scheibe, R. J. (1997). Functional interference between retinoic acid or steroid hormone receptors and the oncoprotein Fli-1. *Oncogene*, *15*(25), 3067–3082. http://dx.doi.org/10.1038/sj.onc.1201503.

Darnell, J. E., Jr. (2002). Transcription factors as targets for cancer therapy. *Nature Reviews Cancer*, *2*(10), 740–749.

Dawson, M. A., & Kouzarides, T. (2012). Cancer epigenetics: From mechanism to therapy. *Cell*, *150*(1), 12–27. http://dx.doi.org/10.1016/j.cell.2012.06.013.

De Carvalho, D. D., You, J. S., & Jones, P. A. (2010). DNA methylation and cellular reprogramming. *Trends in Cell Biology*, *20*(10), 609–617. http://dx.doi.org/10.1016/j.tcb.2010.08.003.

Delmore, J. E., Issa, G. C., Lemieux, M. E., Rahl, P. B., Shi, J. W., Jacobs, H. M., ... Mitsiades, C. S. (2011). BET bromodomain inhibition as a therapeutic strategy to target c-Myc. *Cell*, *146*(6), 903–916. http://dx.doi.org/10.1016/j.cell.2011.08.017.

Deng, T., Liu, J. C., Chung, P. E., Uehling, D., Aman, A., Joseph, B., ... Zacksenhaus, E. (2014). shRNA kinome screen identifies TBK1 as a therapeutic target for HER2+ breast cancer. *Cancer Research*, *74*(7), 2119–2130. http://dx.doi.org/10.1158/0008-5472.CAN-13-2138.

Di Como, C. J., Gaiddon, C., & Prives, C. (1999). p73 function is inhibited by tumor-derived p53 mutants in mammalian cells. *Molecular and Cellular Biology*, *19*(2), 1438–1449.

Dick, J. E. (2003). Breast cancer stem cells revealed. *Proceedings of the National Academy of Sciences of the United States of America*, *100*(7), 3547–3549.

Downward, J. (2015). RAS synthetic lethal screens revisited: Still seeking the elusive prize? *Clinical Cancer Research*, *21*(8), 1802–1809. http://dx.doi.org/10.1158/1078-0432.CCR-14-2180.

Dvinge, H., Git, A., Graf, S., Salmon-Divon, M., Curtis, C., Sottoriva, A., ... Caldas, C. (2013). The shaping and functional consequences of the microRNA landscape in breast cancer. *Nature*, *497*(7449), 378–382. http://dx.doi.org/10.1038/nature12108.

Dyson, N. J. (2016). RB1: A prototype tumor suppressor and an enigma. *Genes & Development*, *30*(13), 1492–1502. http://dx.doi.org/10.1101/gad.282145.116.

Eeckhoute, J., Keeton, E. K., Lupien, M., Krum, S. A., Carroll, J. S., & Brown, M. (2007). Positive cross-regulatory loop ties GATA-3 to estrogen receptor alpha expression in breast cancer. *Cancer Research*, *67*(13), 6477–6483.

Ellis, M. J., Ding, L., Shen, D., Luo, J., Suman, V. J., Wallis, J. W., … Mardis, E. R. (2012). Whole-genome analysis informs breast cancer response to aromatase inhibition. *Nature*, *486*(7403), 353–360. http://dx.doi.org/10.1038/nature11143.

Emami, K. H., Nguyen, C., Ma, H., Kim, D. H., Jeong, K. W., Eguchi, M., … Kahn, M. (2004). A small molecule inhibitor of beta-catenin/CREB-binding protein transcription [corrected]. *Proceedings of the National Academy of Sciences of the United States of America*, *101*(34), 12682–12687. http://dx.doi.org/10.1073/pnas.0404875101.

Farrell, A. S., & Sears, R. C. (2014). MYC degradation. *Cold Spring Harbor Perspectives in Medicine*, *4*(3), 15 pp. http://dx.doi.org/10.1101/cshperspect.a014365.

Fece de la Cruz, F., Gapp, B. V., & Nijman, S. M. (2015). Synthetic lethal vulnerabilities of cancer. *Annual Review of Pharmacology and Toxicology*, *55*, 513–531. http://dx.doi.org/10.1146/annurev-pharmtox-010814-124511.

Ferrari, A., Vincent-Salomon, A., Pivot, X., Sertier, A. S., Thomas, E., Tonon, L., … Thomas, G. (2016). A whole-genome sequence and transcriptome perspective on HER2-positive breast cancers. *Nature Communications*, 7, 12222. http://dx.doi.org/10.1038/ncomms12222.

Filippakopoulos, P., Qi, J., Picaud, S., Shen, Y., Smith, W. B., Fedorov, O., … Bradner, J. E. (2010). Selective inhibition of BET bromodomains. *Nature*, *468*(7327), 1067–1073. http://dx.doi.org/10.1038/nature09504.

Finn, R. S., Crown, J. P., Lang, I., Boer, K., Bondarenko, I. M., Kulyk, S. O., … Slamon, D. J. (2015). The cyclin-dependent kinase 4/6 inhibitor palbociclib in combination with letrozole versus letrozole alone as first-line treatment of oestrogen receptor-positive, HER2-negative, advanced breast cancer (PALOMA-1/TRIO-18): A randomised phase 2 study. *The Lancet Oncology*, *16*(1), 25–35. http://dx.doi.org/10.1016/S1470-2045(14)71159-3.

Finn, R. S., Dering, J., Conklin, D., Kalous, O., Cohen, D. J., Desai, A. J., … Slamon, D. J. (2009). PD 0332991, a selective cyclin D kinase 4/6 inhibitor, preferentially inhibits proliferation of luminal estrogen receptor-positive human breast cancer cell lines in vitro. *Breast Cancer Research*, *11*(5), R77. http://dx.doi.org/10.1186/bcr2419.

Fishel, R., Lescoe, M. K., Rao, M. R., Copeland, N. G., Jenkins, N. A., Garber, J., … Kolodner, R. (1993). The human mutator gene homolog MSH2 and its association with hereditary nonpolyposis colon cancer. *Cell*, *75*(5), 1027–1038.

Foulkes, W. D., Stefansson, I. M., Chappuis, P. O., Begin, L. R., Goffin, J. R., Wong, N., … Akslen, L. A. (2003). Germline BRCA1 mutations and a basal epithelial phenotype in breast cancer. *Journal of the National Cancer Institute*, *95*(19), 1482–1485.

Friedland, J. C., Smith, D. L., Sang, J., Acquaviva, J., He, S., Zhang, C., & Proia, D. A. (2014). Targeted inhibition of Hsp90 by ganetespib is effective across a broad spectrum of breast cancer subtypes. *Investigational New Drugs*, *32*(1), 14–24. http://dx.doi.org/10.1007/s10637-013-9971-6.

Fujii, Y., Yada, M., Nishiyama, M., Kamura, T., Takahashi, H., Tsunematsu, R., … Nakayama, K. I. (2006). Fbxw7 contributes to tumor suppression by targeting multiple proteins for ubiquitin-dependent degradation. *Cancer Science*, *97*(8), 729–736. http://dx.doi.org/10.1111/j.1349-7006.2006.00239.x.

Gatza, M. L., Lucas, J. E., Barry, W. T., Kim, J. W., Wang, Q., Crawford, M. D., … Nevins, J. R. (2010). A pathway-based classification of human breast cancer. *Proceedings of the National Academy of Sciences of the United States of America*, *107*(15), 6994–6999.

Gayvert, K. M., Dardenne, E., Cheung, C., Boland, M. R., Lorberbaum, T., Wanjala, J., … Elemento, O. (2016). A computational drug repositioning approach for targeting oncogenic transcription factors. *Cell Reports*, *15*(11), 2348–2356. http://dx.doi.org/10.1016/j.celrep.2016.05.037.

Godar, S., Ince, T. A., Bell, G. W., Feldser, D., Donaher, J. L., Bergh, J., … Weinberg, R. A. (2008). Growth-inhibitory and tumor-suppressive functions of p53 depend on its repression of CD44 expression. *Cell*, *134*(1), 62–73.

Gorrini, C., Baniasadi, P. S., Harris, I. S., Silvester, J., Inoue, S., Snow, B., … Gauthier, M. L. (2013). BRCA1 interacts with Nrf2 to regulate antioxidant signaling and cell survival. *Journal of Experimental Medicine, 210*(8), 1529–1544. http://dx.doi.org/10.1084/jem.20121337.

Gorrini, C., Gang, B. P., Bassi, C., Wakeham, A., Baniasadi, S. P., Hao, Z., … Mak, T. W. (2014). Estrogen controls the survival of BRCA1-deficient cells via a PI3K-NRF2-regulated pathway. *Proceedings of the National Academy of Sciences of the United States of America, 111*(12), 4472–4477. http://dx.doi.org/10.1073/pnas.1324136111.

Goss, P. E., Ingle, J. N., Martino, S., Robert, N. J., Muss, H. B., Piccart, M. J., … Pater, J. L. (2003). A randomized trial of letrozole in postmenopausal women after five years of tamoxifen therapy for early-stage breast cancer. *The New England Journal of Medicine, 349*(19), 1793–1802. http://dx.doi.org/10.1056/NEJMoa032312.

Goss, P. E., Ingle, J. N., Pritchard, K. I., Robert, N. J., Muss, H., Gralow, J., … Parulekar, W. R. (2016). Extending aromatase-inhibitor adjuvant therapy to 10 years. *The New England Journal of Medicine, 375*(3), 209–219. http://dx.doi.org/10.1056/NEJMoa1604700.

Green, D. R., & Kroemer, G. (2009). Cytoplasmic functions of the tumour suppressor p53. *Nature, 458*(7242), 1127–1130. http://dx.doi.org/10.1038/nature07986.

Grieve, S., Gao, Y., Hall, C., Hu, J., & Greer, P. A. (2016). Calpain genetic disruption and HSP90 inhibition combine to attenuate mammary tumorigenesis. *Molecular and Cellular Biology, 36*(15), 2078–2088. http://dx.doi.org/10.1128/MCB.01062-15.

Hakem, R., de la Pompa, J. L., Sirard, C., Mo, R., Woo, M., Hakem, A., … Mak, T. W. (1996). The tumor suppressor gene Brca1 is required for embryonic cellular proliferation in the mouse. *Cell, 85*(7), 1009–1023.

Halder, G., Dupont, S., & Piccolo, S. (2012). Transduction of mechanical and cytoskeletal cues by YAP and TAZ. *Nature Reviews Molecular Cell Biology, 13*(9), 591–600. http://dx.doi.org/10.1038/nrm3416.

Hanahan, D., & Weinberg, R. A. (2011). Hallmarks of cancer: The next generation. *Cell, 144*(5), 646–674. http://dx.doi.org/10.1016/j.cell.2011.02.013.

Handeli, S., & Simon, J. A. (2008). A small-molecule inhibitor of Tcf/beta-catenin signaling down-regulates PPARgamma and PPARdelta activities. *Molecular Cancer Therapeutics, 7*(3), 521–529. http://dx.doi.org/10.1158/1535-7163.MCT-07-2063.

Hayden, M. S., & Ghosh, S. (2012). NF-kappaB, the first quarter-century: Remarkable progress and outstanding questions. *Genes & Development, 26*(3), 203–234. http://dx.doi.org/10.1101/gad.183434.111.

Hayes, D. F., Van Zyl, J. A., Hacking, A., Goedhals, L., Bezwoda, W. R., Mailliard, J. A., … Schoenfelder, J. (1995). Randomized comparison of tamoxifen and two separate doses of toremifene in postmenopausal patients with metastatic breast cancer. *Journal of Clinical Oncology, 13*(10), 2556–2566.

Heise, C., Sampson-Johannes, A., Williams, A., McCormick, F., Von Hoff, D. D., & Kirn, D. H. (1997). ONYX-015, an E1B gene-attenuated adenovirus, causes tumor-specific cytolysis and antitumoral efficacy that can be augmented by standard chemotherapeutic agents. *Nature Medicine, 3*(6), 639–645.

Hensley, C. T., Wasti, A. T., & DeBerardinis, R. J. (2013). Glutamine and cancer: Cell biology, physiology, and clinical opportunities. *Journal of Clinical Investigation, 123*(9), 3678–3684. http://dx.doi.org/10.1172/JCI69600.

Herschkowitz, J. I., Simin, K., Weigman, V. J., Mikaelian, I., Usary, J., Hu, Z., … Perou, C. M. (2007). Identification of conserved gene expression features between murine mammary carcinoma models and human breast tumors. *Genome Biology, 8*(5), R76.

Ho, A. T., Li, Q. H., Hakem, R., Mak, T. W., & Zacksenhaus, E. (2004). Coupling of caspase-9 to Apaf1 in response to loss of pRb or cytotoxic drugs is cell-type-specific. *The EMBO Journal, 23*(2), 460–472. http://dx.doi.org/10.1038/sj.emboj.7600039.

Hsieh, A. L., Walton, Z. E., Altman, B. J., Stine, Z. E., & Dang, C. V. (2015). MYC and metabolism on the path to cancer. *Seminars in Cell & Developmental Biology*, *43*, 11–21. http://dx.doi.org/10.1016/j.semcdb.2015.08.003.

Hsu, T. Y. T., Simon, L. M., Neill, N. J., Marcotte, R., Sayad, A., Bland, C. S., … Westbrook, T. F. (2015). The spliceosome is a therapeutic vulnerability in MYC-driven cancer. *Nature*, *525*(7569), 384–388. http://dx.doi.org/10.1038/nature14985.

Huen, M. S. Y., Sy, S. M. H., & Chen, J. J. (2010). BRCA1 and its toolbox for the maintenance of genome integrity. *Nature Reviews Molecular Cell Biology*, *11*(2), 138–148. http://dx.doi.org/10.1038/nrm2831.

Hurst, C. D., Tomlinson, D. C., Williams, S. V., Platt, F. M., & Knowles, M. A. (2008). Inactivation of the Rb pathway and overexpression of both isoforms of E2F3 are obligate events in bladder tumours with 6p22 amplification. *Oncogene*, *27*(19), 2716–2727. http://dx.doi.org/10.1038/sj.onc.1210934.

Husemann, Y., Geigl, J. B., Schubert, F., Musiani, P., Meyer, M., Burghart, E., … Klein, C. A. (2008). Systemic spread is an early step in breast cancer. *Cancer Cell*, *13*(1), 58–68.

Hydbring, P., Malumbres, M., & Sicinski, P. (2016). Non-canonical functions of cell cycle cyclins and cyclin-dependent kinases. *Nature Reviews Molecular Cell Biology*, *17*(5), 280–292. http://dx.doi.org/10.1038/nrm.2016.27.

Ijichi, N., Shigekawa, T., Ikeda, K., Horie-Inoue, K., Shimizu, C., Saji, S., … Inoue, S. (2012). Association of double-positive FOXA1 and FOXP1 immunoreactivities with favorable prognosis of tamoxifen-treated breast cancer patients. *Hormones & Cancer*, *3*(4), 147–159. http://dx.doi.org/10.1007/s12672-012-0111-0.

Inoue, S., Hao, Z., Elia, A. J., Cescon, D., Zhou, L., Silvester, J., … Mak, T. W. (2013). Mule/Huwe1/Arf-BP1 suppresses Ras-driven tumorigenesis by preventing c-Myc/Miz1-mediated down-regulation of p21 and p15. *Genes & Development*, *27*(10), 1101–1114. http://dx.doi.org/10.1101/gad.214577.113.

Iorio, F., Knijnenburg, T. A., Vis, D. J., Bignell, G. R., Menden, M. P., Schubert, M., … Garnett, M. J. (2016). A landscape of pharmacogenomic interactions in cancer. *Cell*, *166*(3), 740–754. http://dx.doi.org/10.1016/j.cell.2016.06.017.

Issaeva, N., Bozko, P., Enge, M., Protopopova, M., Verhoef, L. G., Masucci, M., … Selivanova, G. (2004). Small molecule RITA binds to p53, blocks p53-HDM-2 interaction and activates p53 function in tumors. *Nature Medicine*, *10*(12), 1321–1328. http://dx.doi.org/10.1038/nm1146.

Jiang, D. D., Brady, C. A., Johnson, T. M., Lee, E. Y., Park, E. J., Scott, M. P., & Attardi, L. D. (2011). Full p53 transcriptional activation potential is dispensable for tumor suppression in diverse lineages. *Proceedings of the National Academy of Sciences of the United States of America*, *108*(41), 17123–17128. http://dx.doi.org/10.1073/pnas.1111245108.

Jiang, Z., Jones, R., Liu, J. C., Deng, T., Robinson, T., Chung, P. E., … Zacksenhaus, E. (2011). RB1 and p53 at the crossroad of EMT and triple-negative breast cancer. *Cell Cycle*, *10*(10), 1563–1570.

Jiang, Z., Liang, P., Leng, R., Guo, Z., Liu, Y., Liu, X., … Zacksenhaus, E. (2000). E2F1 and p53 are dispensable, whereas p21(Waf1/Cip1) cooperates with Rb to restrict endoreduplication and apoptosis during skeletal myogenesis. *Developmental Biology*, *227*(1), 8–41. http://dx.doi.org/10.1006/dbio.2000.9892.

Jiang, J. X., Sun, C. Y., Tian, S., Yu, C., Chen, M. Y., & Zhang, H. (2016). Tumor suppressor Fbxw7 antagonizes WNT signaling by targeting beta-catenin for degradation in pancreatic cancer. *Tumour Biology*, *37*, 13893–13902. http://dx.doi.org/10.1007/s13277-016-5217-5.

Johnson, R., & Halder, G. (2014). The two faces of Hippo: Targeting the Hippo pathway for regenerative medicine and cancer treatment. *Nature Reviews Drug Discovery*, *13*(1), 63–79. http://dx.doi.org/10.1038/nrd4161.

Jones, P. A., & Baylin, S. B. (2002). The fundamental role of epigenetic events in cancer. *Nature Reviews Genetics*, *3*(6), 415–428. http://dx.doi.org/10.1038/nrg816.

Jones, R. A., Robinson, T. J., Liu, J. C., Shrestha, M., Voisin, V., Ju, Y., ... Zacksenhaus, E. (2016). RB1 deficiency in triple-negative breast cancer induces mitochondrial protein translation. *Journal of Clinical Investigation*, *126*, 3739–3757. http://dx.doi.org/10.1172/JCI81568.

Jonkers, I., & Lis, J. T. (2015). Getting up to speed with transcription elongation by RNA polymerase II. *Nature Reviews Molecular Cell Biology*, *16*(3), 167–177. http://dx.doi.org/10.1038/nrm3953.

Jordan, J. J., Menendez, D., Sharav, J., Beno, I., Rosenthal, K., Resnick, M. A., & Haran, T. E. (2012). Low-level p53 expression changes transactivation rules and reveals superactivating sequences. *Proceedings of the National Academy of Sciences of the United States of America*, *109*(36), 14387–14392. http://dx.doi.org/10.1073/pnas.1205971109.

Kasper, M., Jaks, V., Fiaschi, M., & Toftgard, R. (2009). Hedgehog signalling in breast cancer. *Carcinogenesis*, *30*(6), 903–911. http://dx.doi.org/10.1093/carcin/bgp048.

Kawazu, M., Saso, K., Tong, K. I., McQuire, T., Goto, K., Son, D. O., ... Okada, H. (2011). Histone demethylase JMJD2B functions as a co-factor of estrogen receptor in breast cancer proliferation and mammary gland development. *PLoS One*. *6*(3). e17830. http://dx.doi.org/10.1371/journal.pone.0017830.

Khaled, W. T., Choon Lee, S., Stingl, J., Chen, X., Raza Ali, H., Rueda, O. M., ... Liu, P. (2015). BCL11A is a triple-negative breast cancer gene with critical functions in stem and progenitor cells. *Nature Communications*, *6*, 5987. http://dx.doi.org/10.1038/ncomms6987.

Khuri, F. R., Nemunaitis, J., Ganly, I., Arseneau, J., Tannock, I. F., Romel, L., ... Kirn, D. H. (2000). A controlled trial of intratumoral ONYX-015, a selectively-replicating adenovirus, in combination with cisplatin and 5-fluorouracil in patients with recurrent head and neck cancer. *Nature Medicine*, *6*(8), 879–885. http://dx.doi.org/10.1038/78638.

Kirn, D., Hermiston, T., & McCormick, F. (1998). ONYX-015: Clinical data are encouraging. *Nature Medicine*, *4*(12), 1341–1342. http://dx.doi.org/10.1038/3902.

Klein, C. A. (2009). Parallel progression of primary tumours and metastases. *Nature Reviews Cancer*, *9*(4), 302–312.

Koboldt, D. C., Fulton, R. S., McLellan, M. D., Schmidt, H., Kalicki-Veizer, J., McMichael, J. F., ... Palchik, J. D. (2012). Comprehensive molecular portraits of human breast tumours. *Nature*, *490*, 61–70. http://dx.doi.org/10.1038/nature11412.

Kollmann, K., Heller, G., Schneckenleithner, C., Warsch, W., Scheicher, R., Ott, R. G., ... Sexl, V. (2013). A kinase-independent function of CDK6 links the cell cycle to tumor angiogenesis. *Cancer Cell*, *24*(2), 167–181. http://dx.doi.org/10.1016/j.ccr.2013.07.012.

Kramer, A., Mentrup, T., Kleizen, B., Rivera-Milla, E., Reichenbach, D., Enzensperger, C., ... Kaether, C. (2013). Small molecules intercept Notch signaling and the early secretory pathway. *Nature Chemical Biology*, *9*(11), 731–738. http://dx.doi.org/10.1038/nchembio.1356.

Kreso, A., & Dick, J. E. (2014). Evolution of the cancer stem cell model. *Cell Stem Cell*, *14*(3), 275–291. http://dx.doi.org/10.1016/j.stem.2014.02.006.

Kwak, H., & Lis, J. T. (2013). Control of transcriptional elongation. *Annual Review of Genetics*, *47*, 483–508. http://dx.doi.org/10.1146/annurev-genet-110711-155440.

Kwiatkowski, N., Zhang, T., Rahl, P. B., Abraham, B. J., Reddy, J., Ficarro, S. B., ... Gray, N. S. (2014). Targeting transcription regulation in cancer with a covalent CDK7 inhibitor. *Nature*, *511*(7511), 616–620. http://dx.doi.org/10.1038/nature13393.

Lawrence, M. S., Stojanov, P., Mermel, C. H., Robinson, J. T., Garraway, L. A., Golub, T. R., ... Getz, G. (2014). Discovery and saturation analysis of cancer genes

across 21 tumour types. *Nature*, *505*(7484), 495–501. http://dx.doi.org/10.1038/nature12912.
Lehmann, B. D., Bauer, J. A., Chen, X., Sanders, M. E., Chakravarthy, A. B., Shyr, Y., & Pietenpol, J. A. (2011). Identification of human triple-negative breast cancer subtypes and preclinical models for selection of targeted therapies. *Journal of Clinical Investigation*, *121*(7), 2750–2767. http://dx.doi.org/10.1172/JCI45014.
Leone, G., Sears, R., Huang, E., Rempel, R., Nuckolls, F., Park, C. H., … Nevins, J. R. (2001). Myc requires distinct E2F activities to induce S phase and apoptosis. *Molecular Cell*, *8*(1), 105–113.
Lepourcelet, M., Chen, Y. N., France, D. S., Wang, H., Crews, P., Petersen, F., … Shivdasani, R. A. (2004). Small-molecule antagonists of the oncogenic Tcf/beta-catenin protein complex. *Cancer Cell*, *5*(1), 91–102.
Leprivier, G., Remke, M., Rotblat, B., Dubuc, A., Mateo, A. R., Kool, M., … Sorensen, P. H. (2013). The eEF2 kinase confers resistance to nutrient deprivation by blocking translation elongation. *Cell*, *153*(5), 1064–1079. http://dx.doi.org/10.1016/j.cell.2013.04.055.
Levine, A. J., & Oren, M. (2009). The first 30 years of p53: Growing ever more complex. *Nature Reviews Cancer*, *9*(10), 749–758. http://dx.doi.org/10.1038/nrc2723.
Lewis, J. S., & Jordan, V. C. (2005). Selective estrogen receptor modulators (SERMs): Mechanisms of anticarcinogenesis and drug resistance. *Mutation Research*, *591*(1–2), 247–263. http://dx.doi.org/10.1016/j.mrfmmm.2005.02.028.
Lewis, M. T., Ross, S., Strickland, P. A., Sugnet, C. W., Jimenez, E., Scott, M. P., & Daniel, C. W. (1999). Defects in mouse mammary gland development caused by conditional haploinsufficiency of Patched-1. *Development*, *126*(22), 5181–5193.
Li, B., Gordon, G. M., Du, C. H., Xu, J., & Du, W. (2010). Specific killing of Rb mutant cancer cells by inactivating TSC2. *Cancer Cell*, *17*(5), 469–480. http://dx.doi.org/10.1016/j.ccr.2010.03.019.
Li, Y., Luo, H., Liu, T., Zacksenhaus, E., & Ben-David, Y. (2015). The ets transcription factor Fli-1 in development, cancer and disease. *Oncogene*, *34*(16), 2022–2031. http://dx.doi.org/10.1038/onc.2014.162.
Li, Q., Zhang, Y., Fu, J., Han, L., Xue, L., Lv, C., … Tong, T. (2013). FOXA1 mediates p16 (INK4a) activation during cellular senescence. *The EMBO Journal*, *32*(6), 858–873. http://dx.doi.org/10.1038/emboj.2013.35.
Li, Y. J., Zhao, X., Vecchiarelli-Federico, L. M., Li, Y., Datti, A., Cheng, Y., & Ben-David, Y. (2012). Drug-mediated inhibition of Fli-1 for the treatment of leukemia. *Blood Cancer Journal*. *2*(1). e54. http://dx.doi.org/10.1038/bcj.2011.52.
Libermann, T. A., & Zerbini, L. F. (2006). Targeting transcription factors for cancer gene therapy. *Current Gene Therapy*, *6*(1), 17–33.
Lin, C. Y., Loven, J., Rahl, P. B., Paranal, R. M., Burge, C. B., Bradner, J. E., … Young, R. A. (2012). Transcriptional amplification in tumor cells with elevated c-Myc. *Cell*, *151*(1), 56–67. http://dx.doi.org/10.1016/j.cell.2012.08.026.
Lin, C. J., Malina, A., & Pelletier, J. (2009). c-Myc and eIF4F constitute a feedforward loop that regulates cell growth: Implications for anticancer therapy. *Cancer Research*, *69*(19), 7491–7494. http://dx.doi.org/10.1158/0008-5472.CAN-09-0813.
Liu, J. C., Deng, T., Lehal, R. S., Kim, J., & Zacksenhaus, E. (2007). Identification of tumorsphere- and tumor-initiating cells in HER2/Neu-induced mammary tumors. *Cancer Research*, *67*(18), 8671–8681. http://dx.doi.org/10.1158/0008-5472.CAN-07-1486.
Liu, X., Holstege, H., van der Gulden, H., Treur-Mulder, M., Zevenhoven, J., Velds, A., … Jonkers, J. (2007). Somatic loss of BRCA1 and p53 in mice induces mammary tumors with features of human BRCA1-mutated basal-like breast cancer. *Proceedings of the National Academy of Sciences of the United States of America*, *104*(29), 12111–12116.

Liu, H., Tang, X., Srivastava, A., Pecot, T., Daniel, P., Hemmelgarn, B., … Leone, G. (2015). Redeployment of Myc and E2f1-3 drives Rb-deficient cell cycles. *Nature Cell Biology*, *17*(8), 1036–1048. http://dx.doi.org/10.1038/ncb3210.

Liu, J. C., Voisin, V., Bader, G. D., Deng, T., Pusztai, L., Symmans, W. F., … Zacksenhaus, E. (2012). Seventeen-gene signature from enriched Her2/Neu mammary tumor-initiating cells predicts clinical outcome for human HER2+:ERalpha- breast cancer. *Proceedings of the National Academy of Sciences of the United States of America*, *109*(15), 5832–5837. http://dx.doi.org/10.1073/pnas.1201105109.

Liu, J. C., Voisin, V., Wang, S., Wang, D. Y., Jones, R. A., Datti, A., … Zacksenhaus, E. (2014). Combined deletion of Pten and p53 in mammary epithelium accelerates triple-negative breast cancer with dependency on eEF2K. *EMBO Molecular Medicine*, *6*, 1542–1560. http://dx.doi.org/10.15252/emmm.201404402.

Liu, T. X., Zhang, J. W., Tao, J., Zhang, R. B., Zhang, Q. H., Zhao, C. J., … Chen, Z. (2000). Gene expression networks underlying retinoic acid-induced differentiation of acute promyelocytic leukemia cells. *Blood*, *96*(4), 1496–1504.

Lord, C. J., Tutt, A. N., & Ashworth, A. (2015). Synthetic lethality and cancer therapy: Lessons learned from the development of PARP inhibitors. *Annual Review of Medicine*, *66*, 455–470. http://dx.doi.org/10.1146/annurev-med-050913-022545.

Loven, J., Hoke, H. A., Lin, C. Y., Lau, A., Orlando, D. A., Vakoc, C. R., … Young, R. A. (2013). Selective inhibition of tumor oncogenes by disruption of super-enhancers. *Cell*, *153*(2), 320–334. http://dx.doi.org/10.1016/j.cell.2013.03.036.

Lu, H., Xue, Y., Yu, G. K., Arias, C., Lin, J., Fong, S., … Zhou, Q. (2015). Compensatory induction of MYC expression by sustained CDK9 inhibition via a BRD4-dependent mechanism. *eLife*. *4*. e06535. http://dx.doi.org/10.7554/eLife.06535.

Luo, J., Solimini, N. L., & Elledge, S. J. (2009). Principles of cancer therapy: Oncogene and non-oncogene addiction. *Cell*, *136*(5), 823–837.

Macleod, K. F., Hu, Y., & Jacks, T. (1996). Loss of Rb activates both p53-dependent and independent cell death pathways in the developing mouse nervous system. *The EMBO Journal*, *15*, 6178–6188.

Magee, J. A., Piskounova, E., & Morrison, S. J. (2012). Cancer stem cells: Impact, heterogeneity, and uncertainty. *Cancer Cell*, *21*(3), 283–296. http://dx.doi.org/10.1016/j.ccr.2012.03.003.

Maximov, P. Y., Lee, T. M., & Jordan, V. C. (2013). The discovery and development of selective estrogen receptor modulators (SERMs) for clinical practice. *Current Clinical Pharmacology*, *8*(2), 135–155.

McCain, J. (2015). First-in-class CDK4/6 inhibitor palbociclib could usher in a new wave of combination therapies for HR+, HER2- breast cancer. *P & T: A Peer-Reviewed Journal for Formulary Management*, *40*(8), 511–520.

McNeill, H., & Woodgett, J. R. (2010). When pathways collide: Collaboration and connivance among signalling proteins in development. *Nature Reviews Molecular Cell Biology*, *11*(6), 404–413. http://dx.doi.org/10.1038/nrm2902.

Melchor, L., Molyneux, G., Mackay, A., Magnay, F. A., Atienza, M., Kendrick, H., … Smalley, M. J. (2014). Identification of cellular and genetic drivers of breast cancer heterogeneity in genetically engineered mouse tumour models. *Journal of Pathology*, *233*(2), 124–137. http://dx.doi.org/10.1002/path.4345.

Merino, D., Lok, S. W., Visvader, J. E., & Lindeman, G. J. (2016). Targeting BCL-2 to enhance vulnerability to therapy in estrogen receptor-positive breast cancer. *Oncogene*, *35*(15), 1877–1887. http://dx.doi.org/10.1038/onc.2015.287.

Miyashita, T., & Reed, J. C. (1995). Tumor suppressor p53 is a direct transcriptional activator of the human bax gene. *Cell*, *80*(2), 293–299.

Moon, H. G., Hwang, K. T., Kim, J. A., Kim, H. S., Lee, M. J., Jung, E. M., … Noh, D. Y. (2011). NFIB is a potential target for estrogen receptor-negative breast cancers. *Molecular Oncology*, *5*(6), 538–544. http://dx.doi.org/10.1016/j.molonc.2011.08.002.

Morgenbesser, S. D., Williams, B. O., Jacks, T., & DePinho, R. A. (1994). p53-dependent apoptosis produced by Rb-deficiency in the developing mouse lens. *Nature, 371*(6492), 72–74.

Muller, P. A., Caswell, P. T., Doyle, B., Iwanicki, M. P., Tan, E. H., Karim, S., … Vousden, K. H. (2009). Mutant p53 drives invasion by promoting integrin recycling. *Cell, 139*(7), 1327–1341.

Murray-Zmijewski, F., Lane, D. P., & Bourdon, J. C. (2006). p53/p63/p73 isoforms: An orchestra of isoforms to harmonise cell differentiation and response to stress. *Cell Death and Differentiation, 13*(6), 962–972. http://dx.doi.org/10.1038/sj.cdd.4401914.

Narod, S. A. (2010). BRCA mutations in the management of breast cancer: The state of the art. *Nature Reviews. Clinical Oncology*, 7(12), 702–707. http://dx.doi.org/10.1038/nrclinonc.2010.166.

Nguyen, T., Nioi, P., & Pickett, C. B. (2009). The Nrf2-antioxidant response element signaling pathway and its activation by oxidative stress. *Journal of Biological Chemistry, 284*(20), 13291–13295. http://dx.doi.org/10.1074/jbc.R900010200.

Nielsen, F. C., van Overeem Hansen, T., & Sorensen, C. S. (2016). Hereditary breast and ovarian cancer: New genes in confined pathways. *Nature Reviews Cancer, 16*(9), 599–612. http://dx.doi.org/10.1038/nrc.2016.72.

Nik-Zainal, S., Davies, H., Staaf, J., Ramakrishna, M., Glodzik, D., Zou, X., … Stratton, M. R. (2016). Landscape of somatic mutations in 560 breast cancer whole-genome sequences. *Nature, 534*, 47–54. http://dx.doi.org/10.1038/nature17676.

Nilson, K. A., Guo, J., Turek, M. E., Brogie, J. E., Delaney, E., Luse, D. S., & Price, D. H. (2015). THZ1 reveals roles for Cdk7 in co-transcriptional capping and pausing. *Molecular Cell, 59*(4), 576–587. http://dx.doi.org/10.1016/j.molcel.2015.06.032.

Noman, A. S., Uddin, M., Rahman, M. Z., Nayeem, M. J., Alam, S. S., Khatun, Z., … Islam, S. S. (2016). Overexpression of sonic hedgehog in the triple negative breast cancer: Clinicopathological characteristics of high burden breast cancer patients from Bangladesh. *Scientific Reports, 6*, 18830. http://dx.doi.org/10.1038/srep18830.

O'Leary, B., Finn, R. S., & Turner, N. C. (2016). Treating cancer with selective CDK4/6 inhibitors. *Nature Reviews. Clinical Oncology, 13*(7), 417–430. http://dx.doi.org/10.1038/nrclinonc.2016.26.

Pahlman, S., Ruusala, A. I., Abrahamsson, L., Mattsson, M. E., & Esscher, T. (1984). Retinoic acid-induced differentiation of cultured human neuroblastoma cells: A comparison with phorbolester-induced differentiation. *Cell Differentiation, 14*(2), 135–144.

Patnaik, A., Rosen, L. S., Tolaney, S. M., Tolcher, A. W., Goldman, J. W., Gandhi, L., … Shapiro, G. I. (2016). Efficacy and safety of abemaciclib, an inhibitor of CDK4 and CDK6, for patients with breast cancer, non-small cell lung cancer, and other solid tumors. *Cancer Discovery, 6*(7), 740–753. http://dx.doi.org/10.1158/2159-8290.CD-16-0095.

Pattabiraman, D. R., & Weinberg, R. A. (2014). Tackling the cancer stem cells—What challenges do they pose? *Nature Reviews Drug Discovery, 13*(7), 497–512. http://dx.doi.org/10.1038/nrd4253.

Pavlova, N. N., & Thompson, C. B. (2016). The emerging hallmarks of cancer metabolism. *Cell Metabolism, 23*(1), 27–47. http://dx.doi.org/10.1016/j.cmet.2015.12.006.

Pelletier, J., Graff, J., Ruggero, D., & Sonenberg, N. (2015). Targeting the eIF4F translation initiation complex: A critical nexus for cancer development. *Cancer Research, 75*(2), 250–263. http://dx.doi.org/10.1158/0008-5472.CAN-14-2789.

Pereira, B., Chin, S. F., Rueda, O. M., Vollan, H. K. M., Provenzano, E., Bardwell, H. A., … Caldas, C. (2016). The somatic mutation profiles of 2,433 breast cancers refines their genomic and transcriptomic landscapes. *Nature Communications*. 7. 15 pp. ARTN 11479. http://dx.doi.org/10.1038/ncomms11479.

Perkins, N. D. (2012). The diverse and complex roles of NF-kappaB subunits in cancer. *Nature Reviews Cancer, 12*(2), 121–132. http://dx.doi.org/10.1038/nrc3204.

Perou, C. M., Sorlie, T., Eisen, M. B., van de Rijn, M., Jeffrey, S. S., Rees, C. A., … Botstein, D. (2000). Molecular portraits of human breast tumours. *Nature, 406*(6797), 747–752.

Polager, S., & Ginsberg, D. (2009). p53 and E2f: Partners in life and death. *Nature Reviews Cancer, 9*(10), 738–748. http://dx.doi.org/10.1038/nrc2718.

Pott, S., & Lieb, J. D. (2015). What are super-enhancers? *Nature Genetics, 47*(1), 8–12. http://dx.doi.org/10.1038/ng.3167.

Powell, E., Piwnica-Worms, D., & Piwnica-Worms, H. (2014). Contribution of p53 to metastasis. *Cancer Discovery, 4*(4), 405–414. http://dx.doi.org/10.1158/2159-8290.CD-13-0136.

Prat, A., Parker, J. S., Fan, C., & Perou, C. M. (2012). PAM50 assay and the three-gene model for identifying the major and clinically relevant molecular subtypes of breast cancer. *Breast Cancer Research and Treatment, 135*(1), 301–306. http://dx.doi.org/10.1007/s10549-012-2143-0.

Prat, A., Parker, J. S., Karginova, O., Fan, C., Livasy, C., Herschkowitz, J. I., … Perou, C. M. (2010). Phenotypic and molecular characterization of the claudin-low intrinsic subtype of breast cancer. *Breast Cancer Research, 12*(5), R68.

Prat, A., & Perou, C. M. (2009). Mammary development meets cancer genomics. *Nature Medicine, 15*(8), 842–844. http://dx.doi.org/10.1038/nm0809-842.

Prat, A., & Perou, C. M. (2011). Deconstructing the molecular portraits of breast cancer. *Molecular Oncology, 5*(1), 5–23. http://dx.doi.org/10.1016/j.molonc.2010.11.003.

Radtke, F., & Raj, K. (2003). The role of Notch in tumorigenesis: Oncogene or tumour suppressor? *Nature Reviews Cancer, 3*(10), 756–767. http://dx.doi.org/10.1038/nrc1186.

Roberts, S. A., Lawrence, M. S., Klimczak, L. J., Grimm, S. A., Fargo, D., Stojanov, P., … Gordenin, D. A. (2013). An APOBEC cytidine deaminase mutagenesis pattern is widespread in human cancers. *Nature Genetics, 45*(9), 970–976. http://dx.doi.org/10.1038/ng.2702.

Robinson, D. R., Wu, Y. M., Vats, P., Su, F., Lonigro, R. J., Cao, X., … Chinnaiyan, A. M. (2013). Activating ESR1 mutations in hormone-resistant metastatic breast cancer. *Nature Genetics, 45*(12), 1446–1451. http://dx.doi.org/10.1038/ng.2823.

Rocco, J. W., Leong, C. O., Kuperwasser, N., DeYoung, M. P., & Ellisen, L. W. (2006). p63 mediates survival in squamous cell carcinoma by suppression of p73-dependent apoptosis. *Cancer Cell, 9*(1), 45–56. http://dx.doi.org/10.1016/j.ccr.2005.12.013.

Sahai, V., Redig, A. J., Collier, K. A., Eckerdt, F. D., & Munshi, H. G. (2016). Targeting bet bromodomain proteins in solid tumors. *Oncotarget*, 7, 53997–54009. http://dx.doi.org/10.18632/oncotarget.9804.

Sau, A., Lau, R., Cabrita, M. A., Nolan, E., Crooks, P. A., Visvader, J. E., & Pratt, M. A. (2016). Persistent activation of NF-kappaB in BRCA1-deficient mammary progenitors drives aberrant proliferation and accumulation of DNA damage. *Cell Stem Cell, 19*(1), 52–65. http://dx.doi.org/10.1016/j.stem.2016.05.003.

Scheiber, M. N., Watson, P. M., Rumboldt, T., Stanley, C., Wilson, R. C., Findlay, V. J., … Watson, D. K. (2014). FLI1 expression is correlated with breast cancer cellular growth, migration, and invasion and altered gene expression. *Neoplasia, 16*(10), 801–813. http://dx.doi.org/10.1016/j.neo.2014.08.007.

Schimmer, A. D., Welsh, K., Pinilla, C., Wang, Z., Krajewska, M., Bonneau, M. J., … Reed, J. C. (2004). Small-molecule antagonists of apoptosis suppressor XIAP exhibit broad antitumor activity. *Cancer Cell, 5*, 25–35.

Senoo, M., Pinto, F., Crum, C. P., & McKeon, F. (2007). p63 is essential for the proliferative potential of stem cells in stratified epithelia. *Cell, 129*(3), 523–536.

Shackleton, M., Quintana, E., Fearon, E. R., & Morrison, S. J. (2009). Heterogeneity in cancer: Cancer stem cells versus clonal evolution. *Cell*, *138*(5), 822–829.

Shah, S. P., Roth, A., Goya, R., Oloumi, A., Ha, G., Zhao, Y., … Aparicio, S. (2012). The clonal and mutational evolution spectrum of primary triple-negative breast cancers. *Nature*, *486*(7403), 395–399. http://dx.doi.org/10.1038/nature10933.

Shen, R. R., & Hahn, W. C. (2011). Emerging roles for the non-canonical IKKs in cancer. *Oncogene*, *30*(6), 631–641. http://dx.doi.org/10.1038/onc.2010.493.

Shigekawa, T., Ijichi, N., Ikeda, K., Horie-Inoue, K., Shimizu, C., Saji, S., … Inoue, S. (2011). FOXP1, an estrogen-inducible transcription factor, modulates cell proliferation in breast cancer cells and 5-year recurrence-free survival of patients with tamoxifen-treated breast cancer. *Hormones & Cancer*, *2*(5), 286–297. http://dx.doi.org/10.1007/s12672-011-0082-6.

Shlien, A., Raine, K., Fuligni, F., Arnold, R., Nik-Zainal, S., Dronov, S., … Campbell, P. J. (2016). Direct transcriptional consequences of somatic mutation in breast cancer. *Cell Reports*, *16*(7), 2032–2046. http://dx.doi.org/10.1016/j.celrep.2016.07.028.

Shu, S., Lin, C. Y., He, H. H., Witwicki, R. M., Tabassum, D. P., Roberts, J. M., … Polyak, K. (2016). Response and resistance to BET bromodomain inhibitors in triple-negative breast cancer. *Nature*, *529*(7586), 413–417. http://dx.doi.org/10.1038/nature16508.

Sims, R. J., 3rd, Belotserkovskaya, R., & Reinberg, D. (2004). Elongation by RNA polymerase II: The short and long of it. *Genes & Development*, *18*(20), 2437–2468. http://dx.doi.org/10.1101/gad.1235904.

Slattery, M., Zhou, T., Yang, L., Dantas Machado, A. C., Gordan, R., & Rohs, R. (2014). Absence of a simple code: How transcription factors read the genome. *Trends in Biochemical Sciences*, *39*(9), 381–399. http://dx.doi.org/10.1016/j.tibs.2014.07.002.

Song, W., Li, W., Li, L., Zhang, S., Yan, X., Wen, X., … Cui, J. (2015). Friend leukemia virus integration 1 activates the Rho GTPase pathway and is associated with metastasis in breast cancer. *Oncotarget*, *6*(27), 23764–23775. http://dx.doi.org/10.18632/oncotarget.4350.

Spitz, F., & Furlong, E. E. (2012). Transcription factors: From enhancer binding to developmental control. *Nature Reviews Genetics*, *13*(9), 613–626. http://dx.doi.org/10.1038/nrg3207.

Stephens, P. J., Tarpey, P. S., Davies, H., Van Loo, P., Greenman, C., Wedge, D. C., … Stratton, M. R. (2012). The landscape of cancer genes and mutational processes in breast cancer. *Nature*, *486*(7403), 400–404. http://dx.doi.org/10.1038/nature11017.

Stojanova, A., Tu, W. B., Ponzielli, R., Kotlyar, M., Chan, P. K., Boutros, P. C., … Penn, L. Z. (2016). MYC interaction with the tumor suppressive SWI/SNF complex member INI1 regulates transcription and cellular transformation. *Cell Cycle*, *15*(13), 1693–1705. http://dx.doi.org/10.1080/15384101.2016.1146836.

Sur, I., & Taipale, J. (2016). The role of enhancers in cancer. *Nature Reviews Cancer*, *16*(8), 483–493. http://dx.doi.org/10.1038/nrc.2016.62.

Swanton, C., Soria, J. C., Bardelli, A., Biankin, A., Caldas, C., Chandarlapaty, S., … Andre, F. (2016). Consensus on precision medicine for metastatic cancers: A report from the MAP conference. *Annals of Oncology*, *27*(8), 1443–1448. http://dx.doi.org/10.1093/annonc/mdw192.

Takebe, N., Miele, L., Harris, P. J., Jeong, W., Bando, H., Kahn, M., … Ivy, S. P. (2015). Targeting Notch, Hedgehog, and Wnt pathways in cancer stem cells: Clinical update. *Nature Reviews. Clinical Oncology*, *12*(8), 445–464. http://dx.doi.org/10.1038/nrclinonc.2015.61.

Tannock, I. F., & Hickman, J. A. (2016). Limits to personalized cancer medicine. *The New England Journal of Medicine*, *375*(13), 1289–1294. http://dx.doi.org/10.1056/NEJMsb1607705.

Teo, J. L., Ma, H., Nguyen, C., Lam, C., & Kahn, M. (2005). Specific inhibition of CBP/beta-catenin interaction rescues defects in neuronal differentiation caused by a presenilin-1 mutation. *Proceedings of the National Academy of Sciences of the United States of America*, *102*(34), 12171–12176.

Thomas, Z. I., Gibson, W., Sexton, J. Z., Aird, K. M., Ingram, S. M., Aldrich, A., … Williams, K. P. (2011). Targeting GLI1 expression in human inflammatory breast cancer cells enhances apoptosis and attenuates migration. *British Journal of Cancer*, *104*(10), 1575–1586. http://dx.doi.org/10.1038/bjc.2011.133.

Timmerman, L. A., Holton, T., Yuneva, M., Louie, R. J., Padro, M., Daemen, A., … Gray, J. W. (2013). Glutamine sensitivity analysis identifies the xCT antiporter as a common triple-negative breast tumor therapeutic target. *Cancer Cell*, *24*(4), 450–465. http://dx.doi.org/10.1016/j.ccr.2013.08.020.

Tong, A. J., Liu, X., Thomas, B. J., Lissner, M. M., Baker, M. R., Senagolage, M. D., … Smale, S. T. (2016). A stringent systems approach uncovers gene-specific mechanisms regulating inflammation. *Cell*, *165*(1), 165–179. http://dx.doi.org/10.1016/j.cell.2016.01.020.

Truitt, M. L., & Ruggero, D. (2016). New frontiers in translational control of the cancer genome. *Nature Reviews Cancer*, *16*(5), 288–304. http://dx.doi.org/10.1038/nrc.2016.27.

Tseng, Y. Y., Moriarity, B. S., Gong, W., Akiyama, R., Tiwari, A., Kawakami, H., … Bagchi, A. (2014). PVT1 dependence in cancer with MYC copy-number increase. *Nature*, *512*(7512), 82–86. http://dx.doi.org/10.1038/nature13311.

Tu, W. B., Helander, S., Pilstal, R., Hickman, K. A., Lourenco, C., Jurisica, I., … Penn, L. Z. (2015). Myc and its interactors take shape. *Biochimica et Biophysica Acta*, *1849*(5), 469–483. http://dx.doi.org/10.1016/j.bbagrm.2014.06.002.

Turner, N. C., & Reis-Filho, J. S. (2006). Basal-like breast cancer and the BRCA1 phenotype. *Oncogene*, *25*(43), 5846–5853.

Vaillant, F., Asselin-Labat, M. L., Shackleton, M., Forrest, N. C., Lindeman, G. J., & Visvader, J. E. (2008). The mammary progenitor marker CD61/beta3 integrin identifies cancer stem cells in mouse models of mammary tumorigenesis. *Cancer Research*, *68*(19), 7711–7717.

van Geldermalsen, M., Wang, Q., Nagarajah, R., Marshall, A. D., Thoeng, A., Gao, D., … Holst, J. (2016). ASCT2/SLC1A5 controls glutamine uptake and tumour growth in triple-negative basal-like breast cancer. *Oncogene*, *35*(24), 3201–3208. http://dx.doi.org/10.1038/onc.2015.381.

Van Keymeulen, A., Lee, M. Y., Ousset, M., Brohee, S., Rorive, S., Giraddi, R. R., … Blanpain, C. (2015). Reactivation of multipotency by oncogenic PIK3CA induces breast tumour heterogeneity. *Nature*, *525*(7567), 119–123. http://dx.doi.org/10.1038/nature14665.

Vaquerizas, J. M., Kummerfeld, S. K., Teichmann, S. A., & Luscombe, N. M. (2009). A census of human transcription factors: Function, expression and evolution. *Nature Reviews Genetics*, *10*(4), 252–263. http://dx.doi.org/10.1038/nrg2538.

Vassilev, L. T., Vu, B. T., Graves, B., Carvajal, D., Podlaski, F., Filipovic, Z., … Liu, E. A. (2004). In vivo activation of the p53 pathway by small-molecule antagonists of MDM2. *Science*, *303*(5659), 844–848. http://dx.doi.org/10.1126/science.1092472.

Vilchez, V., Turcios, L., Marti, F., & Gedaly, R. (2016). Targeting Wnt/beta-catenin pathway in hepatocellular carcinoma treatment. *World Journal of Gastroenterology*, *22*(2), 823–832. http://dx.doi.org/10.3748/wjg.v22.i2.823.

Visvader, J. E. (2011). Cells of origin in cancer. *Nature*, *469*(7330), 314–322. http://dx.doi.org/10.1038/nature09781.

Visvader, J. E., & Stingl, J. (2014). Mammary stem cells and the differentiation hierarchy: Current status and perspectives. *Genes & Development*, *28*(11), 1143–1158. http://dx.doi.org/10.1101/gad.242511.114.

Voduc, D., Cheang, M., & Nielsen, T. (2008). GATA-3 expression in breast cancer has a strong association with estrogen receptor but lacks independent prognostic value. *Cancer Epidemiology, Biomarkers & Prevention, 17*(2), 365–373. http://dx.doi.org/10.1158/1055-9965.Epi-06-1090.

Vogel, C., & Marcotte, E. M. (2012). Insights into the regulation of protein abundance from proteomic and transcriptomic analyses. *Nature Reviews Genetics, 13*(4), 227–232. http://dx.doi.org/10.1038/nrg3185.

Walker, M. P., Stopford, C. M., Cederlund, M., Fang, F., Jahn, C., Rabinowitz, A. D., … Major, M. B. (2015). FOXP1 potentiates Wnt/beta-catenin signaling in diffuse large B cell lymphoma. *Science Signaling, 8*(362), ra12. http://dx.doi.org/10.1126/scisignal.2005654.

Wang, S., Liu, J. C., Kim, D., Datti, A., & Zacksenhaus, E. (2016). Targeted Pten deletion plus p53-R270H mutation in mouse mammary epithelium induces aggressive claudin-low and basal-like breast cancer. *Breast Cancer Research, 18*(1), 9. http://dx.doi.org/10.1186/s13058-015-0668-y.

Wang, L., & Wang, J. (2012). MicroRNA-mediated breast cancer metastasis: From primary site to distant organs. *Oncogene, 31*(20), 2499–2511. http://dx.doi.org/10.1038/onc.2011.444.

Wang, S. P., Wang, W. L., Chang, Y. L., Wu, C. T., Chao, Y. C., Kao, S. H., … Yang, P. C. (2009). p53 controls cancer cell invasion by inducing the MDM2-mediated degradation of Slug. *Nature Cell Biology, 11*(6), 694–704.

Wang, L., Zeng, X. Z., Chen, S., Ding, L. Y., Zhong, J., Zhao, J. C., … Huang, H. J. (2013). BRCA1 is a negative modulator of the PRC2 complex. *The EMBO Journal, 32*(11), 1584–1597. http://dx.doi.org/10.1038/emboj.2013.95.

Weinstein, I. B., & Joe, A. (2008). Oncogene addiction. *Cancer Research, 68*(9), 3077–3080. discussion 3080.

Weintraub, S. J., Prater, C. A., & Dean, D. C. (1992). Retinoblastoma protein switches the E2F site from positive to negative element. *Nature, 358*, 259–261.

Wise, D. R., & Thompson, C. B. (2010). Glutamine addiction: A new therapeutic target in cancer. *Trends in Biochemical Sciences, 35*(8), 427–433.

Wong, G. T., Manfra, D., Poulet, F. M., Zhang, Q., Josien, H., Bara, T., … Parker, E. M. (2004). Chronic treatment with the gamma-secretase inhibitor LY-411,575 inhibits beta-amyloid peptide production and alters lymphopoiesis and intestinal cell differentiation. *Journal of Biological Chemistry, 279*(13), 12876–12882. http://dx.doi.org/10.1074/jbc.M311652200.

Wu, Y., Cain-Hom, C., Choy, L., Hagenbeek, T. J., de Leon, G. P., Chen, Y., … Siebel, C. W. (2010). Therapeutic antibody targeting of individual Notch receptors. *Nature, 464*(7291), 1052–1057. http://dx.doi.org/10.1038/nature08878.

Wu, J., Lu, L. Y., & Yu, X. (2010). The role of BRCA1 in DNA damage response. *Protein & Cell, 1*(2), 117–123. http://dx.doi.org/10.1007/s13238-010-0010-5.

Wu, L., Timmers, C., Maiti, B., Saavedra, H. I., Sang, L., Chong, G. T., … Leone, G. (2001). The E2F1-3 transcription factors are essential for cellular proliferation. *Nature, 414*(6862), 457–462.

Xie, L., Gazin, C., Park, S. M., Zhu, L. J., Debily, M. A., Kittler, E. L., … Green, M. R. (2012). A synthetic interaction screen identifies factors selectively required for proliferation and TERT transcription in p53-deficient human cancer cells. *PLoS Genetics. 8*(12). e1003151http://dx.doi.org/10.1371/journal.pgen.1003151.

Xie, Q., Wu, Q., Kim, L., Miller, T. E., Liau, B. B., Mack, S. C., … Rich, J. N. (2016). RBPJ maintains brain tumor-initiating cells through CDK9-mediated transcriptional elongation. *Journal of Clinical Investigation, 126*(7), 2757–2772. http://dx.doi.org/10.1172/JCI86114.

Yoshida, H., Matsui, T., Yamamoto, A., Okada, T., & Mori, K. (2001). XBP1 mRNA is induced by ATF6 and spliced by IRE1 in response to ER stress to produce a highly active transcription factor. *Cell, 107*(7), 881–891. http://dx.doi.org/10.1016/S0092-8674(01)00611-0.

Yumimoto, K., Akiyoshi, S., Ueo, H., Sagara, Y., Onoyama, I., Ueo, H., … Nakayama, K. I. (2015). F-box protein FBXW7 inhibits cancer metastasis in a non-cell-autonomous manner. *Journal of Clinical Investigation, 125*(2), 621–635. http://dx.doi.org/10.1172/JCI78782.

Zacksenhaus, E., Jiang, Z., Phillips, R. A., & Gallie, B. L. (1996). Dual mechanisms of repression of E2F1 activity by the retinoblastoma gene product. *The EMBO Journal, 15*(21), 5917–5927.

Zhang, J. W., Gu, J., Wang, Z. Y., Chen, S. J., & Chen, Z. (2000). Mechanisms of all-trans retinoic acid-induced differentiation of acute promyelocytic leukemia cells. *Journal of Biosciences, 25*(3), 275–284. http://dx.doi.org/10.1007/Bf02703936.

Zirath, H., Frenzel, A., Oliynyk, G., Segerstrom, L., Westermark, U. K., Larsson, K., … Henriksson, M. A. (2013). MYC inhibition induces metabolic changes leading to accumulation of lipid droplets in tumor cells. *Proceedings of the National Academy of Sciences of the United States of America, 110*(25), 10258–10263. http://dx.doi.org/10.1073/pnas.1222404110.

CHAPTER EIGHT

Transcription Factors as a Target for Vaccination Against Ticks and Mites

O.A.E. Sparagano[1]
Coventry University, Coventry, United Kingdom
[1]Corresponding author: e-mail address: olivier.sparagano@coventry.ac.uk

Contents

Abstract

Ticks and mites are well-known ectoparasites as potential vectors for numerous bacteria, viruses, and parasites. Many being blood feeders add to physiological deterioration, morbidity, and mortality of their vertebrate hosts. To control them, transcription factors have been identified and studied in their role to sustain such arthropod pests. This paper summarizes some of the work done on those factors involved during blood feeding, reproduction, or when interacting with their pathogens and symbiont populations. Any transcription factor supporting the equilibrium developed by the ticks/mites could become a potential target for new control methods to prevent some of their key physiological functions.

1. INTRODUCTION

Blood feeding ticks and mites are ectoparasites affecting their vertebrate hosts in many ways (anemia, disease transmission, local inflammation,

Advances in Protein Chemistry and Structural Biology, Volume 107
ISSN 1876-1623
http://dx.doi.org/10.1016/bs.apcsb.2016.11.004

and skin damages). Many control methods have been tested from acaricide products to predators and plant-derived products and some had good results but often limited to laboratory settings and for a short time (Sparagano, George, Harrington, & Giangaspero, 2014). Fortunately transcriptomes are now fully sequenced such as the one for the poultry red mite, *Dermanyssus gallinae* (Schicht, Qi, Poveda, & Strube, 2013a, 2013b).

Transcription factors in mites, ticks, their animal hosts, and the pathogens they transmit or those they harbor, such as symbionts, for instance, have been used to reduce or block their deadly effects. This paper highlights the progress toward understanding the transcriptome in these arthropods to better control them as part of new vaccines or other control methods to be used under different environmental settings. Laboratory results and in some cases field ones have proven that understand better and manipulating transcription factors in these Acari can open future control methods.

2. TARGETS FROM MITES AND TICKS

2.1 Blood Feeding-Related Targets

2.1.1 Impact on Hosts

Blood feeding arthropods such as ticks and mites have direct negative effects due to tick bites, skin damages, blood loss, anemia affecting animal health, and productivity. Indirect effects can be related to vector–borne pathogens being transmitted to the host with disease symptoms being seen sometimes only weeks after the ticks have left the host. Morbidity and mortality levels have economic, societal, and environmental impacts on animals and humans making them second to the mosquitoes in terms of negative impact.

2.1.2 Gut Microbiome

A possible way to indirectly control ticks and mites is to identify and subsequently control the microbiome in those arthropod pests.

Interestingly, Zhang et al. (2014) investigated the microbiome before and after feeding in *Ixodes persulcatus* ticks and in rat blood samples after tick bites. They assigned 373 and 289 bacterial genera to unfed and fed ticks, respectively. Two hundred and thirty-seven bacterial genera (82% of the one found in fed ticks) were found transmitted to the vertebrate host. This approach will provide valuable information regarding pathogen transmission, their interaction with the tick gut proteins, between species and adaptation with the host environment. Reducing pathogen transmission would not impact on physical damages caused by tick bites but could tremendously reduce disease transmission.

This has been demonstrated by Narasimhan and Fikrig (2015) showing that altered gut microbiota reduced Borrelia colonization of the gut epithelium in *Ixodes scapularis* ticks.

2.2 Reproduction-Related Targets

Another possibility to affect Acari reproduction is to target the hormones produced by them for blood feeding and reproduction. Tick synganglion-specific transcriptomes are not fully understood yet but some research has proven that in *Dermacentor variabilis* (Donohue et al., 2010). For more information about the mechanisms for the regulation of adult blood feeding in ticks, see Roe, Donohue, Khalil, and Sonenshine (2008). Hormones involved in the recognition between male and female ticks, hormones transferred by the male tick into the genital track of the female and hormones involved in blood seeking, blood digestion, vitellogenesis, and egg production can be potential targeted factors to block tick reproduction.

2.2.1 *Impact on Sex Determination*

Paradoxically it could be important to disrupt sex determination of blood feeding arthropods such as the poultry red mite, *D. gallinae*, for which the complete transcriptome has been sequenced (Schicht et al., 2013a, 2013b) but on the other hand it could be important to support the reproduction and sex determination of predatory mites attacking other arthropod pests. This is an approach done by Pomerantz, Hoy, and Kawahara (2015) working on the western orchard predatory mite *Metaseiulus occidentalis* used in biological control program in agriculture. Their work showed that certain sex determination genes are conserved in arthropods.

2.2.2 *Vitellogenesis*

GATA factor mRNA is accumulated before blood feeding in the tick gut but translation is only activated by blood feeding. RNA interference-mediated knockdown of GATA factor (synthesized after a blood meal) and S6 kinase in *Haemaphysalis longicornis* ticks resulted in a decrease of the major yolk protein vitellogenin, affecting egg development in female ticks after a blood meal (Boldbaatar et al., 2010). *H. longicornis* GATA factor (HlGATA) is well known now as a downstream regulator of the target of rapamycin (TOR) signaling pathway. Such approach would prevent, if possibly applicable in field programs, female ticks laying the next generation and transmitting pathogens to their eggs (not all tick species can transmit pathogens to their eggs).

2.3 Tick Physiology

2.3.1 Defense Mechanisms

When ticks are blood feeding they have to suppress the local immune response of the host as some ticks will stay attached for days. Umemiya et al. (2007) identified a homologue of the macrophage migration inhibitory factor (MIF) in *H. longicornis* ticks (HlMIF). Such factor is upregulated during blood feeding and immunohistochemistry showed that the endogenous HlMIF is localized to the midgut and epidermal cells. Such factor may facilitate blood feeding by preventing host macrophage migration to the feed.

One way to control ectoparasites is to understand how themselves are blocking insecticides, pesticides, and toxins from affecting them. One example is the work of Cancino-Rodezno et al. (2010) on the mitogen-activated protein kinase p38 (MAPK p38) in *Manduca sexta* (Lepidopteran) and *Aedes aegypti* (Dipteran). When silencing such pathway both insect species became hypersensitive to Cry toxins from *Bacillus thuringiensis*.

Another way for ticks to survive is to produce clotting mechanisms, well known as the limulus clotting factor C (LCFC) is horseshoe crabs but very limited information is available for other invertebrates. However, Urbanova et al. (2014) demonstrated that the clotting factor C in *Ixodes ricinus* (IrFC) could act as an ortholog of the LCFC and would play a role in the tick primordial complement system.

2.3.2 Subolesin-Related Transcription Factors

One of the greatest achievements against Acari has been the development of tick vaccines. Although this has been developed decades ago in Australia against the *Boophilus microplus* ticks (BM86- and bm95-derived vaccines) there are not many antitick/mite vaccines on the market yet. "Concealed antigens" mainly in the gut and salivary glands have been targeted by researchers for many years.

Discovered in *I. scapularis* (Almazan et al., 2003) protective antigens such as the subolesin protein has given hope for more vaccines to be soon considered in the field. Subolesin gene and protein sequences are highly conserved in ticks and are the orthologs of the akirins in insects and vertebrates (Galindo et al., 2009). However, recently Almazán et al. (2012) observed that it is the unstructured region of the subolesin that need to be recognized by vaccine-induced specific antibodies to disrupt its biological activity as observed in the reduction in Ornithodoros tick oviposition and fertility. Crossreactive epitopes on other tick proteins expressed in the midgut and ovaries could explain such protective effects from subolesin/akiri-based vaccines.

Similar work using a subolesin recombinant antigen from ticks against the *D. gallinae* mite has produced a very high mortality rate for an ortholog antigen (Harrington et al., 2009).

3. TARGETING PATHOGENS AND SYMBIONTS

3.1 Controls in Place From the Arthropods

Ticks and mites have also to defend themselves against pathogens invading them.

Therefore, ticks are, for instance, mounting defence mechanisms to detect *Borrelia burgdorferi* during blood meals and produce antimicrobial peptides. Recently, Smith et al. (2016) showed that *I. scapularis* ticks feeding on infected mice upregulated a Rho-like GTPase (IGTPase) which in turns induced antimicrobial peptides with IFNγ acquisition during blood meals necessary for IGTPase induction.

Anaplasma phagocytophilum (agent for human granulocytic anaplasmosis) seems to be controlled by its vector *I. scapularis* through the Janus kinase (JAK)-signaling transducer activator of transcription (STAT) pathways as such control is suppressed by RNA interference (Liu et al., 2012). The authors showed that a 5.3-kDa antimicrobial peptide is highly induced when *A. phagocytophilum* infection reaches tick salivary glands regulated by tick STAT.

3.2 Targeting the Symbionts

Understanding which symbionts are living with which ticks or mites could bring another way to affect those by destabilizing the symbiotic partnership. *Francisella*-like endosymbionts have been detected in *Ornithodoros* and *Dermacentor* ticks, while *Wolbachia* has been known to develop a symbiotic relationship with *I. ricinus* ticks. Those symbiotic bacteria can either reduce pathogen transmission as seen with *Coxiella* endosymbionts found in the salivary glands of *Amblyomma americanum* decreasing the transmission of *Ehrlichia chaffeensis* or the presence of the mitochondrial endosymbiont *Candidatus Midichloria mitochondrii* linked to the vector fitness in terms of transovarial transmission.

3.3 Learning From the Pathogens Themselves

Understanding regulations and counter-regulations between vector–borne pathogens such a *B. burgdorferi* and its arthropod vectors can bring some information of the key mechanisms to be targeted. For instance, the *Borrelia* host adaptation Regulator (BadR) is known to regulate rpoS. BadR-deficient strains showed upregulation of rpoS, important factor for virulence in vertebrate hosts (Miller, Rajasekhar Karna, & Seshu, 2013). Furthermore and more amazingly, *B. burdorferi* as to adjust its transcriptome to adapt to either *Ixodes* ticks or vertebrate hosts with the Lyme disease pathogen requiring to survive and proliferate in very different environments. Dulebohn, Hayes, and Rosa (2014) showed that a novel protein called BBD18 was involved in the regulation of the virulence-associated sigma factor RpoS, which acts at posttranscriptional regulation level with a possible mechanism destabilizing proteins rather than translational control.

4. CONCLUSIONS

This paper has been exclusively looking at transcription factors in ticks and mites used for controlling their development, reproduction, vector capacity, and survival. However, there is much more work done with insect models, which could not be presented in this paper on Acari. Lessons have been learned from the tick vaccine investigations when proteins found in *B. microplus* ticks in Australia were not all present in their counterparts in Cuba for instance explaining why BM86-based vaccines performed less than those using Bm95 a more conserved antigen. It is important to keep in mind species/strains variability when control methods are put in place to avoid regional successes and worldwide failures. Transcriptome sequencing progresses have open more potential to understand regulatory processes to identify more effective control methods. This approach would allow a reduction of the negative impacts of such arthropod pests and would subsequently reduce the transmission of many vector–borne diseases.

ACKNOWLEDGMENTS

This chapter is based upon work from COST action (FA1404-COREMI) and supported by COST (European Cooperation in Science and Technology).

REFERENCES

Almazan, C., Kocan, K. M., Bergman, D. K., Garcia-Garcia, J. C., Blouin, E. F., & de la Fuente, J. (2003). *Ixodes scapularis*, infestations using cDNA expression library immunization. *Vaccine*, *21*, 1492–1501.

Almazán, C., Moreno-Cantú, O., Moreno-Cid, J. A., Galindo, R. C., Canales, M., Villar, M., et al. (2012). Control of tick infestations in cattle vaccinated with bacterial membranes containing surface-exposed tick protective antigens. *Vaccine*, *30*(2), 265–272.

Boldbaatar, D., Battur, B., Umemiya-Shirafuji, R., Liao, M., Tanaka, T., & Fujisaki, K. (2010). GATA transcription, translation and regulation in *Haemaphysalis longicornis* tick: Analysis of the cDNA and an essential role for vitellogenesis. *Insect Biochemistry and Molecular Biology*, *40*, 49–57.

Cancino-Rodezno, A., Alexander, C., Villasenor, R., Pacheco, S., Porta, H., Pauchet, Y., et al. (2010). The mitogen-activated protein kinase p38 involved in insect defense against Cry toxins from *Bacillus thuringiensis*. *Insect Biochemistry and Molecular Biology*, *40*, 58–63.

Donohue, K. V., Khalil, S. M. S., Ross, E., Grozinger, C. M., Sonenshine, D. E., & Roe, R. M. (2010). Neuropeptide signaling sequences identified by pyrosequencing of the American dog tick synganglion transcriptome during blood feeding and reproduction. *Insect Biochemistry and Molecular Biology*, *40*, 79–90.

Dulebohn, D. P., Hayes, B. M., & Rosa, P. A. (2014). Global repression of host-associated genes of the Lyme disease spirochete through post-transcriptional modulation of the alternative sigma factor RpoS. *PLoS One*, *36*, e93141.

Galindo, R. C., Doncel-Perez, E., Zivkovic, Z., Naranjo, V., Gortazar, C., Mangold, A. J., et al. (2009). Tick subolesin is an ortholog of the akirins described in insects and vertebrates. *Developmental and Comparative Immunology*, *33*, 612–617.

Harrington, D., Canales, M., de la Fuente, J., Robinson, K., Guy, J., & Sparagano, O. (2009). Immunisation with recombinant proteins subolesin and Bm86 for the control of *Dermanyssus gallinae* in poultry. *Vaccine*, *27*(30), 4056–4063.

Liu, L., Dai, J., Zhao, Y. O., Narasimhan, S., Yang, Y., Zhang, L., et al. (2012). *Ixodes scapularis* JAK-STAT pathway regulates tick antimicrobial peptides, thereby controlling the agent of human granulocytic anaplasmosis. *Journal of Infectious Diseases*, *206*, 1233–1241.

Miller, C. L., Rajasekhar Karna, S. L., & Seshu, J. (2013). Borrelia host adaptation regulator (BadR) regulates rpoS to modulate host adaptation and virulence factors in *Borrelia burgdorferi*. *Molecular Microbiology*, *88*(1), 105–124.

Narasimhan, S., & Fikrig, E. (2015). Tick microbiome: The force within. *Trends in Parasitology*, *31*(7), 315–323.

Pomerantz, A. F., Hoy, M. A., & Kawahara, A. Y. (2015). Molecular characterization and evolutionary insights into potential sex-determination genes in the western orchard predatory mite *Metaseiulus occidentalis* (Chelicerata: Arachnida: Acari: Phytoseiidae). *Journal of Biomolecular Structure and Dynamics*, *33*(6), 1239–1253.

Roe, R. M., Donohue, K. V., Khalil, S. M. S., & Sonenshine, D. E. (2008). Hormonal regulation of metamorphosis and reproduction in ticks. *Frontiers in Bioscience: A Journal and Virtual Library*, *13*, 7250–7268.

Schicht, S., Qi, W., Poveda, L., & Strube, C. (2013a). The predicted secretome and transmembrane of the poultry red mite Dermanyssus gallinae. *Parasites & Vectors*, *6*(1), 259.

Schicht, S., Qi, W., Poveda, L., & Strube, C. (2013b). Whole transcriptome analysis of the poultry red mite Dermanyssus gallinae (De Geer, 1778). *Parasitology*, *141*, 336–346.

Smith, A. A., Navasa, N., Yang, X., Wilder, C. N., Buyuktani, O., Marques, A., et al. (2016). Cross-species interferon signaling boosts microbicidal activity with the tick vector. *Cell Host & Microbe*, *20*(1), 91–98.

Sparagano, O. A. E., George, D. R., Harrington, D., & Giangaspero, A. (2014). Biology, epidemiology, management and risk related to the poultry red mite, *Dermanyssus gallinae* (de Geer, 1778). *Annual Review of Entomology*, *59*, 447–466.

Umemiya, R., Hatta, T., Liao, M., Tanaka, M., Zhou, J., Inoue, N., et al. (2007). *Haemaphysalis longicornis*: Molecular characterization of a homologue of the macrophage migration inhibitory factor from the partially fed ticks. *Experimental Parasitology, 115*, 135–142.

Urbanova, V., Hartmann, D., Grunclova, L., Sima, R., Flemming, T., Hajdusek, O., et al. (2014). IrFC—An *Ixodes ricinus* injury-responsive molecule related to Limulus Factor C. *Developmental and Comparative Immunology, 46*, 439–447.

Zhang, X.-C., Yang, Z.-N., Lu, B., Ma, X.-F., Zhang, C.-X., & Xu, H.-J. (2014). The composition and transmission of microbiome in hard ticks, *Ixodes persulcatus*, during blood meal. *Ticks and Tick-Borne Diseases, 5*, 864–870.

CHAPTER NINE

Elucidating the Mutational Landscape in Hepatocyte Nuclear Factor 1β (HNF1B) by Computational Approach

P. Sneha, C.G.P. Doss[1]
School of Biosciences and Technology, VIT University, Vellore, Tamil Nadu, India
[1]Corresponding author: e-mail addresses: georgecp77@yahoo.co.in; georgepriyadoss@vit.ac.in

Contents

Abstract

Transcription factors are the major gene-regulatory proteins that recognize specific nucleotide sequences and bind to them. Missense mutations in transcription factors play a significant role in misregulation of gene expression contributing to various diseases and disorders. Understanding their structural and functional impact of the disease-causing mutations becomes prime importance in treating a disease. Commonly associated defect with the mutations of hepatocyte nuclear factor 1 beta (HNF1B) protein, a transcription factor results in maturity-onset diabetes of the young-5 (MODY-5) leading to loss of function. In the study presented, we applied a series of computational strategies to analyze the effect of mutations on protein

Advances in Protein Chemistry and Structural Biology, Volume 107
ISSN 1876-1623
http://dx.doi.org/10.1016/bs.apcsb.2016.11.005

structure or function in protein–DNA complex. The mutations from publicly available databases were retrieved and subjected to an array of in silico prediction methods. Key implementation of the present study consists of a pipeline drawn using well established in silico prediction methods of different algorithms to explain the biochemical changes impaired upon mutations in the binding sites of protein–DNA complex using HNF1B. Prediction scores obtained from the in silico tools suggested H153N and A241T as the major nsSNPs involved in destabilizing the protein. Further, high-end microscopic computational study, such as molecular dynamics simulations was utilized to relate the structural and functional effects upon mutations. Although, both the mutations exhibited similar structural differences, we observed A241T with higher destabilizing effect on the protein. The presented work is a step toward understanding the genotype–phenotype relationships in transcription factor genes using fast and accurate computational approach.

1. INTRODUCTION

Transcriptional control and its key concepts were well established about a half century ago in wide ranges of prokaryotic systems (Jacob & Monod, 1961). This revolutionary study recognized that DNA-binding proteins, i.e., the transcription factors (TFs) occupy specific DNA sequences that control *cis*-element and regulate the transcription apparatus. Extensive study of specific transcription factors and their cofactors along with various chromatin regulators have been made in eukaryotic systems (Adelman & Lis, 2012; Bonasio, Tu, & Reinberg, 2010; Conaway & Conaway, 2011; Fuda, Ardehali, & Lis, 2009; Kozak, 1992; McAdams & Arkin, 1997; Roeder, 2005; Taatjes, 2010; Walter & Newport, 2000). Nonfunctional transcription factors contribute to the pathogenesis of many human diseases, ranging from diabetes, inflammatory disorders, cardiovascular disease, and various cancers (Papavassiliou & Papavassiliou, 2016).

Maturity-onset diabetes of the young-5 (MODY5) is a manifestation of missense mutations in HNF1B or a chromosome 17q12 deletion encompassing the *HNF1B* (TCF2) gene (Barbacci et al., 2004; Raile et al., 2008). HNF1B is a transcription factor with 557 amino acids which binds to specific regions of nucleic acid (Fig. 1). Mutations in HNF1B were first described in patients with MODY5 (Horikawa et al., 1997). A recent study designed by Cuff et al. (2013), reported HNF1B may act as a marker of clear cell phenotype, and supports a mechanistic link between glycogen accumulation and thrombosis (Cuff et al., 2013). HNF1B plays a significant role in early development and organogenesis of several tissues namely

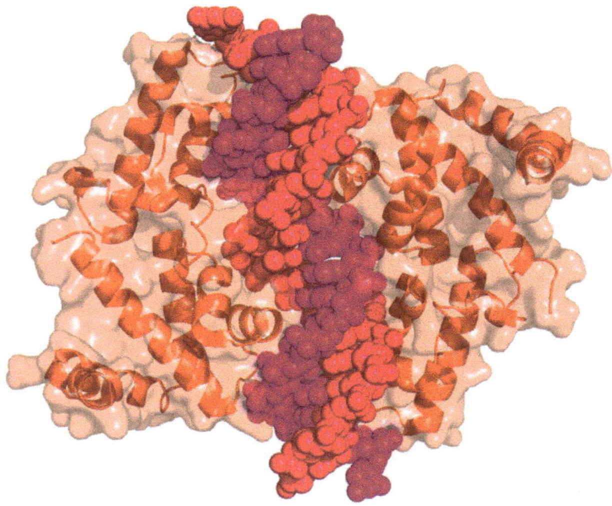

Fig. 1 HNF1B protein–DNA complex. The *purple* and *pink* represent the two strands of the DNA, while the *wheat-colored cartoon structure* represents the protein.

pancreas, liver, lung, and kidney (Barbacci et al., 1999). The gene is also transiently expressed in the neural tube and the epididymis, vas deferens, seminal vesicle, prostate, uterus, and oviduct (Barbacci et al., 1999; Reber & Cereghini, 2001). HNF1B was shown to bind directly to the DNA elements that regulate the expression of several kidney and liver-specific genes where associated mutations are responsible for causing cystic kidney diseases (Alvelos et al., 2015; Gresh et al., 2004), MODY 5 (Horikawa et al., 1997; Raile et al., 2008), hypomagnesemia (Adalat et al., 2009), and renal hypodysplasia (Thomas et al., 2011). HNF1B is a POU (Pit-1Oct-1/2-UNC-86) domain transcription factor, with highly conserved DNA-binding domain composed of a POU-specific (POU_S) and POU homeo (POU_H) domain. POU domain is enough for DNA-specific binding of the protein to the DNA (Sturm & Herr, 1988; Wegner, Drolet, & Rosenfeld, 1993). Mutation-associated MODY5 results in either truncation of codons or an internal deletion within the POU_S domain and thereby hindering the DNA-binding mechanism.

Analyzing each disease-causing mutations especially nonsynonymous (nsSNPs) mutations present in the protein through experimental analysis may be labor intense and time-consuming procedure, for which advanced computational methods have now become a preferred alternative strategy. In the past decade, computational tools and algorithms have drastically evolved (Gnad, Baucom, Mukhyala, Manning, & Zhang, 2013;

Liu, Watson, & Zhang, 2015; Mort et al., 2014) to analyze the various structural and functional changes caused by the disease-causing mutations (George Priya Doss, Chakraborty, Narayan, & Thirumal Kumar, 2014; George et al., 2014; Gray, Kukurba, & Kumar, 2012; Nagasundaram et al., 2015; Thirumal Kumar et al., in press). In the current study, we have conducted a three-stage molecular structural analysis using various in silico prediction methods to identify the biochemical changes occurred upon mutation. As an initial step, the nsSNPs from publicly available databases were retrieved and subjected to in silico prediction tools to determine their pathogenic effect using SIFT (Ng & Henikoff, 2003), PredictSNP (Bendl et al., 2014), PhD-SNP (Capriotti, Calabrese, & Casadio, 2006), PolyPhen-1 (Ramensky, Bork, & Sunyaev, 2002), PolyPhen-2 (Adzhubei, Jordan, & Sunyaev, 2013), MAPP (Stone & Sidow, 2005), and SNAP (Johnson et al., 2008). Subsequently, the pathogenic mutations were subjected to analyze their effect on the stability of the protein using CUPSAT (Parthiban, Gromiha, & Schomburg, 2006), MUpro (Cheng, Randall, & Baldi, 2006), I-Mutant 2.0 (Capriotti, Fariselli, & Casadio, 2005), iStable (Chen, Lin, & Chu, 2013), SNP4.0 effect (Baets et al., 2012), and INPS (Savojardo, Fariselli, Martelli, & Casadio, 2016). Finally, a molecular dynamics simulation study of 40 ns (nanoseconds) was performed to understand the micro changes occurred upon mutation at the structural level.

2. MATERIALS AND METHODS

2.1 Retrieval of Mutations

nsSNP information search related to HNF1B was conducted using HGMD (Stenson et al., 2003) and dbSNP (Sherry et al., 2001) databases. The retrieved nsSNPs were subjected to literature search in PubMed to understand the biological function of the mutations leading to MODY5.

2.2 Disease-Related Mutations

nsSNPs information obtained from the different databases was subjected to various sequence and structure based on silico prediction tools with diverse algorithms to analyze their impact on protein structure and function. The pathogenicity of the mutations were evaluated using SIFT (Ng & Henikoff, 2003), PredictSNP (Bendl et al., 2014), PhD-SNP (Capriotti et al., 2006), PolyPhen-1 (Ramensky et al., 2002), PolyPhen-2 (Adzhubei et al., 2013), MAPP (Stone & Sidow, 2005), and SNAP (Johnson et al., 2008).

Based on the obtained scores (deleterious/neutral) from the in silico prediction methods, nsSNP mutations were prioritized using ranking strategy to suggest how far the mutation might have an effect on the protein. Furthermore, the mutations categorized as Rank1 were selected and subjected to stability assessing tools such as CUPSAT (Parthiban et al., 2006), MUpro (Cheng et al., 2006), I-Mutant 2.0 (Capriotti et al., 2005), iStable (Chen et al., 2013), and SNP4.0 effect (Baets et al., 2012). INPS (Savojardo et al., 2016) server predicts the Gibbs free energy change and an exclusive graph is plotted by comparing the standard energy of the protein.

2.3 Modeling of Native and Mutant Protein

The protein molecule of HNF1B with a PDB ID: 2H8R (Lu, Rha, & Chi, 2007) was retrieved from Protein Data Bank. Using ELASPIC server (Witvliet et al., 2016), we mapped the mutations onto the protein sequence at their corresponding positions. During the preparation of three-dimensional (3D) structures for mutational analysis using SWISS-PDB Viewer (Guex & Peitsch, 1997), the protein molecule was separated from the DNA. Mutations (H153N and A241T) were introduced at their corresponding positions along with corresponding mutant amino acids. HADDOCK server (van Zundert et al., 2016) was used to perform docking analysis and the obtained native and two mutant protein–DNA complexes were subjected to further analysis.

2.4 Molecular Dynamics

Molecular dynamics simulation for the native and mutant complexes was carried out using Gromacs 4.6.3 package (Pronk et al., 2013). The structure of native protein–DNA complex, A241T–DNA complex, and H153N–DNA complex was used as the starting step for the MD simulation analysis. The complexes were solvated in separate systems in a cubic box of 1.0 nm marginal radius. The default water molecules (SPC—simple point charge) were used as the solvent by using water space tool available in the Gromacs package. As a next step, "genion" tool was used to neutralize the negatively charged system by adding 10 sodium ions (Na+) to the simulation box. After neutralization, the whole system was subjected to energy minimization for 50,000 iterations by steepest descent algorithm implementing CHARMM 36 (Huang & MacKerell, 2013) force field. Primarily, the solvent atoms were relaxed while the solute atoms were harmonically restrained with original positions having a force constant of 1000 kcal/mol. The temperature of the system was regulated using Berendsen temperature coupling method (Berendsen, Postma, van Gunsteren, DiNola, & Haak, 1984), and the electrostatic interactions were then computed using PME (particle mesh Ewald) method (Cheatham, Miller, Fox, Darden, & Kollman, 2002). The ionization states of all the residues were set appropriate to a neutral pH 7. Along with this, SHAKE algorithm was also used to constrain the bond lengths involving hydrogen, Van der Waals forces, and Coulomb interactions. After successful minimization, the system was further subjected to three different steps: canonical ensemble, isothermal–isobaric ensembles, and production MD. An NPT ensemble or the canonical ensembles was performed for 50,000 ps (picoseconds) at 300 K followed by an NVT ensemble/isothermal–isobaric ensembles that were carried out at 300 K. Finally, the production molecular dynamics simulation was performed at 300 K for 40 ns. The resultant files of molecular dynamics simulation are trajectory files that are subjected to various analysis to understand the structural and functional changes in the protein–DNA complexes.

2.5 Result Analysis

The resultant trajectory files of 40 ns MD simulation were further subjected to utilities available in Gromacs package to understand the various structural changes occurred during the simulation period. g_rms was used to plot the root-mean-square deviations (RMSDs) that occur at the end of 40 ns (4000 ps) simulation. This analysis provides an overview on the structural stability of the protein–DNA complexes. g_hbond was used to calculate the number of hydrogen bonds formed between the protein and DNA (native and mutant complexes). g_gyrate explains the compactness of the protein to the DNA. g_densmap was used to plot the partial density at the end of 40 ns simulation. Ramachandran plot analysis was performed for the native and mutant using g_rama utility of the Gromacs package. All the graphs were plotted with the aid of GRACE tool (Turner, 2005) and images were visualized using VMD (Humphrey, Dalke, & Schulten, 1996).

2.6 Essential Dynamics

A principal component analysis (PCA) or essential dynamics of the trajectory yields a set of vectors that describe the direction of motion about the average conformations and also provides information about correlations between the movements. Initially, a covariance matrix was constructed with the help of g_covar command. The covariance matrix is based on the eigenvectors and eigenvalues which imply the transition act of the linear points. With this covariance matrix as a starting point, g_anaeig command was used to obtain the PCA of the protein–DNA complexes. Following steps were performed for all the three complexes to analyze the motion changes that have occurred during the simulation period of 40 ns.

3. RESULTS

3.1 In Silico Tool Prediction of nsSNPs

A total of 50 nsSNPs were retrieved from HGMD and dbSNP that may have an association with MODY5. We first conducted a pathogenic analysis for all the nsSNPs using SIFT, PredictSNP, PhD-SNP, PolyPhen-1, PolyPhen-2, MAPP, and SNAP to understand the functional impact a mutation may pose on the protein. Of the 50 nsSNPs, MAPP and PredictSNP predicted highest number of nsSNPs to be deleterious with 39 and 38 nsSNPs, respectively. PolyPhen-1 and PolyPhen-2 identified 20 and 19 nsSNPs to affect the functionality of the protein. Whereas SIFT, PhD-SNP, and SNAP predicted 20, 14, and 18 nsSNPs to cause a deleterious effect, respectively (Table 1). Furthermore, a ranking of the nsSNPs was performed to understand the severity of each nsSNP and classify the most deleterious one in a pool. Ranking assigned ranges between Rank1 to Rank5 (Table 1) where nsSNP predicted to be deleterious by all tools was categorized as Rank1-nsSNP and with none of the tool predicting deleterious categorized as Rank5. The obtained results from the pathogenic tools predicted 16 nsSNPs as RANK1. The stability of the protein directly affects the function of the protein. In this context, we used stability prediction tools to determine the change in the Gibbs free energy and understand the stability change upon mutation (Table 2). Obtained results from the in silico tools, predicted only two nsSNPs (H153N and A241T) to be highly pathogenic and affecting the stability of the protein. INPS server was used to compare the increase in the overall Gibbs free energy between the native and mutant proteins. In Fig. 2, the red line indicates the average change in the Gibbs free energy upon mutation, the blue line indicates a standard deviation, and the dotted line indicates the mean (or actual) energy of the protein in its native state (standard line). In Fig. 2A, H153N mutation does not show a drastic change from the standard line, whereas A241T (Fig. 2B) shows an overall deviation from the standard line.

Table 1 Pathogenic and Ranking Results for the nsSNPs Obtained From dbSNP and HGMD databases

IDs	nsSNPs	PredictSNP	MAPP	PhD-SNP	PolyPhen-1	PolyPhen-2	SIFT	SNAP	Ranks
CM104563	M1I	Deleterious	Deleterious	Neutral	Deleterious	Deleterious	Deleterious	Deleterious	Rank2
rs139107479	V25L	Neutral	Neutral	Neutral	Neutral	Neutral	Neutral	Neutral	Rank5
CM020188	S36F	Neutral	Neutral	Neutral	Deleterious	Neutral	Deleterious	Neutral	Rank4
rs193922483	P47L	Deleterious	Neutral	Neutral	Deleterious	Deleterious	Deleterious	Deleterious	Rank2
CM060492	V61G	Neutral	Neutral	Neutral	Neutral	Neutral	Neutral	Neutral	Rank5
rs144425830/CM055551	G76C	Deleterious	Deleterious	Neutral	Deleterious	Deleterious	Deleterious	Neutral	Rank2
rs140562402	D82N	Deleterious	Neutral	Neutral	Deleterious	Deleterious	Deleterious	Neutral	Rank3
rs121918671	E101Q	Deleterious	Deleterious	Neutral	Deleterious	Deleterious	Deleterious	Neutral	Rank2
CM060493	V110G	Deleterious	Deleterious	Deleterious	Deleterious	Deleterious	Deleterious	Deleterious	Rank1
CM042485	R112P	Deleterious	Deleterious	Neutral	Deleterious	Deleterious	Deleterious	Deleterious	Rank2
rs193922487	S115N	Neutral	Deleterious	Neutral	Neutral	Neutral	Neutral	Neutral	Rank4
CM042486	Q136E	Deleterious	Deleterious	Deleterious	Deleterious	Deleterious	Deleterious	Deleterious	Rank1
rs121918674/CM041863	S148W	Deleterious	Deleterious	Deleterious	Deleterious	Deleterious	Deleterious	Deleterious	Rank1
CM060494	S148l	Deleterious	Deleterious	Deleterious	Deleterious	Deleterious	Deleterious	Deleterious	Rank1
CM022465	S151P	Deleterious	Deleterious	Deleterious	Deleterious	Deleterious	Deleterious	Deleterious	Rank1
CM104565	S151C	Deleterious	Deleterious	Neutral	Deleterious	Deleterious	Deleterious	Deleterious	Rank2
CM040805	H153N	Deleterious	Deleterious	Deleterious	Deleterious	Deleterious	Deleterious	Deleterious	Rank1
CM060487	L156E	Deleterious	Deleterious	Deleterious	Deleterious	Deleterious	Deleterious	Deleterious	Rank1

CM104566	P159L	Deleterious	Deleterious	Deleterious	Deleterious	Deleterious	Deleterious	Deleterious	Rank1
CM042488	L164Q	Deleterious	Deleterious	Deleterious	Deleterious	Deleterious	Deleterious	Deleterious	Rank1
rs121918675/CM042489	R165H	Deleterious	Deleterious	Deleterious	Deleterious	Deleterious	Deleterious	Deleterious	Rank1
CM104567	R165P	Deleterious	Deleterious	Deleterious	Deleterious	Deleterious	Deleterious	Deleterious	Rank1
rs193922490	W171R	Deleterious	Deleterious	Deleterious	Deleterious	Deleterious	Deleterious	Deleterious	Rank1
CM119363	D221V	Neutral	Neutral	Neutral	Deleterious	Neutral	Deleterious	Neutral	Rank4
rs193922491	R235W	Deleterious	Deleterious	Neutral	Deleterious	Deleterious	Deleterious	Deleterious	Rank2
CM055550	R235Q	Deleterious	Deleterious	Deleterious	Deleterious	Deleterious	Deleterious	Deleterious	Rank1
CM118474	G239A	Neutral	Deleterious	Neutral	Neutral	Deleterious	Deleterious	Neutral	Rank3
rs761415487	A241T	Deleterious	Deleterious	Deleterious	Deleterious	Deleterious	Deleterious	Deleterious	Rank1
CM088210	Q243P	Deleterious	Deleterious	Deleterious	Deleterious	Deleterious	Deleterious	Deleterious	Rank1
CM076231	Q253P	Deleterious	Deleterious	Deleterious	Deleterious	Deleterious	Deleterious	Deleterious	Rank1
CM104569	P256S	Deleterious	Deleterious	Deleterious	Deleterious	Deleterious	Deleterious	Deleterious	Rank1
CM034453	E260D	Deleterious	Deleterious	Neutral	Deleterious	Deleterious	Deleterious	Deleterious	Rank2
CM067048	C273Y	Deleterious	Deleterious	Deleterious	Deleterious	Deleterious	Deleterious	Deleterious	Rank1
CM060488	R276Q	Deleterious	Deleterious	Deleterious	Neutral	Deleterious	Deleterious	Neutral	Rank2
CM055552	R276G	Deleterious	Deleterious	Deleterious	Deleterious	Deleterious	Deleterious	Deleterious	Rank1

Continued

Table 1 Pathogenic and Ranking Results for the nsSNPs Obtained From dbSNP and HGMD databases—cont'd

IDs	nsSNPs	PredictSNP	MAPP	PhD-SNP	PolyPhen-1	PolyPhen-2	SIFT	SNAP	Ranks
CM055553	G285D	Deleterious	Deleterious	Deleterious	Deleterious	Deleterious	Deleterious	Deleterious	Rank1
CM117493	L286V	Deleterious	Deleterious	Neutral	Deleterious	Deleterious	Deleterious	Deleterious	Rank2
CM117494	N289D	Deleterious	Deleterious	Deleterious	Neutral	Neutral	Neutral	Deleterious	Rank3
CM042492	R295H	Deleterious	Deleterious	Deleterious	Deleterious	Deleterious	Deleterious	Deleterious	Rank1
CM060489	R295P	Deleterious	Deleterious	Deleterious	Deleterious	Deleterious	Deleterious	Deleterious	Rank1
CM055549	R295C	Deleterious	Deleterious	Deleterious	Deleterious	Deleterious	Deleterious	Deleterious	Rank1
CM104570	W299G	Deleterious	Deleterious	Deleterious	Deleterious	Deleterious	Deleterious	Deleterious	Rank1
CM104910	N302L	Deleterious	Deleterious	Deleterious	Neutral	Deleterious	Deleterious	Deleterious	Rank2
rs193922492	A317S	Neutral	Neutral	Neutral	Neutral	Deleterious	Neutral	Neutral	Rank4
rs193922493	N321S	Neutral	Neutral	Neutral	Neutral	Neutral	Neutral	Neutral	Rank5
rs794727624	N325S	Neutral	Neutral	Neutral	Neutral	Neutral	Neutral	Neutral	Rank5
CM067046	H336D	Neutral	Deleterious	Neutral	Neutral	Neutral	Neutral	Deleterious	Rank4
CM055554	G370S	Neutral	Neutral	Neutral	Neutral	Neutral	Neutral	Neutral	Rank5
CM001351	G492S	Deleterious	Deleterious	Deleterious	Deleterious	Deleterious	Deleterious	Deleterious	Rank1

Ranking Scheme: Rank1-nsSNPs predicted deleterious by all the tools; Rank2-nsSNPs predicted neutral by 1 or 2 tools; Rank3-nsSNPs predicted neutral by 3 or 4 tools; Rank4-nsSNPs predicted neutral by 5 or 6 tools; Rank5—all the tools predicted as neutral.

Table 2 Analyzing the Stabilization Effect Using In Silico Tools

nsSNPs	CUPSAT	MUpro	I-Mutant2.0	iStable	SNP4.0effect
V110G	Decrease	Increase	Increase	Decrease	Increase
Q136E	Increase	Decrease	Increase	Decrease	Increase
S148W	Increase	Increase	Decrease	Increase	Increase
S148l	Increase	Decrease	Increase	Decrease	Increase
S151P	Decrease	Increase	Decrease	Increase	Increase
H153N	Decrease	Decrease	Decrease	Decrease	Decrease
L156E	Increase	Increase	Decrease	Increase	Increase
P159L	Increase	Increase	Decrease	Decrease	Decrease
L164Q	Increase	Decrease	Decrease	Increase	Increase
R165H	Increase	Increase	Decrease	Decrease	Increase
R165P	Increase	Decrease	Decrease	Increase	Decrease
W171R	Decrease	Increase	Increase	Increase	Increase
R235Q	Increase	Decrease	Decrease	Increase	Decrease
A241T	Decrease	Increase	Decrease	Decrease	Decrease
Q243P	Increase	Increase	Increase	Decrease	Increase
Q253P	Increase	Decrease	Decrease	Increase	Increase
P256S	Increase	Decrease	Decrease	Increase	Decrease
C273Y	Decrease	Decrease	Increase	Increase	Increase
R276G	Increase	Decrease	Decrease	Increase	Decrease
G285D	Decrease	Increase	Decrease	Decrease	Increase
R295H	Increase	Decrease	Increase	Increase	Increase
R295P	Decrease	Increase	Decrease	Increase	Decrease
R295C	Increase	Decrease	Decrease	Decrease	Increase
W299G	Decrease	Increase	Decrease	Increase	Increase
G492S	Decrease	Increase	Decrease	Decrease	Increase

3.2 Mutation Mapping and Molecule Preparation

The mutants were mapped on the protein sequence using ELASPIC server. The server marked H153N on the N-terminal end of the protein, while

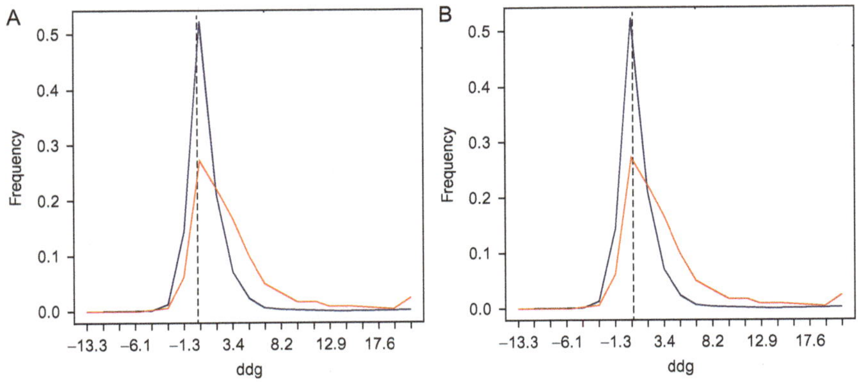

Fig. 2 INPS tool results of mutants (A) H153N and (B) A241T. The *red line* indicates the average change in the Gibbs free energy due to the mutation, the *blue line* indicates a standard deviation, and the *dotted line* indicates the mean (or actual) energy of the protein in its native state (standard line).

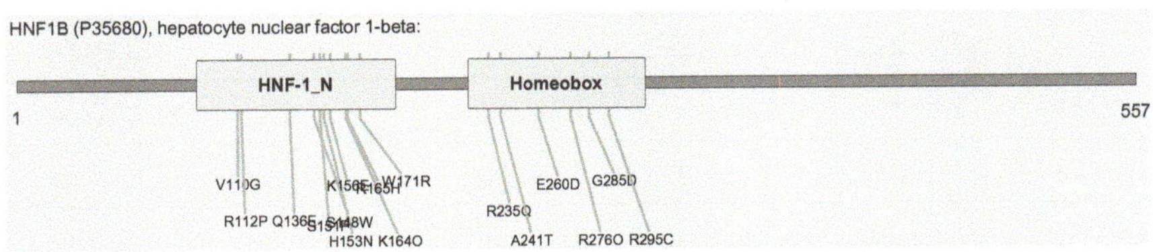

Fig. 3 A few known mutations marked on the different domains of HNF1B protein by ELASPIC tool.

A241T on the Homeobox of the protein (Fig. 3) and mutation was induced in the 3D protein structure. Since the binding of the protein to DNA is a crucial process of a transcription factor, we have mutated the protein and then performed docking analysis. Docked complexes showed the number of interactions observed between the protein (native) and the DNA in comparison to the mutant complexes. Less binding affinity efficiency pattern was seen in A241T mutant complex when compared to native and other mutant complexes (Table 3). Furthermore, energy minimization was carried out for all the three complexes and the native complex was observed with least energy.

3.3 Molecular Dynamics Analysis

The molecular dynamics analysis predicts the structural insights at the atomistic level were used to analyze the possible biochemical change that could

Table 3 Docking Results of DNA–Protein (Native and Mutants) Using HADDOCK

Protein Complex	HADDOCK Binding Score	Van der Waals Energy (kcal/mol)	Electrostatic Energy (kcal/mol)	Desolvation Energy (kcal/mol)
Native	−279.34	−154.1	−657	3.4
A241T	−256.45	−150.6	−654	−1.22
H153N	−263.56	−152.1	−654	−1.32

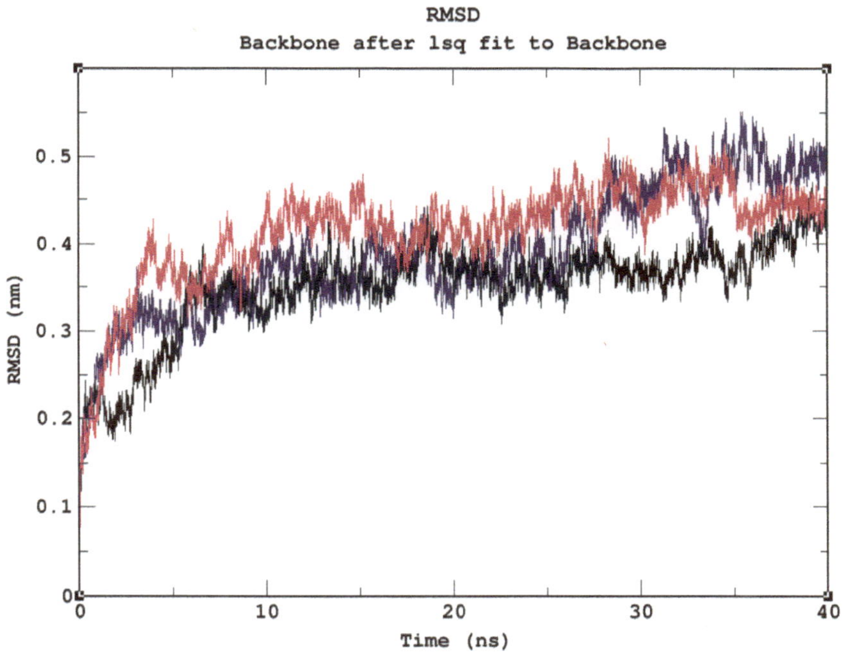

Fig. 4 Root-mean-square deviation of the protein–DNA complexes; color scheme: black, native; magenta, H153N; and blue, A241T.

have taken place upon mutation in the complex. In the beginning of the simulation (RMSD; Fig. 4), huge deviations were observed in all the complexes, which might be due to the observed kinetic shock. At 40 ns, we found that all the three complexes attained convergence. With this, the simulation was terminated at the end of 40 ns. Hydrogen bond analysis, suggested that, there was lesser number of average hydrogen bonds formed in A241T mutant (between DNA and the protein) in comparison to other molecules (Fig. 5). Additionally, radius of gyration (Rg) was estimated for all the three complex molecules to understand the proximity of the protein toward DNA (Fig. 6). We observed higher Rg values in the mutant

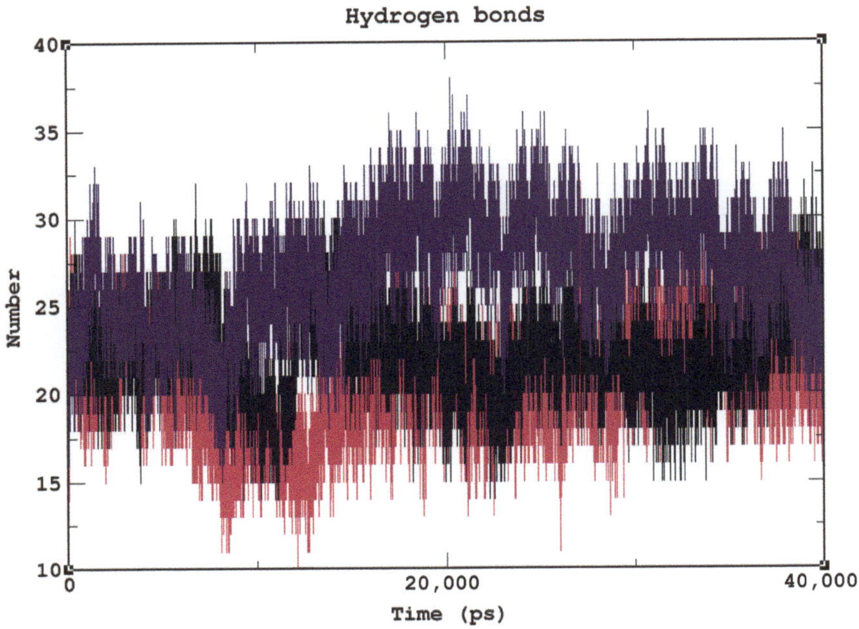

Fig. 5 Number of hydrogen bonds formed in protein–DNA complexes throughout 40 ns simulation; color scheme: black, native; magenta, H153N; and blue, A241T.

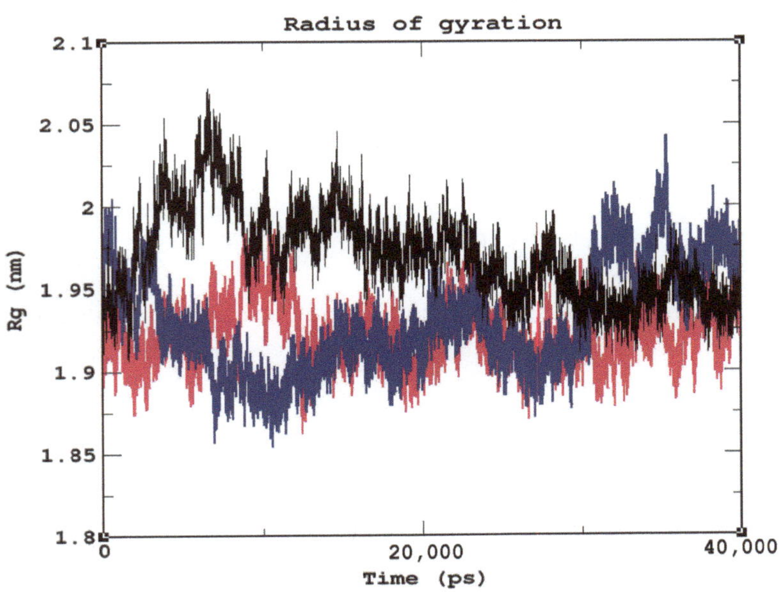

Fig. 6 Radius of gyration of the protein–DNA complexes; color scheme: black, native; magenta, H153N; and blue, A241T.

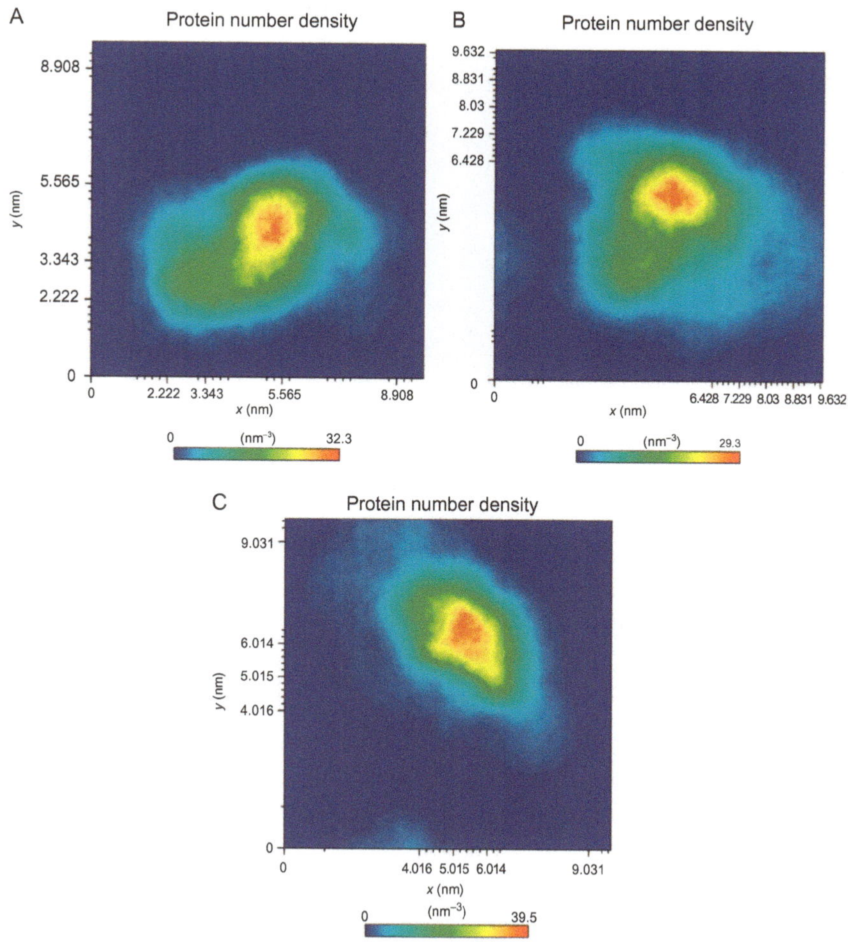

Fig. 7 Protein density mapped for the protein–DNA complexes. (A) Native, (B) H153N, (C) A241T.

complexes A241T and H153N relates to loss of closeness with the DNA molecule. The partial density of the protein was calculated using the densmap analysis available in the Gromacs package (Fig. 7). We observed increase in partial density area in the native complex whereas very least density area was seen in A241T mutant complex (Fig. 7A–C).

As a next step, essential dynamics was carried out to understand the movement changes that had occurred during the 40 ns simulations. In PCA, primarily a covariance matrix is constructed using the eigenvectors and eigenvalues. The eigenvectors are axes along which a linear transformation is explained, thereby a covariance matrix helps in analyzing the PCA

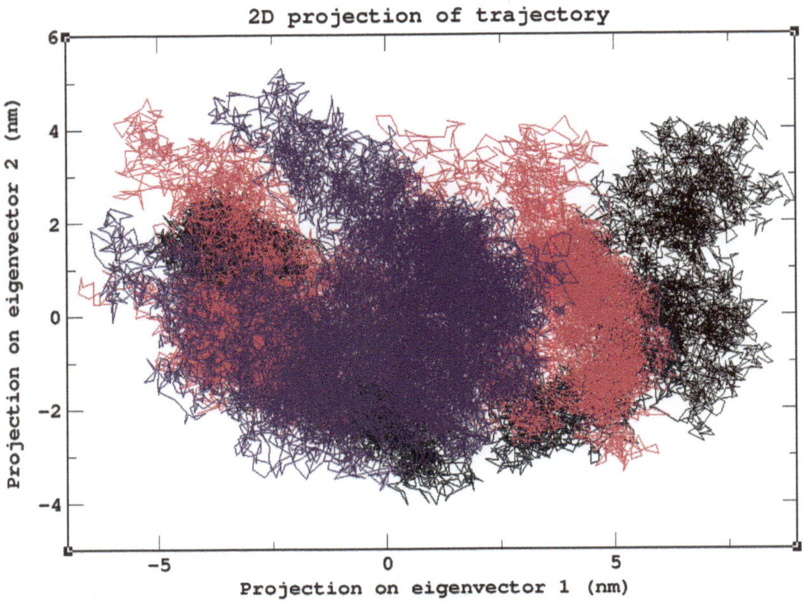

Fig. 8 Principle component analysis of the protein–DNA complexes; color scheme: black, native; magenta, H153N; and blue, A241T.

which further depicts the motional changes in the protein. With the help of PCA, we observed that the native complex covered a larger space of the box in the 40 ns simulation in comparison to the other two mutant complexes (Fig. 8). Additionally, a comparison between the three complexes was computed using the Ramachandran plot analysis. The psi and phi changes in the amino acid of the mutant complex were analyzed to suggest any change at their structural level. We observed similar amino acid orientation in the A241T and H153N mutant complexes when compared to native complex with proper amino acid distribution (Fig. 9) which depicts the effect of mutation with global effect on the protein structure.

4. DISCUSSION

The recent insights into the control of cellular gene expression programs have provided a route understanding of misregulation of gene expression in disease. Many different diseases and syndromes, including cancer, autoimmunity, neurological disorders, diabetes, cardiovascular disease, and obesity, were caused by mutations in regulatory sequences and the transcription factors, cofactors, chromatin regulators, and noncoding RNAs can

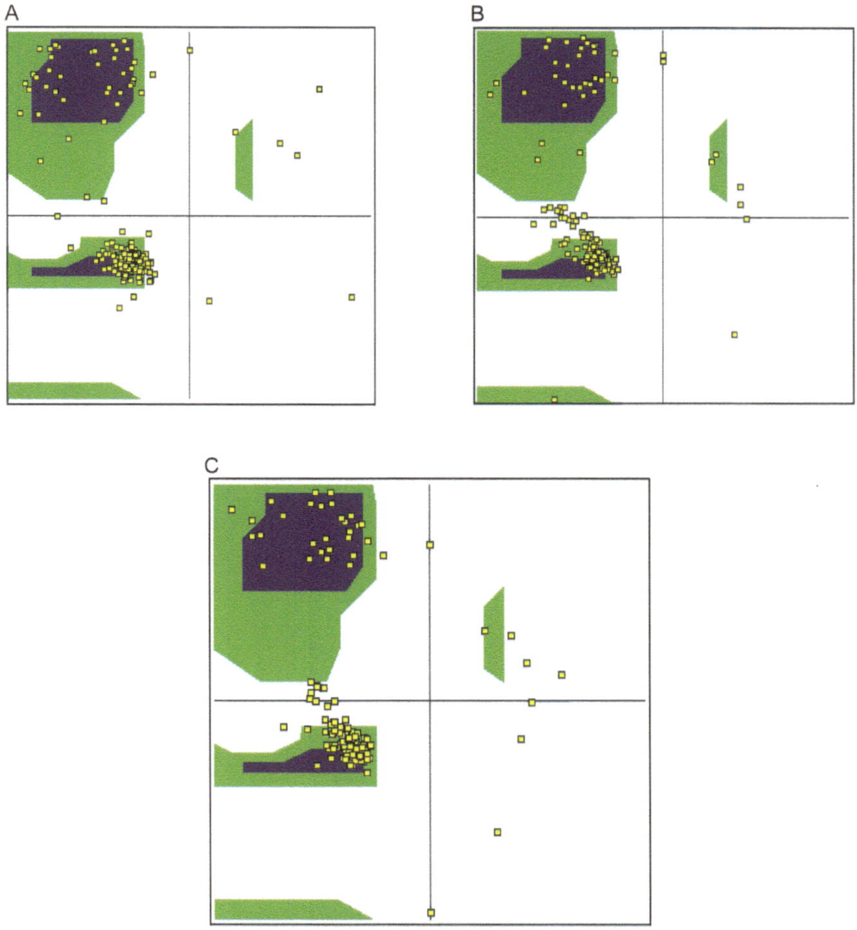

Fig. 9 Ramachandran plot analysis of the protein–DNA complexes at the end of 40 ns simulation. (A) Native, (B) H153N, (C) A241T.

interact in these regions (Lee et al., 2013). nsSNP mutation in transcription factors plays a major role in misregulation of gene expression thereby causing a disease (Lee et al., 2013). A major challenge in current era of human genetics craves to understand the association between the mutants and their corresponding disease expression. Transcription factor HNF1B was proposed to form an integrated regulatory network in b-cells is involved in glucose metabolism and insulin secretion (Barbacci et al., 2004) and mutations are involved in causing MODY5.

MODY is a genetically heterogeneous form of diabetes mellitus with an early onset, abnormal insulin secretion, and autosomal dominant inheritance

(Barbacci et al., 2004). Based on the genes involved, the MODY are classified into different types; unlike other MODY types, patients with MODY5 failed to respond to oral medications such as sulfonylureas (Kim et al., 2014). However, the accurate molecular mechanism behind HNF1B mutations causing a broad spectrum of clinical symptoms remains largely unknown (Barbacci et al., 2004). This encouraged us to investigate the likely changes that could have occurred in the hindering binding of the protein to DNA upon mutation. Analyzing the missense mutations and predicting the possible biochemical changes through experimental analysis may be tedious for which the use of computational tools may be a great alternative. In silico tools in the past decade have developed hugely and evolved to become an efficient strategy in understanding very trivial information at structural level in protein–DNA complex (Doss & NagaSundaram, 2012; Mukherjee, Pandey, & Vidyarthi, 2015; Nasir et al., 2013; Pires, Chen, Blundell, & Ascher, 2015; Sujitha et al., 2016). Thus, in the current study, an array of in silico tools was selected to understand the impact of the mutations on the protein's function or stability. The pathogenic tools predicted 16 nsSNPs to have an effect on the protein function, and stability tools predicted 2 nsSNPs (H153N and A241T) to have a destabilizing effect on the protein. With further investigation using ELASPIC tool, we observed A241T in the Homeobox domain and H153N in the N-terminal domain of the POU domain. The POU domain consists of two distinct domain: POU_H (C-terminal Homeobox) domain and POU_S (N-terminal) specific region (Fig. 3) which is enough for DNA-specific binding of the transcription factor (Sturm & Herr, 1988).

The results obtained from the SNP4.0 effect and INPS suggested that there is structural instability upon mutation. H153N mutation exhibited an energy value of −0.488 kcal/mol, with least destability, whereas A241T exhibited −0.53 kcal/mol, using INPS prediction server. In this current study, an in-depth analysis of the possible mutations in the POU domains was investigated to understand their structural impact. Most of the processes were not well understood at an atomic level; therefore, an adequate description of their MD analysis was required to derive functional impact from structure level (Sneha & George Priya Doss, 2016). Stability changes were observed in the native and the mutant complexes, where A241T and H153N showed higher RMSD values (Fig. 4). A higher RMSD value states a decrease in the stability (Yun & Guy, 2011). Furthermore, the number of the hydrogen bonds formed between the protein and DNA complex was observed using hydrogen bond analysis. Noticeably in Fig. 5

reduction in hydrogen bond formation was observed in A241T mutant complex. The above-observed results provoked us to identify the physiochemical property of native and mutant amino acids in causing stability difference upon mutation. Alanine being a hydrophobic amino acid tends to be in the buried region, wherein HNF1B, it is found to be located in the exposed area. The presence of a hydrophobic amino acid in the exposed region is known to have a stabilizing effect on the protein structure (Strub et al., 2004). A substitution with polar amino acid might invite contact with the surrounding amino acids which may be a cause in affecting the overall protein's stability. In both the mutants, there is an increase in the number of surrounding amino acids with an increase in contact with polar amino acids (Fig. 10). Furthermore, PCA was utilized to explain the

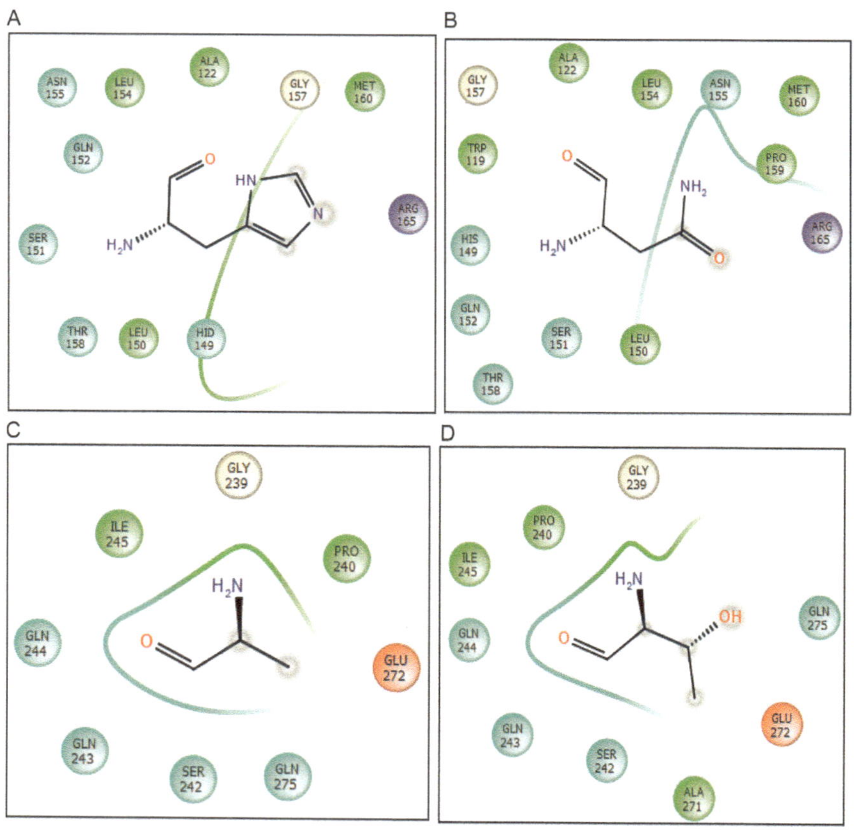

Fig. 10 Variation in surrounding amino acid upon mutation in protein–DNA complexes visualized using Maestro software. (A) Native H153, (B) Mutant H153N, (C) Native A241, (D) Mutant A241T.

direction of movement by constructing a covariance matrix. Reduction in motional change was observed in the case of A241T, whereas native has covered a larger space in the box. When comparing the observed results from partial density with PCA, we predict mutant complex (A241T) was more confined to a position in the box, while the native tend to occupy considerably higher space (Figs. 7 and 8). These conformational changes may be a driving force to hinder the protein–DNA-binding mechanism. The observed findings by computational analysis conclude mutation A241T to be destabilizing the protein was in concurrence with a previously reported study by experimental analysis (Chi, 2005). This illustrates the usefulness of integrating in silico prediction methods and MD analysis in protein–DNA research.

5. CONCLUSION

A combinational use of various computational methods may not only provide better insights on the different biochemical changes caused due to mutation and might also reduce time and cost by experimental methods. Trivial changes in the protein caused upon mutation can drastically modify the protein's function. When these variations occur in the transcription factors, the specificity of the DNA binding is lost leading to disease. Analyzing this variation at structural level can provide better insight into understanding the mechanistic action behind the loss of protein function. This present study is a step toward understanding the genotype–phenotype relationships in HNF1B transcription factor caused by the missense mutations using a computational approach.

ACKNOWLEDGMENT

The authors take this opportunity to thank the management of VIT University for providing the facilities and encouragement to carry out this work.

Conflict of Interest. We declare no competing interests.

REFERENCES

Adalat, S., Woolf, A. S., Johnstone, K. A., Wirsing, A., Harries, L. W., Long, D. A., … Bockenhauer, D. (2009). HNF1B mutations associate with hypomagnesemia and renal magnesium wasting. *Journal of the American Society of Nephrology*, *20*(5), 1123–1131.

Adelman, K., & Lis, J. T. (2012). Promoter-proximal pausing of RNA polymerase II: Emerging roles in metazoans. *Nature Reviews. Genetics*, *13*(10), 720–731.

Adzhubei, I., Jordan, D. M., & Sunyaev, S. R. (2013). Predicting functional effect of human missense mutations using PolyPhen-2. *Current Protocols in Human Genetics*. Chapter 7, Unit 7.20.

Alvelos, M. I., Rodrigues, M., Lobo, L., Medeira, A., Sousa, A. B., Simão, C., & Lemos, M. C. (2015). A novel mutation of the HNF1B gene associated with hypoplastic glomerulocystic kidney disease and neonatal renal failure. *Medicine, 94*(7), e469.

Baets, D. G., Van Durme, J., Reumers, J., Maurer-Stroh, S., Vanhee, P., Dopazo, J., … Rousseau, F. (2012). SNPeffect 4.0: On-line prediction of molecular and structural effects of protein-coding variants. *Nucleic Acids Research, 40*(Database issue), D935–D939.

Barbacci, E., Chalkiadaki, A., Masdeu, C., Haumaitre, C., Lokmane, L., Loirat, C., … Cereghini, S. (2004). HNF1b/TCF2 mutations impair transactivation potential through altered co-regulator recruitment. *Human Molecular Genetics, 13*(24), 3139–3149.

Barbacci, E., Reber, M., Ott, M. O., Breillat, C., Huetz, F., & Cereghini, S. (1999). Variant hepatocyte nuclear factor 1 is required for visceral endoderm specification. *Development (Cambridge, England), 126*(21), 4795–4805.

Bendl, J., Stourac, J., Salanda, O., Pavelka, A., Wieben, E. D., Zendulka, J., … Damborsky, J. (2014). PredictSNP: Robust and accurate consensus classifier for prediction of disease-related mutations. *PLoS Computational Biology, 10*(1), e1003440.

Berendsen, H. J. C., Postma, J. P. M., van Gunsteren, W. F., DiNola, A., & Haak, J. R. (1984). Molecular dynamics with coupling to an external bath. *The Journal of Chemical Physics, 81*, 3684.

Bonasio, R., Tu, S., & Reinberg, D. (2010). Molecular signals of epigenetic states. *Science (New York, N.Y.), 330*(6004), 612–616.

Capriotti, E., Calabrese, R., & Casadio, R. (2006). Predicting the insurgence of human genetic diseases associated to single point protein mutations with support vector machines and evolutionary information. *Bioinformatics (Oxford, England), 22*(22), 2729–2734.

Capriotti, E., Fariselli, P., & Casadio, R. (2005). I-Mutant2.0: Predicting stability changes upon mutation from the protein sequence or structure. *Nucleic Acids Research, 33*(Web Server issue), W306–W310.

Cheatham, T. E. I., Miller, J. L., Fox, T., Darden, T. A., & Kollman, P. A. (2002). Molecular dynamics simulations on solvated biomolecular systems: The particle mesh Ewald method leads to stable trajectories of DNA, RNA, and proteins. *The Journal of Chemical Physics, 81*, 3684.

Chen, C.-W., Lin, J., & Chu, Y.-W. (2013). iStable: Off-the-shelf predictor integration for predicting protein stability changes. *BMC Bioinformatics, 14*(Suppl. 2), S5.

Cheng, J., Randall, A., & Baldi, P. (2006). Prediction of protein stability changes for single-site mutations using support vector machines. *Proteins, 62*(4), 1125–1132.

Chi, Y. I. (2005). Homeodomain revisited: A lesson from disease-causing mutations. *Human Genetics, 116*(6), 433–444.

Conaway, R. C., & Conaway, J. W. (2011). Function and regulation of the mediator complex. *Current Opinion in Genetics & Development, 21*(2), 225–230.

Cuff, J., Salari, K., Clarke, N., Esheba, G. E., Forster, A. D., Huang, S., … Pollack, J. R. (2013). Integrative bioinformatics links HNF1B with clear cell carcinoma and tumor-associated thrombosis. *PLoS One, 8*(9), e74562.

Doss, C. G. P., & NagaSundaram, N. (2012). Investigating the structural impacts of I64T and P311S mutations in APE1-DNA complex: A molecular dynamics approach. *PLoS One, 7*, 1–11.

Fuda, N. J., Ardehali, M. B., & Lis, J. T. (2009). Defining mechanisms that regulate RNA polymerase II transcription in vivo. *Nature, 461*(7261), 186–192.

George Priya Doss, C., Chakraborty, C., Narayan, V., & Thirumal Kumar, D. (2014). Computational approaches and resources in single amino acid substitutions analysis toward clinical research. *Advances in Protein Chemistry and Structural Biology, 94*, 365–423.

George, D. C. P., Chakraborty, C., Haneef, S. A. S., Nagasundaram, N., Chen, L., & Zhu, H. (2014). Evolution- and structure-based computational strategy reveals the impact of deleterious missense mutations on MODY 2 (maturity-onset diabetes of the young, type 2). *Theranostics*, *4*(4), 366–385.

Gnad, F., Baucom, A., Mukhyala, K., Manning, G., & Zhang, Z. (2013). Assessment of computational methods for predicting the effects of missense mutations in human cancers. *BMC Genomics*, *14*(Suppl. 3), S7.

Gray, V. E., Kukurba, K. R., & Kumar, S. (2012). Performance of computational tools in evaluating the functional impact of laboratory-induced amino acid mutations. *Bioinformatics (Oxford England)*, *28*(16), 2093–2096.

Gresh, L., Fischer, E., Reimann, A., Tanguy, M., Garbay, S., Shao, X., … Pontoglio, M. (2004). A transcriptional network in polycystic kidney disease. *The EMBO Journal*, *23*(7), 1657–1668.

Guex, N., & Peitsch, M. C. (1997). SWISS-MODEL and the Swiss-PdbViewer: An environment for comparative protein modeling. *Electrophoresis*, *18*(15), 2714–2723.

Horikawa, Y., Iwasaki, N., Hara, M., Furuta, H., Hinokio, Y., Cockburn, B. N., … Bell, G. I. (1997). Mutation in hepatocyte nuclear factor–1β gene (TCF2) associated with MODY. *Nature Genetics*, *17*(4), 384–385.

Huang, J., & MacKerell, A. D. (2013). CHARMM36 all-atom additive protein force field: Validation based on comparison to NMR data. *Journal of Computational Chemistry*, *34*(25), 2135–2145.

Humphrey, W., Dalke, A., & Schulten, K. (1996). VMD: Visual molecular dynamics. *Journal of Molecular Graphics*, *14*, 33–38. 7855(October 1995).

Jacob, F., & Monod, J. (1961). Genetic regulatory mechanisms in the synthesis of proteins. *Journal of Molecular Biology*, *3*, 1318–1356.

Johnson, A. D., Handsaker, R. E., Pulit, S. L., Nizzari, M. M., O'Donnell, C. J., & de Bakker, P. I. W. (2008). SNAP: A web-based tool for identification and annotation of proxy SNPs using HapMap. *Bioinformatics (Oxford, England)*, *24*(24), 2938–2939.

Kim, E. K., Lee, J. S., Cheong, H. Il, Chung, S. S., Kwak, S. H., & Park, K. S. (2014). Identification and functional characterization of P159L mutation in HNF1B in a family with maturity-onset diabetes of the young 5 (MODY5). *Genomics & Informatics*, *12*(4), 240–246.

Kozak, M. (1992). Regulation of translation in eukaryotic systems. *Annual Review of Cell Biology*, *8*(1), 197–225.

Lee, T. I., Young, R. A., Abramson, J., Giraud, M., Benoist, C., Mathis, D., & Peterlin, B. M. (2013). Transcriptional regulation and its misregulation in disease. *Cell*, *152*(6), 1237–1251.

Liu, M., Watson, L. T., & Zhang, L. (2015). HMMvar-func: A new method for predicting the functional outcome of genetic variants. *BMC Bioinformatics*, *16*, 351.

Lu, P., Rha, G. B., & Chi, Y.-I. (2007). Structural basis of disease-causing mutations in hepatocyte nuclear factor 1beta. *Biochemistry*, *46*(43), 12071–12080.

McAdams, H. H., & Arkin, A. (1997). Stochastic mechanisms in gene expression. *Proceedings of the National Academy of Sciences of the United States of America*, *94*(3), 814–819.

Mort, M., Sterne-Weiler, T., Li, B., Ball, E. V., Cooper, D. N., Radivojac, P., … Mooney, S. D. (2014). MutPred splice: Machine learning-based prediction of exonic variants that disrupt splicing. *Genome Biology*, *15*(1), R19.

Mukherjee, K., Pandey, D. M., & Vidyarthi, A. S. (2015). In silico characterization and analysis of RTBP1 and NgTRF1 protein through MD simulation and molecular docking—A comparative study. *Interdisciplinary Sciences: Computational Life Sciences*, 7, 1–13.

Nagasundaram, N., Zhu, H., Liu, J., Karthick, V., George Priya Doss, C., Chakraborty, C., & Chen, L. (2015). Analysing the effect of mutation on protein function and discovering potential inhibitors of CDK4: Molecular modelling and dynamics studies. *PLoS One*, *10*(8), e0133969.

Nasir, M., Ahmad, N., Sieber, C. M., Latif, A., Malik, S. A., & Hameed, A. (2013). In silico characterization of a novel pathogenic deletion mutation identified in XPA gene in a Pakistani family with severe xeroderma pigmentosum. *Journal of Biomedical Science*, *20*, 70.
Ng, P. C., & Henikoff, S. (2003). SIFT: Predicting amino acid changes that affect protein function. *Nucleic Acids Research*, *31*(13), 3812–3814.
Papavassiliou, K. A., & Papavassiliou, A. G. (2016). Transcription factor drug targets. *Journal of Cellular Biochemistry*, *117*, 2693–2696.
Parthiban, V., Gromiha, M. M., & Schomburg, D. (2006). CUPSAT: Prediction of protein stability upon point mutations. *Nucleic Acids Research*, *34*(Web Server issue), W239–W242.
Pires, D. E. V., Chen, J., Blundell, T. M., & Ascher, B. D. (2015). In silico functional dissection of saturation mutagenesis: Interpreting the relationship between phenotypes and changes in protein stability, interactions and activity. *Scientific Reports*, *6*, 19848.
Pronk, S., Páll, S., Schulz, R., Larsson, P., Bjelkmar, P., Apostolov, R., … Lindahl, E. (2013). GROMACS 4.5: A high-throughput and highly parallel open source molecular simulation toolkit. *Bioinformatics (Oxford, England)*, *29*(7), 845–854.
Raile, K., Klopocki, E., Wessel, T., Deiss, D., Horn, D., Müller, D., … Grüters, A. (2008). HNF1B abnormality (mature-onset diabetes of the young 5) in children and adolescents: High prevalence in autoantibody-negative type 1 diabetes with kidney defects. *Diabetes Care*, *31*(11), e83.
Ramensky, V., Bork, P., & Sunyaev, S. (2002). Human non-synonymous SNPs: Server and survey. *Nucleic Acids Research*, *30*(17), 3894–3900.
Reber, M., & Cereghini, S. (2001). Variant hepatocyte nuclear factor 1 expression in the mouse genital tract. *Mechanisms of Development*, *100*(1), 75–78.
Roeder, R. G. (2005). Transcriptional regulation and the role of diverse coactivators in animal cells. *FEBS Letters*, *579*(4), 909–915.
Savojardo, C., Fariselli, P., Martelli, P. L., & Casadio, R. (2016). INPS-MD: A web server to predict stability of protein variants from sequence and structure: Table 1. *Bioinformatics*, *32*(16), 2542–2544.
Sherry, S. T., Ward, M. H., Kholodov, M., Baker, J., Phan, L., Smigielski, E. M., & Sirotkin, K. (2001). dbSNP: The NCBI database of genetic variation. *Nucleic Acids Research*, *29*(1), 308–311.
Sneha, P., & George Priya Doss, C. (2016). Chapter seven—Molecular dynamics: New frontier in personalized medicine. *Advances in Protein Chemistry and Structural Biology*, *102*, 181–224.
Stenson, P. D., Ball, E. V., Mort, M., Phillips, A. D., Shiel, J. A., Thomas, N. S. T., … Cooper, D. N. (2003). Human gene mutation database (HGMD): 2003 update. *Human Mutation*, *21*(6), 577–581.
Stone, E. A., & Sidow, A. (2005). Physicochemical constraint violation by missense substitutions mediates impairment of protein function and disease severity. *Genome Research*, *15*(7), 978–986.
Strub, C., Alies, C., Lougarre, A., Ladurantie, C., Czaplicki, J., & Fournier, D. (2004). Mutation of exposed hydrophobic amino acids to arginine to increase protein stability. *BMC Biochemistry*, *5*(1), 9.
Sturm, R. A., & Herr, W. (1988). The POU domain is a bipartite DNA-binding structure. *Nature*, *336*(6199), 601–604.
Sujitha, S. P., Thirumal Kumar, D., George Priya Doss, C., Aavula, K., Ramesh, R., Lakshmanan, S., … Anilkumar, G. (2016). DNA repair gene (XRCC1) polymorphism (Arg399Gln) associated with schizophrenia in south Indian population: A genotypic and molecular dynamics study. *PLoS One*, *11*(1), e0147348.
Taatjes, D. J. (2010). The human mediator complex: A versatile, genome-wide regulator of transcription. *Trends in Biochemical Sciences*, *35*(6), 315–322.

Thirumal Kumar D., George Priya Doss C., Sneha P., Tayubi I.A., Siva R., Chakraborty C. and Magesh R., Influence of V54M mutation in giant muscle protein titin: A computational screening and molecular dynamics approach, *Journal of Biomolecular Structure & Dynamics*, 1–12, in press.

Thomas, R., Sanna-Cherchi, S., Warady, B. A., Furth, S. L., Kaskel, F. J., & Gharavi, A. G. (2011). HNF1B and PAX2 mutations are a common cause of renal hypodysplasia in the CKiD cohort. *Pediatric Nephrology*, *26*(6), 897–903.

Turner, P. J. (2005). *XMGRACE (version 5.1.19)*. Beaverton: Center for Coastal and Land-Margin Research, Oregon Graduate Institute of Science and Technology.

van Zundert, G. C. P., Rodrigues, J. P. G. L. M., Trellet, M., Schmitz, C., Kastritis, P. L., Karaca, E., … Bonvin, A. M. J. J. (2016). The HADDOCK2.2 Web server: User-friendly integrative modeling of biomolecular complexes. *Journal of Molecular Biology*, *428*(4), 720–725.

Walter, J., & Newport, J. (2000). Initiation of eukaryotic DNA replication: Origin unwinding and sequential chromatin association of Cdc45, RPA, and DNA polymerase α. *Molecular Cell*, *5*(4), 617–627.

Wegner, M., Drolet, D. W., & Rosenfeld, M. G. (1993). POU-domain proteins: Structure and function of developmental regulators. *Current Opinion in Cell Biology*, *5*(3), 488–498.

Witvliet, D. K., Strokach, A., Giraldo-Forero, A. F., Teyra, J., Colak, R., & Kim, P. M. (2016). ELASPIC web-server: Proteome-wide structure-based prediction of mutation effects on protein stability and binding affinity. *Bioinformatics*, *32*(10), 1589–1591.

Yun, S., & Guy, H. R. (2011). Stability tests on known and misfolded structures with discrete and all atom molecular dynamics simulations. *Journal of Molecular Graphics & Modelling*, *29*(5), 663–675.

AUTHOR INDEX

Note: Page numbers followed by "*f*" indicate figures.

B

C

D

H

I

J

K

L

N

O

P

Q

R

S

T

U

V

W

X

Y

Z

SUBJECT INDEX

Note: Page numbers followed by "*f*" indicate figures and "*t*" indicate tables.

FSC
www.fsc.org
MIX
Papier | Fördert
gute Waldnutzung
FSC® C083411

Kolibri 360 GmbH
Ferdinand-Jühlke-Straße 7
99095 Erfurt, Deutschland
produktsicherheit@kolibri360.de